THE PCR REVOLUTION

The invention of the polymerase chain reaction (PCR) won the Nobel Prize in Chemistry in 1994 and remains one of the most important scientific discoveries of the twentieth century. More than 50,000 researchers in the United States use PCR, and this is reflected in the thousands of publications using this technology. In this book, Professor Stephen A. Bustin, a world-renowned PCR expert, has gathered contributions that describe in detail the latest innovations and the overall impact of PCR on many areas of molecular research. The book contains personal reflections, opinions, and comments by leading authorities on the many applications of PCR and how this technology has revolutionized their respective areas of interest. This book conveys the ways in which PCR has overcome many obstacles in life science and clinical research and also charts the PCR's development from time-consuming, low throughput, nonquantitative procedure to today's rapid, high throughput, quantitative super method.

Stephen A. Bustin (PhD, Trinity College, University of Dublin) is Professor of Molecular Science, Barts and the London School of Medicine and Dentistry, Queen Mary University of London, United Kingdom.

The PCR Revolution

BASIC TECHNOLOGIES AND APPLICATIONS

Edited by

Stephen A. Bustin

The Institute of Cell and Molecular Science, Barts
and the London School of Medicine and Dentistry
Queen Mary University of London, United Kingdom

CAMBRIDGE
UNIVERSITY PRESS

32 Avenue of the Americas, New York NY 10013-2473, USA

Cambridge University Press is part of the University of Cambridge.

It furthers the University's mission by disseminating knowledge in the pursuit of education, learning and research at the highest international levels of excellence.

www.cambridge.org
Information on this title: www.cambridge.org/9781107423589

First published 2010
First paperback edition 2014

A catalogue record for this publication is available from the British Library

Library of Congress Cataloguing in Publication data

The PCR revolution : basic technologies and applications / edited by Stephen A. Bustin.
p. ; cm.
Includes bibliographical references and index.
ISBN 978-0-521-88231-6 (hardback)
1. Polymerase chain reaction – Diagnostic use. 2. Polymerase chain reaction.
I. Bustin, Stephen A., 1954– II. Title.
[DNLM: 1. Polymerase Chain Reaction – methods. 2. Diagnostic Techniques and Procedures. 3. Nucleic Acids – analysis. QU 450 P3485 2009]
RB43.8.P64P376 2009
572'.43 – dc22 2009013026

ISBN 978-0-521-88231-6 Hardback
ISBN 978-1-107-42358-9 Paperback

Contents

Contributors

Vladimir Benes, PhD
Genomics Core Facility
European Molecular Biology Laboratory
Heidelberg, Germany

Susan A. Burchill, MD
Candlelighter's Children's Cancer Research Group
Cancer Research UK Clinical Centre
Leeds Institute of Molecular Medicine
St. James's University Hospital
Leeds, United Kingdom

Stephen A. Bustin, PhD
The Institute of Cell and Molecular Science, Barts and the London School of Medicine and Dentistry
Queen Mary University of London
United Kingdom

Mirco Castoldi, PhD
University of Heidelberg
Heidelberg, Germany

Lin Chen
Institute for Analytical Sciences
Dortmund, Germany

Weijun Chen, PhD
Beijing Genomics Institute
Chinese Academy of Sciences
Beijing, China

Philip J. Day, PhD
Centre for Integrated Genomic Medical Research
Manchester Interdisciplinary Biocentre
The University of Manchester
Manchester, United Kingdom

Russell Higuchi
Cepheid Fellow, R&D
Cepheid
Sunnyvale, California

Michael Hofreiter, PhD
Junior Group Molecular Ecology
Max Planck Institute for Evolutionary Anthropology
Leipzig, Germany

Yang Huanming, PhD
Beijing Genomics Institute
Chinese Academy of Sciences
Beijing, China

Jim Huggett, PhD, BSc
Centre for Infectious Diseases and International Health
Windeyer Institute of Medical Research
London, United Kingdom

David Ibberson
Genomics Core Facility
European Molecular Biology Laboratory
Heidelberg, Germany

Fred Russell Kramer
Department of Molecular Genetics
Public Health Research Institute
Newark, New Jersey

Pui-Yan Kwok, MD, PhD
University of California, San Francisco
San Francisco, California

Ulrich Lehmann, PhD
Institute of Pathology
Medizinische Hochschule Hannover
Hannover, Germany

Y. M. Dennis Lo
Li Ka Shing Institute of Health Sciences and
Department of Chemical Pathology
The Chinese University of Hong Kong
Prince of Wales Hospital
Hong Kong SAR, China

Ian M. Mackay, MD
Queensland Pediatric Infectious Diseases Laboratory
Queensland Children's Medical Research Institute
Clinical Medical Virology Centre
University of Queensland and Royal Children's Hospital
Herston, Queensland, Australia

Melissa Mariani
Institute for Analytical Sciences
Dortmund, Germany

Salvatore A. E. Marras, PhD
Public Health Research Institute Center and
Department of Microbiology and Molecular Genetics
New Jersey Medical School
University of Medicine and Dentistry of New Jersey
Newark, New Jersey

Martina Muckenthaler
University of Heidelberg
Heidelberg, Germany

Tania Nolan, PhD
Sigma Life Science Custom Products
Haverhill, United Kingdom

Tanya Novak
Centre for Infectious Diseases and International Health
Windeyer Institute for Medical Sciences
University College London
London, United Kingdom

Michael W. Pfaffl
Technical University of Munich
Freising, Germany

Sudip K. Rakshit
Food Engineering and Bioprocess Technology Program
Asian Institute of Technology
Pathumthani, Thailand

Randy P. Rasmussen
Idaho Technology
Salt Lake City, Utah

Kirk M. Ririe
Idaho Technology
Salt Lake City, Utah

Holger Römpler
Rudolf-Boehm-Institute of Pharmacology and Toxicology
University of Leipzig
Leipzig, Germany

Ben Sowers
Biosearch Technologies, Inc.
Novato, California

Jens Stolte
Genomics Core Facility
European Molecular Biology Laboratory
Heidelberg, Germany

Sanjay Tyagi, PhD
Public Health Research Institute Center
University of Medicine and Dentistry of New Jersey
Newark, New Jersey

Mickey Williams, PhD
Genomics and Oncology Department
Roche Molecular Systems
Pleasanton, California

Carl T. Wittwer, MD, PhD
Department of Pathology
University of Utah Medical Center
Salt Lake City, Utah

Elisa Wurmbach, PhD
Office of Chief Medical Examiner
Department of Forensic Biology
New York, New York

Foreword

Russell Higuchi

Advances in science and technology come at an ever increasing pace. What was state of the art can be obsolete in just a few years. The old saw "The work that earns a Nobel prize today will be a graduate thesis ten years from now" may underestimate the rate of change. The hard-earned accomplishments of scientists and technologists can be made to seem trivial well within the time frame of a career. Faced with this, what do scientists celebrate as their accomplishments?

In the Stephen Sondheim play *Sunday in the Park with George*, about the pioneering artist, George Seurat, the character Dot says that the only things worth "passing on" to posterity are "children and art." I agree wholeheartedly with the children part but I wondered about how broadly art could be defined in this context. As evidenced by university colleges still organized around "Arts and Sciences" – a holdover I believe from classical uses of the terms – science and art did not used to be so far apart. I believe that beautiful, inventive thinking can be art, and good science is full of beautiful, inventive thinking – such as PCR.

When I first heard of PCR, I thought – art. When I later joined Cetus and heard from Kary Mullis his idea to use a thermostable enzyme – art. When Kary proposed the Hot Start (a bit arcane but to someone who was now an aficionado) – art.

Nonetheless, I did find myself thinking, "PCR – five years and something even better will come along." More than twenty years later, however, that has not proven to be the case, as PCR has matured and, as evidenced by the chapters in this book, is still increasingly useful. Part of that is due to real-time PCR, which I was the first to put into practice, and which is well covered in this book.

However, something better will definitely come along. We already have highly parallel sequencing of clonally amplified single DNA molecules with a throughput of a human genome a week. We are close to true single-molecule sequencing of a human genome a day. If ways can be found to parse this enormous throughput cost effectively among large numbers of samples, why use PCR and a probe to guess at what sequences might be in a sample when you can know everything that is there and at what frequency?

So the question: If its use is so transient, can it be art? I think so. With art in general, "usefulness" is not a measure of its import. How and why it comes into existence is. Hence the import of books like this, in which, unlike in journal articles, the how and why are told.

Lastly, I was pleased to see a chapter here on ancient DNA and PCR. The late Allan Wilson and I reported the first recombinant DNA cloning and sequencing of ancient DNA.[1] It was through trying to make this more efficient that I first learned PCR.[2] In referring to this work, a hard-core biochemist I knew said, "this is not science, it's art." Now I like to think it was a bit of both.

REFERENCES

1. Higuchi R, Bowman B, Freiberger M, Ryder OA, Wilson AC. DNA sequences from the quagga, an extinct member of the horse family. *Nature* 1984;**312**: 282–284.
2. Pääbo S, Higuchi RG, Wilson AC. Ancient DNA and the polymerase chain reaction. The emerging field of molecular archaeology. *Journal of Biological Chemistry* 1989;**264**: 9709–9712.

Preface

We live in an age in which hyperbole has become so pervasive that it is difficult to find apt expressions for something truly exceptional. Furthermore, impatience, haste, and short attention span seem to be added hallmarks of our times, inviting technological bandwagon effects that briefly promise the earth, but then cannot deliver because the technologies were either conceived in haste without proper regard for technical and biological concerns or are superseded by the next technological "revolution."

The polymerase chain reaction (PCR) has been around a long time now: certainly as US Patent 4,683,202 since 1987, as a practical tool since 1985,[1] and as a theoretical proposition since 1971.[2] A Google search for "polymerase chain reaction" throws up more than 1.3×10^7 results, roughly the same number as a search for "monoclonal antibody," the other wonder technology in the molecular arsenal. Its conceptual clarity, practical accessibility, and ubiquitous applicability have made PCR the defining technology of our molecular age, with a three-letter abbreviation as distinctive as that of deoxyribonucleic acid (DNA). It has even made it to Hollywood, where the re-creation of dinosaurs in *Jurassic Park* was accomplished using PCR technology. The concept is so perfectly simple that the elemental scheme remains unchanged since its inception: two oligonucleotide primers that define converging sequences on opposite strands of a DNA molecule, a DNA polymerase, dNTP building blocks, and a series of heating and cooling cycles. This prompts the selection and enormous amplification of specific DNA sequences; consequently, the needle-in-a-haystack stumbling block is magically recast as a solution that creates a haystack made up of needles.

In contrast, the detection of amplification products has undergone, and continues to undergo, pronounced changes that have led the technology into new contexts and uses. The most dramatic, and dare I say revolutionary, innovation has been the invention of real-time quantitative PCR (qPCR), which in a flash has addressed many of the practical limitations associated with legacy, gel-based PCR.[3] This adaptation allows detection of the accumulation of amplified DNA in real time after each amplification cycle. At its simplest, a qPCR experiment uses legacy PCR protocols, with the simple addition of a DNA intercalating dye, most commonly SYBR Green.[4] When complexed with double-stranded DNA, the dye absorbs blue light (λ_{max} = 488 nm) and emits green light (λ_{max} = 522 nm) that is easily detected using a qPCR instrument that combines a thermal cycler with a fluorimeter.[5] At its most complex, qPCR can use a number of fluorescent dye–labeled probes to detect and quantify multiple targets in the same tube.[6] Certainly, qPCR has been a requisite for the translation of PCR from pervasive research technology to practical process and has propelled progress in every branch of the life sciences, from agriculture to zoology; indeed it has created and sustains whole new sectors.

So, how to describe this technology and do it justice? Luckily, PCR speaks for itself through the vast numbers of applications, settings, and achievements that are unthinkable without this simple technique. This book attempts to tell the story of the PCR and to shine the light on some of the scientific advances that would never have happened without it. It presents personal views of authors from a wide range of backgrounds, pursuing an eclectic mixture of interests but united in their appreciation of the key role played by the PCR in their individual pursuits. Contributors include giants of the PCR field: Carl Wittwer, the "father" of qPCR instrumentation[5,7–9] as well as the pacesetter behind numerous practical qPCR innovations[4,10–39]; Mickey Williams, one of the original "Taqmen" and major contributor to further modifications[40–42]; Fred Kramer and Sanjay Tyagi, inventors of the ingenious molecular beacons and variants, as well as practical applications for them[43–50]; and Michael Pfaffl, major contributor and instigator of the drive toward more reliable and appropriate target quantification.[51–59] The essays of these exceptional individuals are complemented by a range of contributions that focus on practical applications of the technology. These range from Susan Burchill, who has explored the clinical potential of this technology,[60–73] to the wonderful world of ancient DNA research.[74–83] Their stories are all about the impact PCR has had on these and many other areas of molecular research. They aim to convey a flavor of the obstacles faced by life science and clinical researchers and how the unique properties of the PCR have been instrumental in overcoming these limitations. The book also aspires to chart the development of this technique from a time-consuming, low throughput, nonquantitative procedure to today's rapid, high throughput quantitative super method. Reading through these chapters clarifies just how phenomenally powerful this technology is.

Sadly, something as commanding as this technology is also open to abuse and inappropriate use – which is also addressed within these pages. This is particularly distressing when it involves peoples' health and highlights the need for constant vigilance when interpreting PCR-derived data. Opportunely, the year 2009 sees the publication of the first set of guidelines for researchers publishing PCR-based results.[84]

Tempus fugit and PCR continues to evolve. I hope that this snapshot and the associated reflections of a group of individuals who have lived this technology will help to inspire others, and make them reflect that while a technology might be great, it is the individuals who craft and struggle with it that make it truly extraordinary.

REFERENCES

1. Saiki RK, Scharf S, Faloona F, Mullis KB, Horn GT, Erlich HA, Arnheim N. Enzymatic amplification of beta-globin genomic sequences and restriction site analysis for diagnosis of sickle cell anemia. *Science* 1985;**230**:1350–1354.
2. Kleppe K, Ohtsuka E, Kleppe R, Molineux I, Khorana HG. Studies on polynucleotides. XCVI. Repair replications of short synthetic DNAs as catalyzed by DNA polymerases. *J Mol Biol* 1971;**56**:341–361.
3. Higuchi R, Fockler C, Dollinger G, Watson R. Kinetic PCR analysis: real-time monitoring of DNA amplification reactions. *Biotechnology (NY)* 1993;**11**:1026–1030.
4. Wittwer CT, Herrmann MG, Moss AA, Rasmussen RP. Continuous fluorescence monitoring of rapid cycle DNA amplification. *Biotechniques* 1997;**22**:130–138.
5. Wittwer CT, Ririe KM, Andrew RV, David DA, Gundry RA, Balis UJ. The LightCycler: a microvolume multisample fluorimeter with rapid temperature control. *Biotechniques* 1997;**22**:176–181.
6. Weller SA, Elphinstone JG, Smith NC, Boonham N, Stead DE. Detection of *Ralstonia solanacearum* strains with a quantitative, multiplex, real-time, fluorogenic PCR (TaqMan) assay. *Appl Environ Microbiol* 2000;**66**:2853–2858.
7. Wittwer CT, Fillmore GC, Hillyard DR. Automated polymerase chain reaction in capillary tubes with hot air. 1989;**17**:4353–4357.
8. Wittwer CT, Fillmore GC, Garling DJ. Minimizing the time required for DNA amplification by efficient heat transfer to small samples. *Anal Biochem* 1990;**186**:328–331.
9. Wittwer CT, Garling DJ. Rapid cycle DNA amplification: time and temperature optimization. *Biotechniques* 1991;**10**:76–83.
10. Weis JH, Tan SS, Martin BK, Wittwer CT. Detection of rare mRNAs via quantitative RT-PCR. 1992;**8**:263–264.
11. Ririe KM, Rasmussen RP, Wittwer CT. Product differentiation by analysis of DNA melting curves during the polymerase chain reaction. *Anal Biochem* 1997;**245**:154–160.
12. Lay MJ, Wittwer CT. Real-time fluorescence genotyping of factor V Leiden during rapid-cycle PCR. 1997;**43**:2262–2267.
13. Ririe KM, Rasmussen RP, Wittwer CT. Product differentiation by analysis of DNA melting curves during the polymerase chain reaction. *Anal Biochem* 1997;**245**:154–160.
14. Wittwer CT, Herrmann MG, Moss AA, Rasmussen RP. Continuous fluorescence monitoring of rapid cycle DNA amplification. *Biotechniques* 1997;**22**:130–131, 4–8.
15. Bernard PS, Ajioka RS, Kushner JP, Wittwer CT. Homogeneous multiplex genotyping of hemochromatosis mutations with fluorescent hybridization probes. 1998;**153**:1055–1061.
16. Bernard PS, Lay MJ, Wittwer CT. Integrated amplification and detection of the C677T point mutation in the methylenetetrahydrofolate reductase gene by fluorescence resonance energy transfer and probe melting curves. *Anal Biochem* 1998;**255**:101–107.
17. Morrison TB, Weis JJ, Wittwer CT. Quantification of low-copy transcripts by continuous SYBR Green I monitoring during amplification. *Biotechniques* 1998;**24**:954–962.
18. Bernard PS, Ajioka RS, Kushner JP, Wittwer CT. Homogeneous multiplex genotyping of hemochromatosis mutations with fluorescent hybridization probes. *Am J Pathol* 1998;**153**:1055–1061.
19. Bernard PS, Pritham GH, Wittwer CT. Color multiplexing hybridization probes using the apolipoprotein E locus as a model system for genotyping. 1999;**273**:221–228.
20. Elenitoba-Johnson KS, Bohling SD, Wittwer CT, King TC. Multiplex PCR by multicolor fluorimetry and fluorescence melting curve analysis. *Nat Med* 2001;**7**:249–253.
21. Wittwer CT, Herrmann MG, Gundry CN, Elenitoba-Johnson KS. Real-time multiplex PCR assays. *Methods* 2001;**25**:430–442.

22. Bernard PS, Wittwer CT. Real-time PCR technology for cancer diagnostics. *Clin Chem* 2002;**48**:1178–1185.
23. Wittwer CT, Reed GH, Gundry CN, Vandersteen JG, Pryor RJ. High-resolution genotyping by amplicon melting analysis using LCGreen. *Clin Chem* 2003;**49**:853–860.
24. Gundry CN, Vandersteen JG, Reed GH, Pryor RJ, Chen J, Wittwer CT. Amplicon melting analysis with labeled primers: a closed-tube method for differentiating homozygotes and heterozygotes. *Clin Chem* 2003;**49**:396–406.
25. Palais R, Wittwer CT. Mathematical algorithms for high-resolution DNA melting analysis. *Methods Enzymol* 2009;**454**:323–343.
26. Zhou L, Errigo RJ, Lu H, Poritz MA, Seipp MT, Wittwer CT. Snapback primer genotyping with saturating DNA dye and melting analysis. *Clin Chem* 2008;**54**:1648–1656.
27. Seipp MT, Pattison D, Durtschi JD, Jama M, Voelkerding KV, Wittwer CT. Quadruplex genotyping of F5, F2, and MTHFR variants in a single closed tube by high-resolution amplicon melting. *Clin Chem* 2008;**54**:108–115.
28. Gundry CN, Dobrowolski SF, Martin YR, Robbins TC, Nay LM, Boyd N, et al. Base-pair neutral homozygotes can be discriminated by calibrated high-resolution melting of small amplicons. *Nucleic Acids Res* 2008;**36**:3401–3408.
29. Elenitoba-Johnson O, David D, Crews N, Wittwer CT. Plastic versus glass capillaries for rapid-cycle PCR. *Biotechniques* 2008;**44**:487–488, 90, 92.
30. Sundberg SO, Wittwer CT, Greer J, Pryor RJ, Elenitoba-Johnson O, Gale BK. Solution-phase DNA mutation scanning and SNP genotyping by nanoliter melting analysis. *Biomed Microdevices* 2007;**9**:159–166.
31. Reed GH, Kent JO, Wittwer CT. High-resolution DNA melting analysis for simple and efficient molecular diagnostics. *Pharmacogenomics* 2007;**8**:597–608.
32. Dames S, Pattison DC, Bromley LK, Wittwer CT, Voelkerding KV. Unlabeled probes for the detection and typing of herpes simplex virus. *Clin Chem* 2007;**53**:1847–1854.
33. Dames S, Margraf RL, Pattison DC, Wittwer CT, Voelkerding KV. Characterization of aberrant melting peaks in unlabeled probe assays. *J Mol Diagn* 2007;**9**:290–296.
34. Margraf RL, Mao R, Wittwer CT. Masking selected sequence variation by incorporating mismatches into melting analysis probes. *Hum Mutat* 2006;**27**:269–278.
35. Palais RA, Liew MA, Wittwer CT. Quantitative heteroduplex analysis for single nucleotide polymorphism genotyping. *Anal Biochem* 2005;**346**:167–175.
36. Chou LS, Meadows C, Wittwer CT, Lyon E. Unlabeled oligonucleotide probes modified with locked nucleic acids for improved mismatch discrimination in genotyping by melting analysis. *Biotechniques* 2005;**39**:644, 6, 8 passim.
37. Zhou L, Myers AN, Vandersteen JG, Wang L, Wittwer CT. Closed-tube genotyping with unlabeled oligonucleotide probes and a saturating DNA dye. *Clin Chem* 2004;**50**:1328–1335.
38. Reed GH, Wittwer CT. Sensitivity and specificity of single-nucleotide polymorphism scanning by high-resolution melting analysis. *Clin Chem* 2004;**50**:1748–1754.
39. Margraf RL, Page S, Erali M, Wittwer CT. Single-tube method for nucleic acid extraction, amplification, purification, and sequencing. *Clin Chem* 2004;**50**:1755–1761.
40. Heid CA, Stevens J, Livak KJ, Williams PM. Real time quantitative PCR. *Genome Res* 1996;**6**:986–994.
41. Gibson UE, Heid CA, Williams PM. A novel method for real time quantitative RT-PCR. *Genome Res* 1996;**6**:995–1001.
42. Winer J, Jung CK, Shackel I, Williams PM. Development and validation of real-time quantitative reverse transcriptase-polymerase chain reaction for monitoring gene expression in cardiac myocytes in vitro. *Anal Biochem* 1999;**270**:41–49.
43. Tyagi S, Kramer FR. Molecular beacons: probes that fluoresce upon hybridization. *Nat Biotechnol* 1996;**14**:303–308.
44. Tyagi S, Bratu DP, Kramer FR. Multicolor molecular beacons for allele discrimination. 1998;**16**:49–53.
45. Kostrikis LG, Tyagi S, Mhlanga MM, Ho DD, Kramer FR. Spectral genotyping of human alleles. *Science* 1998;**279**:1228–1229.

46. Vet JA, Majithia AR, Marras SA, Tyagi S, Dube S, Poiesz BJ, Kramer FR. Multiplex detection of four pathogenic retroviruses using molecular beacons. 1999;**96**:6394–6399.
47. Bonnet G, Tyagi S, Libchaber A, Kramer FR. Thermodynamic basis of the enhanced specificity of structured DNA probes. 1999;**96**:6171–6176.
48. Marras SA, Kramer FR, Tyagi S. Multiplex detection of single-nucleotide variations using molecular beacons. 1999;**14**:151–156.
49. Tyagi S, Marras SA, Kramer FR. Wavelength-shifting molecular beacons. *Nat Med* 2000;**18**:1191–1196.
50. El-Hajj HH, Marras SA, Tyagi S, Shashkina E, Kamboj M, Kiehn TE, et al. Identification of mycobacterial species utilizing sloppy molecular beacon probes. *J Clin Microbiol* 2009 (in press).
51. Fleige S, Pfaffl MW. RNA integrity and the effect on the real-time qRT-PCR performance. *Mol Aspects Med* 2006;**27**:126–139.
52. Fleige S, Walf V, Huch S, Prgomet C, Sehm J, Pfaffl MW. Comparison of relative mRNA quantification models and the impact of RNA integrity in quantitative real-time RT-PCR. *Biotechnol Lett* 2006;**28**:1601–1613.
53. Pfaffl MW, Tichopad A, Prgomet C, Neuvians TP. Determination of stable housekeeping genes, differentially regulated target genes and sample integrity: BestKeeper – Excel-based tool using pair-wise correlations. *Biotechnol Lett* 2004;**26**:509–515.
54. Tichopad A, Didier A, Pfaffl MW. Inhibition of real-time RT-PCR quantification due to tissue-specific contaminants. *Mol Cell Probes* 2004;**18**:45–50.
55. Stahlberg A, Kubista M, Pfaffl M. Comparison of reverse transcriptases in gene expression analysis. *Clin Chem* 2004;**50**:1678–1680.
56. Tichopad A, Dilger M, Schwarz G, Pfaffl MW. Standardized determination of real-time PCR efficiency from a single reaction set-up. *Nucleic Acids Res* 2003;**31**:e122.
57. Pfaffl MW, Horgan GW, Dempfle L. Relative expression software tool (REST) for group-wise comparison and statistical analysis of relative expression results in real-time PCR. *Nucleic Acids Res* 2002;**30**:e36.
58. Pfaffl MW. A new mathematical model for relative quantification in real-time RT-PCR. *Nucleic Acids Res* 2001;**29**:e45.
59. Pfaffl MW, Hageleit M. Validities of mRNA quantification using recombinant RNA and recombinant DNA external calibration curves in real-time RT-PCR. *Biotechnol Lett* 2001;**23**:275–282.
60. Viprey VF, Lastowska MA, Corrias MV, Swerts K, Jackson MS, Burchill SA. Minimal disease monitoring by QRT-PCR: guidelines for identification and systematic validation of molecular markers prior to evaluation in prospective clinical trials. *J Pathol* 2008;**216**:245–252.
61. Viprey VF, Corrias MV, Kagedal B, Oltra S, Swerts K, Vicha A, et al. Standardisation of operating procedures for the detection of minimal disease by QRT-PCR in children with neuroblastoma: quality assurance on behalf of SIOPEN-R-NET. *Eur J Cancer* 2007;**43**:341–350.
62. Beiske K, Ambros PF, Burchill SA, Cheung IY, Swerts K. Detecting minimal residual disease in neuroblastoma patients – the present state of the art. *Cancer Lett* 2005;**228**:229–240.
63. Patel K, Whelan PJ, Prescott S, Brownhill SC, Johnston CF, Selby PJ, Burchill SA. The use of real-time reverse transcription-PCR for prostate-specific antigen mRNA to discriminate between blood samples from healthy volunteers and from patients with metastatic prostate cancer. *Clin Cancer Res* 2004;**10**:7511–7519.
64. Riley RD, Heney D, Jones DR, Sutton AJ, Lambert PC, Abrams KR, et al. A systematic review of molecular and biological tumor markers in neuroblastoma. *Clin Cancer Res* 2004;**10**:4–12.
65. Burchill SA, Perebolte L, Johnston C, Top B, Selby P. Comparison of the RNA-amplification based methods RT-PCR and NASBA for the detection of circulating tumour cells. *Br J Cancer* 2002;**86**:102–109.

66. Burchill SA, Kinsey SE, Picton S, Roberts P, Pinkerton CR, Selby P, Lewis IJ. Minimal residual disease at the time of peripheral blood stem cell harvest in patients with advanced neuroblastoma. *Med Pediatr Oncol* 2001;**36**:213–219.
67. Burchill SA, Selby PJ. Molecular detection of low-level disease in patients with cancer. *J Pathol* 2000;**190**:6–14.
68. Keilholz U, Willhauck M, Rimoldi D, Brasseur F, Dummer W, Rass K, et al. Reliability of reverse transcription-polymerase chain reaction (RT-PCR)-based assays for the detection of circulating tumour cells: a quality-assurance initiative of the EORTC Melanoma Cooperative Group. *Eur J Cancer* 1998;**34**:750–753.
69. Wyld DK, Selby P, Perren TJ, Jonas SK, Allen-Mersh TG, Wheeldon J, Burchill SA. Detection of colorectal cancer cells in peripheral blood by reverse-transcriptase polymerase chain reaction for cytokeratin 20. *Int J Cancer* 1998;**79**:288–293.
70. Keilholz U, Willhauck M, Scheibenbogen C, de Vries TJ, Burchill S. Polymerase chain reaction detection of circulating tumour cells. EORTC Melanoma Cooperative Group, Immunotherapy Subgroup. *Melanoma Res* 1997;7 Suppl **2**:S133–141.
71. Pittman K, Burchill S, Smith B, Southgate J, Joffe J, Gore M, Selby P. Reverse transcriptase–polymerase chain reaction for expression of tyrosinase to identify malignant melanoma cells in peripheral blood. *Ann Oncol* 1996;**7**:297–301.
72. Johnson PW, Burchill SA, Selby PJ. The molecular detection of circulating tumour cells. *Br J Cancer* 1995;**72**:268–276.
73. Burchill SA, Bradbury MF, Pittman K, Southgate J, Smith B, Selby P. Detection of epithelial cancer cells in peripheral blood by reverse transcriptase-polymerase chain reaction. *Br J Cancer* 1995;**71**:278–281.
74. Knapp M, Rohland N, Weinstock J, Baryshnikov G, Sher A, Nagel D, et al. First DNA sequences from Asian cave bear fossils reveal deep divergences and complex phylogeographic patterns. *Mol Ecol* 2009 (in press).
75. Hofreiter M. DNA sequencing: mammoth genomics. *Nature* 2008;**456**:330–331.
76. Krause J, Unger T, Nocon A, Malaspinas AS, Kolokotronis SO, Stiller M, et al. Mitochondrial genomes reveal an explosive radiation of extinct and extant bears near the Miocene-Pliocene boundary. *BMC Evol Biol* 2008;**8**:220.
77. Gilbert MT, Jenkins DL, Gotherstrom A, Naveran N, Sanchez JJ, Hofreiter M, et al. DNA from pre-Clovis human coprolites in Oregon, North America. *Science* 2008;**320**:786–789.
78. Meyer M, Briggs AW, Maricic T, Hober B, Hoffner B, Krause J, et al. From micrograms to picograms: quantitative PCR reduces the material demands of high-throughput sequencing. *Nucleic Acids Res* 2008;**36**:e5.
79. Rohland N, Hofreiter M. Ancient DNA extraction from bones and teeth. *Nat Protoc* 2007;**2**:1756–1762.
80. Rompler H, Dear PH, Krause J, Meyer M, Rohland N, Schoneberg T, et al. Multiplex amplification of ancient DNA. *Nat Protoc* 2006;**1**:720–728.
81. Leonard JA, Rohland N, Glaberman S, Fleischer RC, Caccone A, Hofreiter M. A rapid loss of stripes: the evolutionary history of the extinct quagga. *Biol Lett* 2005;**1**:291–295.
82. Hofreiter M, Lister A. Mammoths. *Curr Biol* 2006;**16**:R347–348.
83. Hofreiter M, Poinar HN, Spaulding WG, Bauer K, Martin PS, Possnert G, Pääbo S. A molecular analysis of ground sloth diet through the last glaciation. *Mol Ecol* 2000;**9**:1975–1984.
84. Bustin SA, Benes V, Garson JA, Hellemans J, Huggett J, Kubista M, et al. The MIQE guidelines: minimum information for publication of quantitative real-time PCR experiments. *Clinical Chemistry* 2009;**55**:611–622.

I Basic technologies

1 Real-time polymerase chain reaction

Mickey Williams

The invention and successful practice of the polymerase chain reaction (PCR) by Kary Mullis and colleagues in 1983 set the stage for a scientific revolution. PCR established a base technology from which many specific and diverse applications have grown. PCR has played a crucial underlying technological role in many aspects of the genomic age that we experience today. The power to assess complete genomic sequences starting with minuscule amounts of target molecules entrenched PCR as the backbone of many subsequent analytical techniques. The sequencing of the genomes of many diverse species and the ability to discriminate individuals within a species have relied on PCR as an instrumental component.

The knowledge of genomes has led to the ability to identify sequences representing the coding genes that carry the blueprints for the construction of proteins. It is of great scientific interest to study the regulation of these gene-encoding messenger ribonucleic (mRNA) molecules. The study of gene expression has led to a better understanding of different biological states that exist within different tissue types, reflecting their different functions. Gene expression changes provide insight into underlying molecular and functional differences that exist between diseased and normal tissues. PCR has had a profound impact on gene expression studies as well. In 1991, while I was a junior scientist at Genentech, my scientific life was intensely affected by PCR. I was part of a team charged with developing assays to assess clinical outcomes of a vaccine treatment for human immunodeficiency virus (HIV) infections. I became aware of PCR and reverse transcriptase (RT)–PCR as means of quantifying specific sequences found in biological samples and was fortunate to meet some of the best and brightest PCR gurus at Roche Molecular Systems, born from Cetus Corporation, where Kary Mullis had worked. I also was introduced to another equally brilliant group of scientists from Applied BioSystems Inc. (ABI), which had gained research rights to the PCR patents. ABI had some exciting research instruments under development. My introduction to these companies and scientists occurred during the period when PCR became a powerful tool for the quantitative assessment of gene expression. Following are some of my recollections of the time preceding and leading to the introduction of real-time quantitative PCR (qPCR).

EARLY DAYS

During the mid-1980s, I was a part of the growing number of postdoctoral trainees and scientists exploring gene expression as a means to gain insight into differentiation of tissues and the functional workings of cells. Specifically, I was studying the mechanisms by which embryonic fibroblasts could differentiate into fat (adipose) cells. The art of gene expression analysis was laborious, relying on such techniques as construction and screening of complementary deoxyribonucleic acid (cDNA) libraries, subtractive library screening, northern blots, and of course lots of cell culture. Considerable effort was needed to establish the infra-break;structure for gene expression experiments. My end goal was to correlate changes in gene expression with biological changes within the cell during the differentiation into adipocytes. Many scientists were using similar approaches for their biological quests. Experiments were laborious and time consuming, often taking weeks to obtain results – provided that all of the technical aspects worked well. Oftentimes a flawed reagent or careless mistake meant weeks lost. An integral part of this effort was the use of copious amounts of radioactive labeling compounds needed to detect sequences of interest. I remember that one of the key elements of our experimental planning was the shipment schedule of the radiolabeled nucleotides. We all wanted to have the freshest batch of ^{32}P-deoxyribonucleotide triphosphate (NTP) for our "important" experiments. These were indeed fun times!

Molecular biologists, using the tools of nucleic acid hybridization, cloning, and sequencing, were busy discovering new transcripts and gene sequences. This was the time when discovery of a new gene or its transcript was often the serendipitous result of an unexpected band on a northern blot or a colony detected in a screening experiment. The use of nonstringent hybridization and washing conditions would permit related but nonidentical sequences to "light up" with radioactively labeled hybridization probes. Researchers who followed up and identified these new transcripts (or genomic sequences) often immediately wanted to learn the tissues of action for these uncharacterized genes. Beyond discovering the cells and tissues of expression of these newly discovered genes, efforts were launched to understand the regulation of this gene's expression as it correlated with such biological state changes as differentiation, cell stimulation with growth factors, and disease. Another active endeavor was the understanding of gene transcriptional regulation. These were the days of promoter bashing (deleting various DNA segments upstream of a gene to determine the impact on transcriptional regulation). These were also the times during which the gel-shift assay was used for identifying the transcriptional regulatory proteins.

Interestingly, these were also the times when nothing more than the detailed description of unexpected bands on a gel led to assured acceptance of a manuscript. Northern blots were the workhorse technology for these activities. Originally most northern blots used agarose gels that included formaldehyde and lots of meticulously and painstakingly prepared RNA (10 to 20 μg), ethidium bromide, nitrocellulose filters, copious amounts of precisely cut paper towels, seal-a-meal bags with a radioactive seal-a-meal instrument (left behind Plexiglas shields), radioactive probes, x-ray film, boxes of latex gloves, and weeks of time. It was not uncommon to develop a film after weeks of effort to realize that one of many possible glitches had impaired the results. I vividly remember one experience from which I learned that plastic wrap sloppily left sticking out of an x-ray film cassette permits light to leak onto the film. This event ruined weeks of waiting for the perfect film image. So it was back to the beginning of the experiment oftentimes after learning such lessons.

THE STORY UNFOLDS

To obtain quality data it was *very important* to begin with the best quality RNA. Many of us remember the first time we were trained in the art of RNA preparation. Many laboratory rules were devised and often posted to prevent degradation of the much-sought-after prize of high-quality full-length RNA. We were taught to use only oven-baked glassware and diethyl pyrocarbonate (DEPC)-treated water, always wear sterile gloves, use only pipettes dedicated to RNA, never open a tube of the dreaded RNase enzyme on a bench where RNA would be purified, never use a pipetteman used for dispersing RNase for purification of RNA, and so on. Although we still follow strict protocols, the introduction of many commercial kits for purification of total RNA has made this a more reliable and less stressful aspect of routine laboratory practice. I do not think many scientists today experience the anxiety that many of us “senior” scientists felt prior to the isolation of RNA from a large and important experiment.

As the study of gene expression continued to be a focus for many experiments, many technical improvements came into the picture. Soon we replaced the messy seal-a-meal bags with glass tubes and hybridization ovens. Brittle and flaky nitrocellulose found competition from more flexible nylon membranes; radioactive labeling techniques were challenged by nonradioactive chemiluminescent approaches. Many new enzymes and tools were harnessed for amplification and labeling of probes. The introduction of riboprobes added a means of producing high-specific-activity RNA probes, which permitted sensitive detection of low-level transcripts. The forerunner of microarrays, the dot blot, was introduced. Dot blots did not permit the visualization of transcript size as did northern blots, but dot blots afforded easy multiwell experiments in which many samples or probes could be analyzed simultaneously. Ninety-six-well dot blots did not quite parallel the massive standards of today's microarrays, but they were a long step beyond 12-well northern blots.

One of the most prevalent dilemmas of the era of northern and dot blots involved the aspect of quantitative assessment of transcript expression levels. It was of great interest to document the expression changes in mRNA levels resulting from biological state changes. Many issues needed to be overcome to permit meaningful quantitative assessment: (1) the painstaking task of preparing good-quality RNA, (2) how to detect and quantify the amount of a given transcript, and (3) how to normalize the load from well to well. This was an era of scientific art during which many creative attempts were made to address these topics. For detection and quantitation of mRNA transcripts, many researchers relied on densitometer analysis of x-ray film images as a means of adding quantitative values to the intensity of the northern blot bands or dot blots. This technique required efforts to ensure that all measured quantities were within the dynamic range of accurate measurement. A saturated image would obviously lead to lack of quantitative results. I believe that some did a much better job than others at attempting to understand and apply these techniques in the best manner. For sample-to-sample comparisons, utilization of additional transcripts termed "housekeeping genes" permitted sample-to-sample normalization for the amount of total RNA loaded. These housekeeping genes represented a class of genes needed for essential metabolic functions in all cell types. It was believed that expression of this class of genes would remain constant as their functions were essential for basic functions of all cells. During this period, many data were published that relied on such housekeeping genes as β-actin and glyceraldehyde 3-phosphate dehydrogenase (GAPDH) for sample normalization. Often the bands on a northern blot were so bulbous that one could be assured that the image was saturated and accurate measurement was not feasible. Such saturated bands are clearly not proof of equal sample loading. Nevertheless, this was the state of affairs during these days. The art of quantification produced many valuable insights that kept the science moving forward. It was in this backdrop that PCR entered our lives.

UNDERSTANDING THE PCR REACTION

Shortly after the introduction of PCR for the amplification of DNA, the addition of reverse transcription added the ability to amplify RNA via a cDNA intermediate. RT–PCR was born. The original methods used two independent steps, whereas reverse transcription was done prior to PCR amplification. Now we have been provided with blended enzyme mixtures or even single, dual-activity enzymes that permit the process of RT–PCR to proceed in a single unopened tube. It was not long after scientists began applying PCR and RT–PCR when the logical extension was made to use this technology for the quantitative assessment of the starting template. Many researchers made early attempts at quantifying the starting target by running PCR products onto a gel and using techniques to measure the amount of product generated. The notion that more initial target would generate more product (as evidenced by a darker stained band on a gel) was true, to an extent. However, it was soon made clear by the PCR gurus that PCR assays eventually cease to produce exponential gains with each successive cycle. The understanding of the PCR plateau put an end to the simple gel-based methods of qPCR. Simply put, the initial excesses of enzymes and primers that exist at the beginning cycles of PCR soon become limiting as excess amounts of product are generated. At this point there is more PCR product than reagents available in the tube for the next round of amplification. When this occurs in later PCR cycles, the exponential increase in product per cycle is lost. Eventually, as the PCR continues, a cycle is reached at which little or no product is generated. It is during this initial plateau phase and final plateau that quantitative measurements are confounded. One fix to this phenomenon was short lived: One used a serial dilution of input target and measured the correlation with dilution of target and the target accumulation. One could use linear analysis to choose a range of input targets that resulted in linear output of product and estimate the input target quantity across samples. A similar approach used stopping the PCRs every couple of cycles throughout the assay. A small sample would be removed for gel analysis, and, again, early preplateau products could be compared for a semiquantitative assessment of input target quantity. These approaches required normalization, and housekeeping gene analysis was used for each sample. The excessive sample manipulation required with these early approaches was conducive to the dreaded PCR contamination, which could be nearly impossible to stop once started. Because such excessive amounts of product are generated during the exponential amplification, it was easy to contaminate clean reagents and also samples that should not contain the target. Tracking the source of PCR product contamination is often difficult and usually results in destroying all reagents and sometimes even changing labs. These early attempts at qPCR or semiquantitative PCR clearly resulted in cumbersome experiments that were almost not worth the effort. Northern blots were still frequently used as the method of choice.

If it were not for the exquisite sensitivity and rapid commercialization of PCR, this could have been the end of the story for qPCR applications. Researchers in the area of HIV and other infectious disease specialties realized the power of the sensitivity and impact that PCR would have on their fields of study. It was in this arena of infectious disease quantitation (especially HIV) that the next major improvements were made to bring qPCR and RT–PCR into everyday practice.

QUANTIFICATION PROSPECTS

My postdoctoral studies ended, and I accepted a job offer at Genentech, where my first project was to help develop a quantitative assay for measurement of HIV infections. Genentech had launched a two-pronged clinical effort to study the effects of HIV vaccines in preventative and therapeutic applications. The field of HIV research was entrenched in the use of CD4 cell counts as a surrogate end point of disease status, but interest was growing in the use of quantitative viral load as another possible surrogate. It was clear to many researchers that the application of RT–PCR brought exquisite sensitivity to detect and quantitatively measure viral load. I was fortunate that our project leader, Jack Nunberg, had a connection to Roche Molecular Systems as he had come to Genentech from Cetus. It was a new beginning for me as I was introduced to many of the PCR gurus who had developed and commercialized this technology. It was my good fortune to have had the opportunity to work and learn from such people as David Gelfand, Russ Higuchi, Shirley Kwok, John Sninsky, and Bob Watson (to name a few). As it turns out, these scientists had already initiated a development of a U.S. Food and Drug Administration (FDA)-approved test for quantitative RT–PCR analysis of HIV viral burden. Their approach used a technology termed Amplicor™. This assay proved to be a workhorse in support of a multitude of clinical trials searching for therapeutically efficacious treatment of HIV, and has proven to provide a sensitive and accurate means to detect HIV in blood. Although these early trials did not yield a successful treatment, we had witnessed the birth of a powerful new assay tool, quantitative RT–PCR. We all were encouraged by the improvements in this technology and were motivated to help establish these techniques in routine laboratory research as well as in clinical research.

At the same time we were working with the Amplicor methodology, others (i.e., Jeffrey Lifson and Michael Piatak) described a gel-based approach.[1] It was called, among many names, "quantitative competitive" RT–PCR. The central component of this approach was the design and use of a competitor molecule that was spiked into the sample at known serial dilutions. The critical aspect of the competitor design was that it included the sequence for the same primers used for the target molecule of interest. The resultant competitor product amplicon needed to be a different length (or internal sequence) such that it could be differentiated from product in gel electrophoresis. The most critical aspect of this competitor was the demonstration that the PCR efficiency (i.e., how much product is generated with each successive cycle) was identical for both the target sequence of interest and the competitor. Demonstration of equivalent PCR efficiency required some assay development during which a series of mixed concentrations of competitor and target were tested to demonstrate expected ratios. The advantage of this approach when compared to all previous methods was the ability to run reactions to any end cycle and still obtain quantitative results. Even if a reaction was run into the plateau stage of PCR, the ratio of products from the target of interest to the competitor product was maintained from the starting sample throughout the entire assay into plateau. The assays were simple to run and interpret. A serial dilution of known amounts of competitor was put into replicates of the sample of interest or vice versa where the competitor concentration was held constant and the sample diluted. The PCR assay was performed, and results could be read from a gel. Line equations for target and competitor were made, and the quantity of the target was determined by comparison with the known amounts of competitor. After the technique was published, this approach was used by many researchers beyond the infectious disease arena. This approach was quickly adopted by scientists studying cell-based gene expression.

The application of PCR technology during this time was growing and impinged on science in multiple arenas. As a result of a strong PCR patent portfolio protecting the rights of the inventors, many PCR work-around techniques began to blossom. Techniques such as strand-displacement amplification, self-sustained sequence replication, and so on began to flourish in the literature and at conferences. Many of these alternative methods had clear potential, but only a few of these competing technologies are still used to any extent for routine research applications.

REAL-TIME PCR

Several hallmark studies began to lay the groundwork for the soon-to-be-described real-time PCR. It was during this time in 1991 when David Gelfand and his colleagues described TaqMan™ methodology.[2] In that article, the use of a radiolabeled hybridization probe designed to hybridize to a sequence within the amplicon was introduced. During the reaction, the DNA polymerase would displace the radiolabeled probe and nucleolytic activity in the DNA polymerase would cleave the probe. In the conclusions the authors stated that the amount of probe cleavage correlates with the amount of product accumulation and hence correlates with the starting target amount. Another major event occurred in 1993 when Russ Higuchi, working with Bob Watson, demonstrated the quantitative accumulation of PCR product with a simple cycle-by-cycle ultraviolet (UV) box visualization of reactions containing ethidium bromide.[3] As more product was generated, the tubes accumulated more ethidium-derived fluorescence. A remarkable photograph of tubes on a UV box clearly demonstrated the concept. The final piece of the puzzle came from efforts by Ken Livak and colleagues at Applied Biosystems. They were making hybridization probes that contained two fluorescent dyes. One dye was a reporter dye, which was quenched in the intact oligonucleotide by a second dye (quencher dye) that by fluorescence resonance energy transfer (FRET) accepted the energy from the reporter, preventing reporter emission of light. Upon polymerase cleavage of the oligonucleotide probe, the quencher was no longer spatially in close proximity to the reporter and the reporter fluorescent light was now detectable.[4] It was the collaborative efforts of these scientists that led to the birth of real-time qPCR. I was fortunate to be collaborating with both groups during this period and was permitted to be the beta-test site for the first ABI real-time PCR instrument. This was a very exciting time. I think we all realized how important this technology would become. We had lots of fun during this period. Virtually every experiment gave us insight into the technique. A tremendous additional advantage of this technique was the closed-tube format. When a reaction was prepared and the tube was sealed, there was not a need to open the tube after the reaction. This technique reduced the potential for product contamination that was prevalent with competitive PCR gel formats. Our first assays were painstakingly developed as we often used a dilution series of probe and primer concentrations to optimize the reaction and obtain the most robust results. I remember the very first real-time experiment I ever ran; about halfway through the run, the power was interrupted and the instrument crashed. The experiment ended without results. After that we soon invested in an uninterruptible power supply (UPS) unit. We also were lucky to have an engineer from ABI, Bob Grossman, personally on call for unexplained phenomena. Eventually, with the efforts of many, we began mastering the technology. These initial efforts resulted in two publications by Chris Heid et al.[5] and Ursula Gibson et al.,[6] which contained the first descriptions of real-time PCR and RT–PCR, respectively.

After the commercial launch of the ABI Model 7700 we began to add more instruments to our research group. This platform soon became a workhorse for many research projects. I was given an opportunity to speak at many conferences during this time period and was truly inspired by the potential power that this new tool would contribute to scientific research. I was in awe of the quantitative dynamic range, which was close to 7 logs of input target. The precision of well-designed assays was astonishing. Additionally, because this was based on PCR, the sensitivity was excellent. Soon there was an explosion of the use of this technology, with many others making significant contributions to its use and expansion. As with any new technology there was a learning curve, but soon a community of experts began to grow. Many other companies came into the arena as suppliers of real-time instruments, reagents, and kits. One of the more important applications came in clinical studies. Many researchers used the enormous sensitivity and quantitative data to study a variety of medically related topics. Quantitative pathogen detection and monitoring comprised many of the early clinical uses. One application that made a lasting impression on me was the first description of real-time PCR monitoring of minimal residual disease. The sensitivity of this technique clearly added to the ability to detect cancer-related chromosome translocations in the blood or bone marrow of leukemia patients.

It had always been a goal for the developers of this technology to use multiplex capabilities of instruments and dyes to add more genes to the analysis in the same tube. Although multiplexing is still not routine, many researchers have taken advantage of this aspect to include normalization genes in the same tube as the gene of interest. Another important advance in real-time instruments came as high-density thermal blocks were introduced. Today it is common to find 384-well blocks available. Some companies have moved to microfluidic devices and have increased the number of individual assay chambers to 1,536.

Today, real-time PCR is in routine use for research and clinical applications. Recently a breast cancer recurrence prediction assay was introduced that uses real-time RT–PCR (OncoType Dx; Genomic Health). This test is recognized by many oncologists as a valid tool to assist in patient management. Research efforts in oncology often rely on formalin-fixed paraffin-embedded (FFPE) archival samples. The processes of fixing and archiving contribute to degradation of RNA quality, often resulting in fragmented RNA of an average size of 150 to 200 bases. A strength of RT–PCR is that small amplicons can be designed such that even poor-quality fragmented FFPET samples are amendable to quantitative assessment. Real-time PCR is an accepted standard for many projects and has been approved for use in in vitro diagnostic assays by the FDA. In the early days of gene expression analysis with microarrays, it was common to verify microarray results with follow-up real-time PCR assays. It is well known that the dynamic range of real-time PCR is much greater than that of microarrays. Real-time PCR has become the gold standard of quantitative nucleic acid analysis.

SOME FINAL THOUGHTS

As I reflect on the power of real-time PCR, I remind myself that good assay development is critical to success. It is important that the normalization genes are carefully chosen and validated. As we have learned, there is probably no one gene that is invariant in all biological situations. Therefore, selection of the best gene or genes is critical for data interpretation. Assays should be assessed for precision of technical replicates. If one would like to demonstrate that a twofold difference in gene expression is meaningful, the assay should have sufficient precision to statistically discriminate this difference. The linear dynamic range of quantitation should be explored. A sample is best analyzed for quantitation if it falls within this range. The impact of biological matrices is of great importance. It is known that such things as heme found in blood can inhibit PCR polymerases. Hence, methods of nucleic acid sample preparation should be robust. I often prefer analyzing a dilution series of a sample. This permits analysis of linear dilution data to the expected dilution slope. A sample with a slope too far removed from the expected may be problematic for quantitation and should be closely examined before conclusions are drawn. Although real-time PCR is a powerful tool, it still requires a sound understanding of the basics of the technology and the assumptions that are made to draw valid scientific conclusions from the results.

As we move forward in this era of genomic exploration, real-time qPCR will continue to play a central role in this effort. I am certain that continual improvement in instruments, reagents, and techniques will aid this effort. I am happy to have been a part of this story!

REFERENCES

1. Kappes JC, Saag MS, Shaw GM, Hahn BH, Chopra P, Chen S, et al. (1995) Assessment of antiretroviral therapy by plasma viral load testing: standard and ICD HIV-1 p24 antigen and viral RNA (QC-PCR) assays compared. *Journal of Acquired Immune Deficiency Syndromes and Human Retrovirology* **10**: 139–149.
2. Holland PM, Abramson RD, Watson R, Gelfand DH (1991) Detection of specific polymerase chain reaction product by utilizing the 5′–3′ exonuclease activity of Thermus aquaticus DNA polymerase. *Proceedings of the National Academy of Sciences of the United States of America* **88**: 7276–7280.
3. Higuchi R, Fockler C, Dollinger G, Watson R (1993) Kinetic PCR analysis: real-time monitoring of DNA amplification reactions. *Biotechnology (NY)* **11**: 1026–1030.
4. Livak KJ, Flood SJ, Marmaro J, Giusti W, Deetz K (1995) Oligonucleotides with fluorescent dyes at opposite ends provide a quenched probe system useful for detecting PCR product and nucleic acid hybridization. *PCR Methods and Applications* **4**: 357–362.
5. Heid CA, Stevens J, Livak KJ, Williams PM (1996) Real time quantitative PCR. *Genome Research* **6**: 986–994.
6. Gibson UE, Heid CA, Williams PM (1996) A novel method for real time quantitative RT-PCR. *Genome Research* **6**: 995–1001.

2 Thermostable enzymes used in polymerase chain reaction

Sudip K. Rakshit

THERMOSTABLE ENZYMES

Many living organisms have been found in most challenging environments that most of us would not believe possible. These microorganisms, known as extremophiles, are found in most extreme conditions of salinity, alkalinity, pressure, temperature, and so on. Most of these microorganisms have been identified as members of domain archae,[1] which are ancient living organisms. The utilization of these organisms and their components (including enzymes) has been studied for a number of applications.

To live in these extreme conditions, those microorganisms have cellular components including biocatalysts – enzyme proteins – that are active in such conditions. Many studies have been carried out to enrich knowledge about such enzymes by identifying and characterizing them. Thermostable enzymes present in microorganisms living in extremely high temperatures are the most extensively studied enzymes and have a number of industrial applications.[2]

Organisms that grow at high temperatures are called thermophiles when they grow optimally between 50 °C and 80 °C and hyperthermophiles when they grow optimally between 80 °C and 110 °C. There are some organisms that can grow under extremely hot conditions up to 113 °C.[3] Such enzymes are typically not active if the temperature is less than 40 °C. These enzymes are also useful in understanding enzyme evolution and molecular mechanisms for thermal stability of proteins and identification of upper temperature limits for enzyme functions.

Whereas the turnover (conversion rate) of most biocatalysts is high at low temperatures (e.g., 37 °C), they denature at high temperatures. Chemical catalysts, which are usually metals, in contrast have high turnover numbers at high temperatures (e.g., 300 °C). The thermally stable enzymes, identified and extracted from thermally tolerant microorganisms, enable high productivity of industrial processes by making possible the required specific conversion at high temperature, without the protein catalyst getting denatured. Thermally stable enzymes are also preferred in biotechnological applications because the high temperature used effectively reduces the risk of contamination. Amylolytic enzymes extracted from thermophilic microorganisms such as pullulanases, β-amylases, and the commercially available Termamyl and Fungamyl have been used in the starch industry for hydrolysis and modification of useful raw materials. This has led to immense cost and time savings as the hydrolysis reaction can be carried out at temperatures at which starch is gelatinized. Xylanases have been effectively used in pulp and paper technology, and thermally stable cellulases extracted from many microorganisms are useful in the textile industry. The latter enzymes have to be active at temperatures as high as 100 °C for the bio-polishing process of cotton. Among all thermally stable enzymes, thermally stable deoxyribonucleic acid (DNA) polymerases, used in the polymerase chain reaction (PCR), are among the most extensively studied and used enzymes.

Many types of thermally stable DNA polymerases have been identified. Taq DNA polymerase extracted from *Thermus aquaticus* – one of the most important DNA polymerases – is the first thermally stable DNA polymerase used in PCR technology. In addition to Taq DNA polymerase, several other thermostable DNA polymerases have been isolated. Pfu DNA polymerase isolated from *Pyrococcus furiosus* and Vent DNA polymerase isolated from *Thermococcus litoralis* are some other commonly used thermostable DNA polymerases.

PCR

Studies of the genetic materials of organisms and cells are important in the identification of organisms, disease diagnosis, characterization of organisms, and many other applications. Earlier such genetic studies depended on obtaining large amounts of pure DNA from such sources. To obtain such large amounts of DNA, numerous cells also had to be used. This process was laborious, time consuming, and (in some cases, as in forensic samples) nearly impossible to obtain. Scientists tried to find a solution to this problem.

The discovery of the PCR method by Kary Mullis resulted in the appearance of a good solution to this problem.[4] The PCR technique is used for in vitro amplification of nucleic acids. In the last two decades, PCR has become an integral resource for most biotechnology laboratories throughout the world. It is a powerful tool because it is sensitive and specific in amplification of DNA. This method consists of a cyclic process of three steps – namely, denaturation, primer annealing, and extension of the DNA fragment. In the first step, double-stranded DNA is denatured into single strands by heating to 95 °C; then specific short DNA fragments called primers are annealed to these DNA strands at 35 °C to 40 °C, in the second step. In the third step, primers are extended by DNA polymerase at 72 °C by adding complementary nucleotides to the 3′ end of the primers. Starting from a single target DNA or ribonucleic acid (RNA) sequence, theoretically more than one billion product sequences can routinely be synthesized by a PCR in one run. This amplification eliminates the need for extraction of large amount of pure DNA for molecular studies. A key requirement for the amplification of the DNA in a PCR is the availability of effective DNA polymerase enzymes.

USE OF THERMOSTABLE POLYMERASE ENZYMES FOR PCR

DNA polymerase enzymes that catalyze the formation and repair of DNA occur naturally in all organisms. These enzymes bind to single DNA strands and create new double-stranded DNA by making complementary DNA (cDNA) strands. The accurate replication of all living matter and transfer of genetic information from one generation to the next generation depend on this process. The PCR technique duplicates part of this method in an in vitro environment. In the third step of a PCR cycle, the DNA polymerase is used for extension of primers that were previously annealed to single-stranded DNA to get double-stranded new DNA.

In the early stages of PCR development, the DNA polymerases used were DNA polymerase extracted from the bacterium *Escherichia coli*. This DNA polymerase has many disadvantages, although it was an invaluable tool for the early pioneers of the PCR. In every PCR cycle, the double-stranded DNA present in the PCR mixture must be denatured by heating to 95 °C to separate cDNA strands that then serve as templates for the synthesis of two double-stranded nucleic acid molecules. Unfortunately, this heating irreversibly inactivated the *E. coli* DNA polymerase. Because of this fresh DNA polymerase enzyme had to be added manually after the annealing step of each PCR cycle. Because a PCR run typically consists of 30 to 40 cycles, this was a labor-intensive, time-consuming, and boring task. Besides, it also required large amounts of DNA polymerase and continual attention throughout the process.

There was thus a need for a DNA polymerase that remained stable during the DNA denaturation step of PCR and retained good activity at high temperatures. The bacterium *Thermophilus aquaticus* isolated from water hot springs provided a solution to this problem.[5] The DNA polymerase isolated from this bacterium, called Taq polymerase, was not rapidly inactivated at high temperatures. This thermally stable Taq polymerase was introduced in place of the *E. coli* DNA polymerase.[6] This DNA polymerase and this change enabled the automation of the PCR process. This new introduction allowed the performance of 30 to 40 cycles of PCR amplification without the need for opening the PCR tube and adding fresh DNA polymerase after each cycle. An additional advantage is that contamination, which can occur when the polymerase enzyme is added manually more than 20 to 30 times in the PCRs, is considerably reduced. The nonspecific products that were obtained in the earlier PCR technique were also reduced by the application of other thermostable DNA polymerases discovered later, which had additional proofreading (fidelity) characteristics. Because the primer annealing could be performed at higher temperatures, higher stringency and higher fidelity were obtained.

DIFFERENT THERMOSTABLE ENZYMES USED IN PCR

Enzymes that replicate DNA using a DNA template are called DNA polymerases. Other enzymes that synthesize DNA include reverse transcriptases (RTs), which make use of an RNA as a template (to produce cDNA), and terminal transferases, which make DNA without using a template. Many organisms have more than one type of DNA polymerase, but all follow the same basic rules. The organism *E. coli*, for example, has five DNA polymerases.

The DNA polymerase I from *E. coli*, usually referred to as pol I, is a single large protein with a molecular weight of approximately 103 kDa (103,000 grams/mole) and requires a divalent cation (Mg^{2+}) for activity. The main activities associated with the enzyme include 5′-to-3′ DNA polymerase activity and the 3′-to-5′ exonuclease proofreading activity (fidelity).

The rate of DNA synthesis by pol I during the polymerase activity is 20 nucleotides/second. The role of the 3′-to-5′ exonuclease is to edit DNA. It removes incorrectly polymerized nucleotides while extension of the primer takes place. In general, the 3′-to-5′ exonuclease increases the accuracy or fidelity of DNA synthesis by a factor of 10 to 1000. Thus, when the 5′-to-3′ polymerization of the enzyme accidentally puts the wrong base into DNA, the 3′-to-5′ exonuclease proofreading activity immediately removes it. Thus, errors due to incorporation of the wrong bases by the DNA polymerase are low because of the base-pairing rules followed and the 3′-to-5′ exonuclease proofreading activity.

DNA polymerase cannot synthesize DNA in the 3′-to-5′ direction because in such a case the nucleotides would add to the primer terminus, which has a 5′-triphosphate. The 3′-OH of each incoming deoxyribonucleoside triphosphate would attack the 5′-triphosphate of the growing chain. This attack would be followed by the removal of an incorrect 5′-terminal nucleoside triphosphate by edition. The removal prevents the DNA chain from being further extended because the primer terminus would now be a 5′-monophosphate, not a 5′-triphosphate.

DNA polymerases have various characteristics, some of which are advantageous and some are disadvantageous. The proofreading 3′-to-5′ exonuclease activity and 5′-to-3′ exonuclease activity are examples of this. Some polymerases possess both of these activities, but some of them have only one. The 5′-to-3′ polymerase activity is important in DNA amplification for producing double-stranded DNA.

The enzymes that have proofreading activity are normally less active and slower to give products, but thermostable DNA polymerases like Taq polymerase, which amplifies faster than others, have less proofreading activity. To optimize the PCR procedure, some researchers use a mix of thermostable DNA polymerases having high activity and high fidelity. The characteristics of some widely used DNA polymerases are given later in this chapter.

As mentioned earlier, Brock and colleagues[5] found one important microorganism that could survive at high temperatures and called it *Thermus aquaticus*. Chien and colleagues extracted thermostable DNA polymerase Taq from this microorganism.[7] This enzyme is one of the most important and the first thermostable DNA polymerases used in PCR work. The characteristics of the enzyme include an optimum temperature of 75 °C to 80 °C and half-lives of 9 minutes and 40 minutes at 97.5 °C and 95 °C, respectively. It can amplify a 1-kb strand in roughly 30 to 60 seconds. One of the major disadvantages is that it has low fidelity because it does not have the 3′-to-5′ exonuclease proofreading capacity. As a result, it has one error per 9,000 nucleotides.[8]

The Pfu polymerase is a thermostable DNA polymerase found in the hyperthermophilic archaeon *Pyrococcus furiosus*. Compared to other thermostable enzymes used in PCR, this enzyme has superior thermostability and proofreading properties because it possesses 3′-to-5′ exonuclease activity. The error rate of Pfu polymerase is 1 in 1.3 million base pairs. The disadvantage of this polymerase as compared to the Taq polymerase is the high time requirement, as it needs 1 to 2 minutes to amplify 1 kb of DNA at 72 °C.

The Pwo DNA polymerase is a thermostable polymerase isolated from the hyperthermophilic archaebacterium *Pyrococcus woese*, which has 3′-to-5′ exonuclease, proofreading activity only, and a high thermal stability with a half-life of greater than 2 hours at 100 °C. Because of the 3′-to-5′ exonuclease proofreading activity, Pwo has 18-fold higher fidelity of DNA synthesis compared to Taq polymerase. The rTth DNA polymerase extracted from the recombinant organism *Thermus thermophilus* (rTth) is used in RT to synthesize cDNA efficiently. Similarly, other such enzymes from a number of other microbial sources are used depending on their polymerization rate, proofreading capacities, and the priorities of the user.

FUTURE TRENDS: TRANSGENIC ORGANISMS FOR PRODUCING THERMOSTABLE ENZYMES

The thermostable enzymes are naturally present in organisms living in high-temperature conditions. Extraction of these thermostable enzymes from such microorganisms requires growing pure cultures in high-temperature reactors. An additional problem is that these organisms produce the polymerase enzymes at low levels. Hence higher separation costs are involved. This problem has been overcome by screening for mutants that produce these enzymes at enhanced levels due to mutation[8] and by genetic engineering.[9] Considerable work has been done using recombinant methods to produce transgenic organisms that have a high potential for producing thermostable enzymes efficiently under normal laboratory conditions.

HELICASE-DEPENDENT AMPLIFICATION

The need for thermally stable enzymes arises because of the denaturation of the polymerases during the DNA denaturation step in PCR. In nature, for organisms growing in normal conditions, replication occurs without the need for heat to separate the DNA strands during replication, hence without the need for thermostable DNA polymerase enzymes. In such natural processes, other accessory enzymes are used in different steps of replication. DNA helicase, for example, is an enzyme protein that is used to separate double-stranded DNA into single strands in this process.[10] A new DNA amplification method has been proposed based on the action of this helicase enzyme, which is called helicase-dependent amplification (HDA). In HDA, the DNA helicase is used to generate single-stranded DNA templates to anneal with primers in the next step of amplification. The primer annealing is followed by subsequent extension of primers to produce new double-stranded DNA. This process can then be catalyzed by DNA polymerases that are not necessarily thermostable. This new method of DNA amplification eliminates the need for costly thermostable DNA polymerase and also the need for expensive thermocyclers.

The time requirement for the HDA method is less than that for the normal PCR-based amplification method. This process has a simpler reaction scheme as it can carry out the whole process cycle at normal laboratory temperatures. Simpler and more portable diagnostic devices also may be possible as no thermocycling will be required. Although this new method has these advantages, it has its own disadvantages as well. One major disadvantage of this method is the need for large quantities of DNA. Because sensitivity is a major concern in molecular diagnostics, this method cannot be used for such applications. Another hurdle is the need to develop recombinant organisms that can produce higher quantities of efficient helicase enzymes.

CONCLUSIONS

Heat-stable enzymes are the key ingredients of the PCR; a wide range of enzymes has been identified and modified to match the different requirements demanded of this adaptable technology. Use of the appropriate enzyme can maximize sensitivity and yield as well as minimize polymerization errors and background amplification. Specialized enzymes are available for longer PCRs – templates with high guanine–cytosine content – and for real-time PCR. This constant development is an essential contributing factor to the continued popularity of the PCR. Novel thermophilic bacteria continue to be detected, and it is likely that additional DNA polymerases with novel and useful features for the PCR will be identified and introduced.

REFERENCES

1. Rothschild L, Manicinelli R (2001) Life in extreme environments. *Nature* **409**: 1092–1101.
2. Haki GD, Rakshit SK (2003) Developments in industrially important thermostable enzymes: a review. *Bioresources Technology* **89**(1): 17–34.
3. Blochl E, Rachel R, Burggraf S, Hafenbradl D, Jannasch HW, Stetter KO (1997) *Pyrolobus fumarii*, gen. and sp. *Nov.*, represents a novel group of archaea, extending the upper temperature limit for life to 113 °C. *Extremophiles* **1**: 14–21.
4. Mullis K (1990) The unusual origin of the polymerase chain reaction.*Scientific American* **262**(4): 56–61, 64–65.
5. Brock TD, Freeze H (1969) *Thermus aquaticus* gen. n. and sp. n., a nonsporulating extreme thermophile. *Journal of Bacteriology* **98**: 289–297.
6. Kogan SC, Doherty M, Gitschier J (1987) An improved method for prenatal diagnosis of genetic diseases by analysis of amplified DNA sequences. Application to hemophilia A. *New England Journal of Medicine* **317**: 985–990.
7. Chien A, Edgar DB, Trela JM (1976) Deoxyribonucleic acid polymerase from the extreme thermophile *Thermus aquaticus*. *Journal of Bacteriology* **127**: 1550–1557.
8. Tindall KR, Kunkel TA (1988) Fidelity of DNA synthesis by the *Thermus aquaticus* DNA polymerase. *Biochemistry* **27**: 6008–6013.
9. Yamamoto T, Kazuhisa M, Hiroshi Y, Michio K, Shigeharu F, Masashi K, et al. (2005) Enhancement of thermostability of kojibiose phosphorylase from *Thermoanaerobacter brockii* ATCC35047 by random mutagenesis. *Journal of Bioscience and Bioengineering* **100**(2): 212–215.
10. Browski SD, Birgitte KA (2003) Cloning, expression, and purification of the His6-tagged hyper-thermostable dUTPase from *Pyrococcus woesei* in *Escherichia coli*: application in PCR. *Protein Expression and Purification* **31**: 72–78.
11. Vincent M, Yan X, Huimin K (2004) Helicase-dependent isothermal DNA amplification. *EMBO Reports* **5**(8): 795–800.

3 Inventing molecular beacons

Fred Russell Kramer, Salvatore A. E. Marras, and Sanjay Tyagi

The invention of molecular beacons followed a rather circuitous route. Our laboratory had been studying the remarkable mechanism of replication of the single-stranded genomic ribonucleic acid (RNA) of bacteriophage Qβ, a virus that infects *Escherichia coli*. When a few molecules of Qβ RNA are incubated in a test tube with the viral RNA-directed RNA polymerase, Qβ replicase, millions of copies of each Qβ RNA molecule are generated in only a few minutes by exponential amplification,[1] without primers and without thermal cycling. Unfortunately, Qβ replicase is so specific for the particular sequences and structures present in Qβ RNA that it ignores almost all other nucleic acid molecules, disappointing scientists who would use its extraordinary amplification characteristics to generate large amounts of any desired RNA in vitro. However, our laboratory discovered that if a heterologous RNA sequence is inserted into an appropriate site within Midivariant RNA (MDV-1), which is a naturally occurring small RNA isolated from Qβ-infected *E. coli*[2] that possesses the sequences and structures required for replication,[3] the resulting "recombinant RNA" can be amplified exponentially by incubation with Qβ replicase.[4] This discovery enabled the design of recombinant RNAs that contained inserted hybridization probe sequences,[5] which were employed in the earliest real-time exponential amplification assays, and the use of which, paradoxically, led to the invention of molecular beacons.

Spurred by the emergence of the pernicious infectious agent human immunodeficiency virus (HIV)-1, which is present in as few as 1 in 100,000 peripheral blood mononuclear cells in infected asymptomatic individuals, we developed an assay that was designed to use the exponential amplification of recombinant RNA hybridization probes to measure the number of HIV-1 target molecules present in clinical samples.[6] The basic idea was to insert an HIV-1 probe sequence into the sequence of MDV-1 RNA. The resulting recombinant RNAs were bifunctional in that they served as hybridization probes, but after washing away the RNAs that are not hybridized to target sequences, the remaining recombinant RNAs served as templates for exponential amplification by Qβ replicase. The expectation was that the large number of RNA copies that are generated from each hybridized probe would enable the detection of extremely rare targets.[7]

FIRST REAL-TIME EXPONENTIAL AMPLIFICATION ASSAYS

Implicit in the use of these replicatable probes was the realization that the number of RNA molecules doubles at regular intervals (approximately every 15 seconds) as exponential amplification progresses. Consequently, the amount of time that elapses before a preselected measurable quantity of RNA is synthesized is dependent upon the number of replicatable probes that are bound to targets prior to amplification. Put mathematically, the time it takes to synthesize a predetermined number of amplicons is inversely proportional to the logarithm of the number of target molecules initially present in a sample,[8] enabling accurate measurements to be made over a wide range of initial target concentrations. This relationship applies to all exponential amplification assays, including the polymerase chain reaction (PCR), and it is the principle underlying quantitative real-time PCR. Moreover, we proposed the use of ethidium bromide[9] as a means of providing a fluorescence signal that can be measured by a simple instrument during the course of an amplification reaction to determine, in real time, the number of amplicons synthesized as the amplification progresses.[6] Thus, our laboratory was intimately familiar with the advantages of real-time assays long before the invention of real-time PCR.

Our work was incorporated into the design of the first commercial real-time exponential amplification assays by Gene-Trak Systems,[10] and it was Gene-Trak (now a part of Abbott Laboratories) that developed the first kinetic fluorescence reader that continuously monitored the fluorescence of an intercalating dye in ninety-six sealed reaction tubes, permitting precise quantification of rare nucleic acid targets in clinical samples.[11,12]

THE INTRINSIC PROBLEM OF USING AMPLIFIABLE PROBES

Even though these early real-time exponential amplification reactions yielded precise quantitation of target amounts, and had a large dynamic range like their more modern counterparts, they suffered from a serious drawback – reactions without any target also produced a positive signal.[6] The ability to detect rare target molecules was dependent on washing away all of the amplifiable probes that were not hybridized to targets. However, we found that no matter how many wash steps were employed, and no matter whether we used a sophisticated method for separating probe-target hybrids from nonhybridized probes,[13] some nonhybridized probe molecules remained, and they were amplified along with the probes that were bound to targets, thereby limiting the sensitivity of the assay to approximately 10,000 target molecules.

To overcome this debilitating limitation, we decided to explore various designs for "smart probes," which are probes that can be amplified only if they are hybridized to their target sequence. Consequently, persistent nonhybridized probes (which, by definition, are not bound to target sequences) would not be amplified, and the resulting reactions would be extraordinarily sensitive.[8] It took us seven years to find a smart probe design that worked well. In this particular scheme, the recombinant RNA hybridization probes were cleaved into two sections (through the middle of the inserted probe sequence), with neither section possessing all of the sequences and structures required for exponential amplification. However, when these "binary probes" hybridize to adjacent positions on a target, they can be joined to each other by incubation with a template-directed RNA ligase, generating an exponentially amplifiable reporter. Persistent nonhybridized binary probes, however, are not aligned on a target, so they cannot be ligated, and they do not generate a background signal. Consequently, the resulting assays were extraordinarily sensitive, and provided quantitative results for clinical samples containing as few as ten HIV-1 target molecules.[14]

SMART PROBES CONTAINING MOLECULAR SWITCHES

Before working on binary probes, we explored two other smart probe designs. In one design, the smart probe was not itself an amplifiable molecule. Instead, it was a small oligonucleotide that was needed to initiate a series of steps that led to the synthesis of exponentially amplifiable reporters, but these steps could only occur if the smart probe bound to its target and changed its shape. In the other design, the smart probes were recombinant RNA molecules that could hybridize to targets and be amplified exponentially. However, they also possessed sequences that enabled them to be destroyed by incubation with *E. coli* ribonuclease III. When these probes bind to their target, they undergo a change in shape that eliminates the recognition site for ribonuclease III. The probe-target hybrids are then incubated with ribonuclease III to destroy all of the probes that are not bound to targets.

The common aspect of both of these designs was that they contained a "molecular switch," which is an oligonucleotide segment possessing a probe sequence embedded between two arm sequences that are complementary to each other but not complementary to the target sequence.[8] In the absence of targets, molecular switches form a hairpin-shaped stem-and-loop structure, in which the arms are bound to each other to form a stem hybrid, and the probe sequence is in the single-stranded loop. In the presence of targets, the probe sequence binds to its target sequence. Because of the rigidity of double-helical nucleic acids,[15] the probe-target hybrid formed by the loop of the molecular switch cannot coexist with the stem hybrid formed by the arms of the molecular switch. In effect, the molecular switch must "make a choice" to either retain the stem hybrid and not bind to the target or to undergo a conformational reorganization in which the stem hybrid unwinds and the probe sequence in the loop binds to the target sequence to form a probe-target hybrid. Molecular switches are therefore designed to contain loop sequences that are sufficiently long (or that possess nucleotides that will form sufficiently strong hybrids) to favor the conformational reorganization of the molecular switch in the presence of target sequences that enables probe-target hybrids to form.

In our first design, the smart probe was a hairpin-shaped oligodeoxyribonucleotide possessing a probe sequence in its loop. When this probe binds to its target, it undergoes a conformational reorganization that unwinds its arm sequences. These probes were designed so that the unwound 3′ arm sequence could subsequently be hybridized to the 5′ end of a complementary DNA strand, enabling the open arm to serve as a promoter for the synthesis of an RNA copy of the DNA strand. The idea works as follows: The probes are hybridized to target nucleic acids present in a sample. Nonhybridized probes are then washed away (although a few persist despite vigorous washing). Template DNA is then added to the washed probe-target hybrids. The probes that are bound to targets, of necessity, will have undergone a conformational reorganization. Consequently, their 3′ arms are free to bind to the complementary DNA, forming substrates for the synthesis of RNA by incubation with DNA-directed RNA polymerase. The resulting transcripts were to be MDV-1 RNA, which could then be amplified exponentially by incubation with Qβ replicase, enabling the detection of rare target sequences. The beauty of this approach is that persistent unbound probes that are not washed away will retain their hairpin structure, so their 3′ arm sequences are not available to serve as promoters, and these nonhybridized probes cannot generate a background signal. The problem with this approach was that no matter what we did (and we tried many things), a small amount of MDV-1 RNA was synthesized from the template DNA by the DNA-directed RNA polymerase in the absence of targets, and even in the absence of probes possessing promoter sequences. This approach, despite immense effort, was abandoned, and we never published our findings.

In our second design, a molecular switch, the loop of which contained a probe sequence, was inserted into an appropriate site in MDV-1 RNA. The resulting recombinant RNA hybridization probes included a stem hybrid (enclosing the probe sequence) that served as a double-stranded cleavage site for RNase III.[8] The idea works as follows: Recombinant RNAs are hybridized to target nucleic acids present in a sample. Nonhybridized recombinant RNAs are then washed away (although a few persist despite vigorous washing). The resulting probe-target hybrids are then incubated with RNase III, which ignores all probes that are bound to target sequences, because the molecular switch within the recombinant RNA hybridization probe undergoes a conformational reorganization (to enable the probe to bind to the target) that eliminates the double-stranded RNase III cleavage site. Persistent nonhybridized recombinant RNA probes, in contrast, retain the hairpin stem enclosing the probe sequence, so those molecules are cleaved by RNase III and are therefore unable to serve as templates for exponential amplification. The hybridized probes, in contrast, are not cleaved, and they are amplified exponentially by incubation with Qβ replicase, generating a detectable signal. This approach also failed, as we made the discovery that the binding of recombinant RNA probe sequences to RNA target sequences creates unexpected cleavage sites for RNase III that result in the destruction of the target-bound probes.[16]

MISERY LOVES COMPANY

The situation in the summer of 1992 (when molecular beacons were invented) was that we had developed molecular diagnostic assays based on the exponential amplification of recombinant RNA hybridization probes that had the potential to be extraordinarily sensitive, and that were amenable to being carried out rapidly, in real time, by automated instruments. However, the sensitivity of these probe-amplification assays was seriously compromised by the presence of background signals that obscured the presence of rare targets, and we had not yet developed binary probes, which eventually provided a workable solution to this problem.

We were also aware that molecular diagnostic assays based on PCR, in which the targets, rather than the probes, are amplified exponentially,[17] suffered from a similar problem. The PCR primers, which are probes (in the sense that they are designed to hybridize to the target sequence), can occasionally bind to non-target sequences in the sample, generating a background signal that consists of false amplicons. Moreover, primers can bind to each other, generating "primer–dimers." In either case, these background signals (just like the background signals in Qβ amplification assays) occurred in the absence of targets, and their existence obscured the presence of rare targets. It became clear that a way around this problem in PCR assays was to employ a product-recognition probe that could distinguish the intended amplicons from the false amplicons.[18]

Moreover, if the products of a PCR assay are taken out of the reaction tube for further analysis (such as by gel electrophoresis or by hybridization to product-recognition probes), there is a likelihood that some of the amplicons will escape, contaminating as-yet-untested samples, thereby generating false-positive signals in the contaminated samples.[19] Thus, for PCR to be of wide use in clinical diagnostic laboratories, it was essential that product-recognition probes be present in the reaction mixtures prior to the initiation of exponential amplification, and that they have the ability to generate a detectable signal in sealed reaction tubes.[20]

PROBES THAT BECOME FLUORESCENT UPON HYBRIDIZATION

This is where molecular beacons come in. We had spent years trying to develop extraordinarily sensitive exponential amplification assays in which amplicon synthesis was dependent on whether hairpin-shaped probes present in the reaction tubes hybridize to their intended target sequences. However, all of our efforts to use molecular switches for this purpose had failed. We were also familiar with the advantages of carrying out assays in which amplification products were measured without their removal from the reaction tube, as a few years earlier we had proposed the use of ethidium bromide as a means of generating fluorescence signals that would enable amplicons to be measured in real time.[6] Moreover, this technique had recently been applied to PCR assays.[21,22] Thus, we were keenly aware that the extraordinary power of PCR[23] could be fully realized only if (1) product-recognition probes could be included in the reaction mixture to distinguish intended amplicons from false amplicons, (2) those probes could generate a signal indicative of the amount of probe-target hybrid present as amplification occurs, and (3) those signals could be detected in sealed reaction tubes to avoid sample cross-contamination. If all of these elements could be achieved, PCR would be transformed from a novel research tool into a practical and extraordinarily sensitive clinical diagnostic technique. It is at this point that it struck us that hairpin-shaped probes, if labeled in such a manner as to signal their change in conformation upon binding to their target, could fulfill all of these requirements.

To enable measurements to be made in sealed reaction tubes, product-recognition probes need to generate a distinguishable fluorescence signal only when they become hybridized to the intended amplicons. A promising fluorescence signal-generation technique was fluorescence resonance energy transfer (FRET), which is dependent on the distance between a "donor fluorophore" and an "acceptor fluorophore."[24] Years earlier, it had been shown that, when two different probes are hybridized to a target nucleic acid at adjacent positions (one labeled with a donor fluorophore, and the other labeled with a different acceptor fluorophore), the intensity of the characteristic fluorescent color from the donor fluorophore is reduced and the intensity of the characteristic fluorescent color of the acceptor fluorophore is increased, and this occurs only when the two probes are hybridized to the target at adjacent positions.[25,26] This principle is employed in LightCycler® probes.[27] Alternatively, a pair of complementary oligonucleotides is used.[28] One strand, which serves as the probe, is labeled with a donor fluorophore and is hybridized to a complementary oligonucleotide labeled with an acceptor fluorophore. The fluorophores are linked to the oligonucleotides in such a manner that they are close to each other, decreasing the intensity of the fluorescence of the donor fluorophore. However, when target strands are present, a competition occurs, and some of the probes hybridize to the target strands instead of hybridizing to the complementary strands possessing the acceptor fluorophore. Consequently, the intensity of the fluorescence from the donor fluorophore increases, signaling the presence of the target.

These FRET techniques require two separate probe molecules, each labeled with a differently colored fluorophore. Our concept was that a hairpin-shaped probe labeled with a differently colored fluorophore on the end of each arm sequence would accomplish the same task. When such a probe is free in solution, it forms a hairpin structure, causing the two label moieties to interact. However, when the probe hybridizes to its target, it undergoes a conformational reorganization that unwinds the arm sequences. Because of the rigidity of the resulting probe-target hybrid, the arms are kept far apart, preventing the two label moieties from interacting. Consequently, the fluorescence signal from probes that are hybridized to targets would be distinguishable from the fluorescence signal from nonhybridized probes.

There was, however, one additional property that had to be included before these product-recognition probes would become true "molecular beacons." We wanted hairpin-shaped probes that were not fluorescent when free in solution, but that became fluorescent when they bind to their targets. We therefore conceived of using a pair of label moieties in which the donor was a fluorophore and the acceptor was not able to fluoresce at all. In this labeling scheme, the fluorescence of the donor would be significantly reduced when the donor was in close proximity to the nonfluorescent acceptor (i.e., the acceptor would serve as a "quencher" of the donor's fluorescence). However, when the probe is hybridized to its target, the fluorophore and the quencher would be forced apart from each other, generating a measurable fluorescence signal. That is the origin of molecular beacon probes.[29]

The use of just that sort of label pair had been known for several years in a different context. In an effort to identify the first drugs that could serve as therapeutic agents against HIV-1, Abbott Laboratories developed an assay to identify compounds that inhibit the activity of the viral protease, which is essential for the maturation of the virus. The assay was based on the preparation of a short peptide that serves as a substrate for the protease. The peptide was covalently linked to a blue-emitting fluorophore at one end (EDANS) and to a nonfluorescent quencher (dabcyl) at the other end. These two label moieties served as a well-matched FRET pair. However, because dabcyl is not a fluorescent moiety, the energy stored in the donor (EDANS) is transferred to the acceptor (dabcyl) and is then released as heat, rather than as light of a characteristic color. Because of the proximity of the dabcyl to the EDANS, there was very little fluorescence. When HIV-1 protease is incubated with this dual-labeled peptide, it cleaves the peptide, physically separating the EDANS from the suppressive presence of the nearby dabcyl, leading to a fluorescence signal. Thousands of individual assays were carried out, each possessing dual-labeled peptides and HIV-1 protease, and each containing a different test compound, in the hopes of identifying potential protease inhibitors by their ability to prevent the generation of a fluorescence signal.[30] EDANS and dabcyl possessed the properties that we had been looking for. We therefore decided to label our hairpin-shaped product-recognition probes with EDANS and dabcyl, thereby creating probes that were dark when free in solution but fluorescent when hybridized to their targets.

About the same time, it was shown that small, single-stranded oligonucleotide probes can be labeled with a different fluorophore at each end, and that when those probes are free in solution, they form a random-coil configuration that brings the ends of each oligonucleotide so close to one another that the fluorophores can undergo efficient FRET. However, because of the rigidity of probe-target hybrids, when these linear probes hybridize to their target, the ends are forced apart, FRET is disrupted, and the consequent changes in the fluorescence intensity of each fluorophore signal the presence of the target.[31,32] This key feature was incorporated into linear fluorescent probes that were used in the 5′-nuclease assay[33] to convert them into TaqMan® probes.[34,35]

CONTACT QUENCHING ENABLES MULTICOLOR PROBES

The first molecular beacons, which were labeled with EDANS and dabcyl, achieved a signal-to-background ratio of 25, which was an unqualified success. However, molecular beacons would have remained just a scientific curiosity if EDANS had been the only fluorophore that could have been used. Its limitations soon became apparent. It was only one fiftieth as bright as fluorescein (the most commonly used fluorophore), and its emission range coincided with the autofluorescence of the plastics used in reaction tubes. Furthermore, to realize the full promise of real-time amplification, and to detect multiple targets in the same tube, it is necessary to use a set of probes, each specific for a different target, and each possessing spectrally distinguishable fluorophores. Therefore, we explored the use of other fluorophore-quencher pairs.

For efficient FRET, not only must the acceptor and the donor be a short distance from each other, but the emission spectrum of the donor must substantially overlap the absorption spectrum of the acceptor.[36] This limitation, combined with the desirability that the quencher be nonfluorescent, proved to be too restrictive, and no other appropriate FRET pairs could be found. Hoping that a partial spectral overlap might yield a useful degree of quenching, we synthesized a molecular beacon possessing fluorescein as the donor and dabcyl as the acceptor. The results were pleasantly surprising. We obtained a better quenching efficiency than we had reported for EDANS and dabcyl. Encouraged by this, we tried other fluorophores with dabcyl, including tetramethylrhodamine and Texas red. The emission spectra of both fluorophores were farther toward the red end of the spectrum, and therefore had little or no overlap with the absorption spectrum of dabcyl; yet both fluorophores were efficiently quenched. Indeed, when we tried a series of fluorophores, the emission spectra of which ranged from deep blue to far red, we obtained a uniformly high degree of quenching (signal-to-background ratios between 100 and 1,000), irrespective of the degree to which their emission spectra overlapped the absorption spectrum of dabcyl.[37]

This apparent violation of the cardinal FRET rule prompted us to reexamine the mechanism of fluorescence quenching that takes place in molecular beacons. Fluorescence is a hard-to-achieve property of a molecular moiety, and the introduction of new chemical bonds into that moiety is likely to destroy its ability to fluoresce. Furthermore, we hypothesized that when molecular beacons are in a hairpin conformation, the fluorophore and the quencher are brought so close to each other that they should be able to form chemical bonds, just as nucleotides that are present on complementary strands of a double helix form hydrogen bonds. To test this hypothesis, we compared the visible absorption spectra of molecular beacons in the presence and in the absence of target strands (i.e., we observed whether the combined absorption spectrum of a fluorophore and a quencher, which depends on which chemical bonds are present, changes when the two label moieties are brought close to one another). The results showed that all molecular beacons (even molecular beacons that possess identical fluorophores on either end) have a different visible absorption spectrum, depending upon whether they are "closed" or "open."[37] Thus, the quenching of fluorescence in molecular beacons possessing labels on their 5′ and 3′ ends is not primarily due to FRET, and this realization proved to be extremely useful. Virtually any fluorophore could be used in combination with the same nonfluorescent quencher, and dabcyl was just one example of a "universal quencher."[37] We were therefore able to design extremely sensitive, multiplex, real-time PCR assays containing sets of molecular beacon probes, each of which was specific for a different target sequence, and each of which possessed a differently colored fluorophore in combination with dabcyl.[38,39]

As a consequence of these observations, it became clear to us that other types of probes that were designed to be used in "homogeneous" PCR assays were not limited to the selection of label pairs that interact by FRET. Any probe design that brings a label pair into intimate contact, depending on whether the probe is, or is not, hybridized to its target, can employ label pairs that interact by "contact quenching." It was subsequently confirmed that, when they are free in solution, molecular beacons form a ground-state intramolecular heterodimer, the spectral qualities of which can be accurately described by exciton theory.[40] The quenching moiety need not be a fluorophore or a dye; it can be any moiety that forms transient chemical bonds with a fluorophore – it can even be a guanosine nucleotide.[41,42] Moreover, it does not matter whether the probe design involves single oligonucleotides, such as TaqMan® probes[43] and molecular beacons,[44] or whether two separate oligonucleotides are employed, such as in strand displacement probes[45] and in "molecular zippers."[46] Our results spurred the introduction of novel, highly efficient "dark quenchers" with exotic names, such as "eclipse quenchers" and "black hole quenchers." Most significantly, it was shown that the use of nonfluorescent quenchers that form transient chemical bonds stabilizes the probes, serving the same function as the hairpin stem in molecular beacons, thereby yielding linear (TaqMan®) probes that are almost as well quenched as molecular beacons.[47,48]

Indeed, the widespread use of molecular beacons (and other dual-labeled probes) led to improvements in the materials available for their synthesis. Initially, we prepared molecular beacons by synthesizing oligonucleotides possessing a 5′-terminal sulfhydryl group and a 3′-terminal amino group. Then, in separate reactions, an iodoacetylated fluorophore was covalently linked to the 5′-terminal sulfhydryl group, and a succinimidyl ester of dabcyl was covalently linked to the 3′-terminal amino group. The introduction of controlled-pore glass columns possessing dabcyl as the starting material by Glen Research in 1997, and the parallel introduction of a variety of fluorophore-labeled phosphoramidite precursors, enabled the automated synthesis of molecular beacons.[49]

REAL-TIME PCR ASSAYS WITH MOLECULAR BEACONS

Initially, we had no idea how to design an effective molecular beacon. Our first version had a probe sequence 40 nucleotides long and arm sequences 20 nucleotides long, based on the naïve idea that a probe-target hybrid twice as long as a stem hybrid would drive the opening of the stem. When we tested this construct by the addition of an excess of complementary oligonucleotides, using an ultraviolet view box to see if there was an increase in fluorescence, no fluorescence was detected. However, when we heated the mixture with a hairdryer, the faint blue fluorescence of EDANS appeared in the solution, indicating that, even though the probe did not respond to its target, it was well quenched in its hairpin state. Hypothesizing that the stem might be too strong to open, we tried different stem lengths, and made the astounding discovery that a stem hybrid only 5 base pairs long worked best, responding spontaneously to the target at room temperature. It soon became clear that we had completely missed the fact that the arm sequences, because they are tethered to each other by the probe sequence, are much more likely to bind to each other than they would if they existed separately in solution. Ultimately, we found that molecular beacons that are designed to detect amplicons during the annealing stage of each PCR thermal cycle (which is usually between 55°C and 60°C) should preferably possess probe sequences between 18 and 25 nucleotides in length (depending on the guanosine–cytidine content of the target sequence) and arm sequences between 5 and 7 nucleotides in length (depending on the number of guanosine–cytidine base pairs that they form).

When we carried out our first real-time PCR assays with molecular beacons, no instrument was available that could simultaneously carry out thermal cycling and measure fluorescence. We therefore adopted a strategy in which a cuvette containing the reactants was cyclically transferred from a bath maintained at 95°C to a spectrofluorometer maintained at 37°C, and then to a bath maintained at 72°C. Several hours of "cycling" left us really tired, and only one reaction could be done per day. Later, we purchased a thermal cycler and carried out identical PCR assays containing molecular beacons, terminating each reaction after a different number of thermal cycles, and then measuring the fluorescence intensity in each tube. Although the data were noisy,[29] we could see the key feature of real-time PCR: The number of thermal cycles that needed to be carried out to synthesize sufficient amplicons for the reaction to complete the exponential phase of synthesis and enter the linear phase of synthesis was inversely proportional to the logarithm of the initial number of target strands. These results were particularly gratifying, as the kinetics were the same as those that we had observed years before for the exponential synthesis of RNA by Qβ replicase.[9] Later, in 1996, we formed a research relationship with Applied Biosystems, under which they provided us with the earliest real-time spectrofluorometric thermal cycler (the ABI Prism 7700). The very first PCR assays performed with this instrument yielded smooth amplification curves, enabling us to simultaneously carry out multiplex reactions with differently colored molecular beacons in sealed reaction tubes.[38,39,50,51]

THE EXTRAORDINARY SPECIFICITY OF MOLECULAR BEACONS

We compared the specificity of hairpin-shaped probes to the specificity of corresponding linear probes.[37] We prepared hairpin-shaped probes and hybridized them to target oligonucleotides that were perfectly complementary to the probe sequence in the hairpins. We then measured the stability of the resulting hybrids, as expressed by their melting temperature. We repeated these measurements, using otherwise identical probes, in which the sequence of one of their arms was rearranged so that it could not form a hairpin stem. We found that both the hairpin-shaped probes and the corresponding linear probes formed hybrids that melted apart at the same temperature, indicating that they were equally stable. However, when we repeated the experiments with target oligonucleotides that caused there to be a mismatched base pair in the middle of each hybrid, the results revealed a fundamental and powerful property of molecular beacons: Because both hybrids now possessed a mismatched base pair, they both melted apart at a lower temperature. However, the melting temperature of the hybrids containing the probes that could form a hairpin was much lower than the melting temperature of the corresponding hybrids containing the probes that could not form a hairpin. These results demonstrated that molecular beacons are considerably more "finicky" than corresponding linear probes[52] and are thus ideal for detecting single-nucleotide polymorphisms and other mutations in PCR amplicons.[39,50,53,54]

We carried out an extensive series of experiments to compare the thermodynamic attributes of probe-target hybrids formed by our hairpin-shaped probes to the attributes of probe-target hybrids formed by corresponding linear probes,[55] and the following picture emerged: Probes, just like any other molecules, are most likely to assume the most stable state possible under a given set of conditions. The presence of a mismatched base pair causes probe-target hybrids to be less stable, but has no effect on the stability of the stem hybrid that can form in hairpin-shaped probes if they are no longer hybridized to a target. When we design molecular beacons that are intended to discriminate against single-nucleotide polymorphisms in a target, we select the length (and strength) of the probe sequence and the length (and strength) of the arm sequences with two criteria in mind: under the detection conditions (the annealing temperature) of the PCR assay, the molecular beacon should assume a nonfluorescent, stable hairpin conformation; and in the presence of perfectly complementary targets, the resulting probe-target hybrid should be just a little more stable than the hairpins by themselves. If a mutation is present in the target, thereby lowering the stability of the potential probe-target hybrid, the molecular beacon will prefer to remain in the nonfluorescent hairpin configuration, which is more stable than the mismatched probe-target hybrid that could be formed under those conditions. Our experiments showed that the reason that corresponding linear probes are not as useful for discriminating mutations is that, unlike probes that can form a hairpin stem, they do not have an alternative stable state to assume in the face of the destabilizing presence of a mismatched base pair in the hybrid, so they tend to remain bound to the targets despite the presence of a mismatched base pair. Thus, the discriminatory power of molecular beacons is an example of "stringency clamping,"[56] which is based on the fundamental principle that the specificity of any intermolecular interaction is significantly higher if one or both of the interacting molecules has the possibility of forming an alternative stable structure, rather than forming an intermolecular complex.[55,57]

Of course, linear probes (such as TaqMan® probes) can also be designed so that they distinguish single-nucleotide polymorphisms. However, because there is not such a great difference in stability between a perfectly complementary hybrid formed by a linear probe and a hybrid containing a single mismatched base pair formed by the same probe, it is quite difficult to find conditions under which a set of linear probes, each specific for a different target, can all display the same degree of specificity in the same reaction tube. Molecular beacons, in contrast, because they form alternative stable structures, rather than forming mismatched probe-target hybrids, easily can be designed so that many different molecular beacon probes, each specific for a different target, can be used under the same reaction conditions in a sealed reaction tube; and each molecular beacon will be so specific that it will bind only to a perfectly complementary target sequence. What this means in practice is that four different molecular beacons can be designed to distinguish four different nucleotides that can occur in the same position in an otherwise identical target sequence. In addition, they can be used together in the same PCR assay tube, and each will generate a fluorescence signal (in a color determined by each probe's fluorophore) only when that probe binds to its perfectly complementary target. The other three probes will not generate a signal if their target sequence is not present.[39] Indeed, because it is easy to design highly specific molecular beacons that bind only to complementary target sequences, even though they are used in the same reaction tube, we have been able to develop PCR screening assays that contain fifteen different species-specific probes (each labeled with a distinctive combinatorial color code) that can rapidly identify a sepsis-causing bacterium that is present in a (normally sterile) blood sample.[58] Moreover, we are developing "molecular blood cultures" that use thirty-five different highly specific, color-coded molecular beacons to simultaneously screen for the presence of thirty-five different infectious agents in the same PCR assay tube. Finally, because of the extraordinary specificity of molecular beacons, they are ideal probes to attach to the surface of oligonucleotide hybridization arrays, because many different molecular beacons can all be designed so that they discriminate target sequences under the same set of conditions.

MOLECULAR BEACONS AS BIOSENSORS

When a conventional single-stranded oligonucleotide probe hybridizes to a target nucleic acid sequence, very little in the way of a physical change occurs that enables one to determine that the target is present. Molecular beacons, in contrast, undergo a conformational reorganization when they bind to their target that separates a fluorophore from a quencher, thereby generating a bright, easily detectable fluorescence signal of a characteristic color. Molecular beacons are therefore classic "biosensors," and they can be adapted for use in a variety of different analytical systems. They can be linked to other macromolecules to create multifunctional detectors; they can be linked to surfaces for use in diagnostic arrays; and they can be synthesized from unnatural nucleotides so that they can function without being destroyed in living cells. Here are a few examples.

In addition to their use in conventional PCR assays, molecular beacons are ideal for detecting amplicons in isothermal gene amplification assays, such as those that use nucleic acid sequence-based amplification (NASBA)[59–61] or rolling-circle amplification.[62,63] Moreover, they have been used in digital PCR assays,[64] linear-after-the-exponential (LATE)-PCR assays,[65,66] and for the detection of RNA during transcription.[67] Finally, in a new paradigm termed "multiprobe species typing," sets of differently colored "sloppy molecular beacons" possessing unusually long probe sequences have been used in screening assays, in conjunction with a determination of the thermal melting characteristics of the probe-target hybrids,[68] to generate species-specific signatures that uniquely identify which infectious agent (from a long list) is present in a clinical sample.[69]

Molecular beacons have been covalently linked to other macromolecules to create bifunctional biosensors. A classic example is "Amplifluor® primers," which are molecular beacons linked to the 5′ ends of PCR primers.[70] After an Amplifluor® primer has bound to its target strand and been extended, the resulting amplicon then serves as a template for the synthesis of a complementary strand, causing the molecular beacon to open, generating a fluorescence signal that is detected at the end of the polymerization step in each PCR cycle. In a clever variant, called "LUX™ primers," guanosine nucleotides serve as the quenchers.[71] An even more intriguing combination occurs in "Scorpion® primers," which are molecular beacons linked to primers via a blocker group that prevents the molecular beacon from being copied during PCR.[72] The probe sequence in the molecular beacon segment is designed to hybridize to a target sequence in the amplicon strand created by the extension of the primer segment, causing the molecular beacon to open, thereby generating a fluorescence signal that is detected at the end of the annealing step in each PCR cycle. An advantage of Scorpion® primers, as compared to Amplifluor® primers and LUX™ primers, is that they do not fluoresce when incorporated into false amplicons. Molecular beacons also have been covalently linked to viral peptides, enabling the probes to cross cell membranes to light up target messenger RNAs (mRNAs) in living cells.[73] In addition, molecular beacons have been linked to transfer RNAs to prevent the probes from becoming sequestered in nuclei, thereby enabling target mRNAs to be detected in the cytoplasm of living cells.[74] Finally, molecular beacons have been linked to an additional sequence that serves as a capture probe. The resulting "tentacle probes" combine the high affinity of capture probes with the extraordinary specificity of molecular beacons.[75]

A key feature of molecular beacon probes is that they can be designed so that they cannot fluoresce when they are in their hairpin conformation, because the fluorophore and the quencher interact by contact quenching to form a transient "ground-state" heterodimer that is not fluorescent.[40,47] However, the function of some molecular beacons has been enhanced by the incorporation of additional elements that interact by FRET. For example, some spectrofluorometric thermal cyclers use a blue argon ion laser to stimulate fluorescence, which efficiently excites blue and green fluorophores, but inefficiently excites orange and red fluorophores. To remedy this inefficiency, molecular beacons were designed that include a secondary fluorophore that interacts with the primary fluorophore by FRET. When these "wavelength-shifting molecular beacons" bind to their target, they undergo a conformational reorganization that enables the energy that was efficiently absorbed from the blue laser light by a primary fluorophore to be transferred by FRET to a secondary fluorophore, resulting in the generation of a bright fluorescence signal in orange or red.[76] By combining wavelength-shifting molecular beacons with conventional molecular beacons, highly multiplex clinical diagnostic PCR assays have been developed that contain differently colored probes that span the entire visible spectrum and are designed to be used with spectrofluorometric thermal cyclers that contain powerful monochromatic lasers.[77] An additional advantage of wavelength-shifting molecular beacons is that they have a large Stokes shift, which means that the color of their fluorescence is different from the color of the light that stimulates their fluorescence, thus enabling the detection of the fluorescence signal with little interference from the stimulating light. This principle was incorporated into "dual-FRET" probes, which are pairs of molecular beacons (one possessing a donor fluorophore on its 5′ end, and the other possessing an acceptor fluorophore on its 3′ end), which bind to adjacent target sequences on mRNAs in living cells.[78,79] Although the fluorophore and quencher in each of the molecular beacons in the pair interact by contact quenching, nucleic acid–binding proteins in the cell occasionally unwind nonhybridized probes, and cellular nucleases occasionally cleave nonhybridized probes, creating background signals from the fluorophores on the probes. However, when a pair of dual-FRET molecular beacons are bound to adjacent positions on a target mRNA, the two fluorophores interact by FRET, creating a unique fluorescence signal that can be distinguished from the background fluorescence.

Ideal biosensors are designed to respond to the presence of their targets but should be unaffected by the environment in which they operate. Molecular beacons, however, are usually made of DNA or RNA, and they are often used in reaction mixtures that contain enzymes, the substrates of which are nucleic acids. For example, if molecular beacons composed of deoxyribonucleotides are used to monitor the synthesis of RNA from a double-stranded DNA template by a DNA-directed RNA polymerase, the molecular beacons themselves are copied, generating a background signal that mimics the synthesis of the transcripts.[67] It is therefore desirable for some assays to use molecular beacons that contain unnatural nucleotides, rendering them resistant to enzymatic activity. For example, molecular beacons have been synthesized that contain peptide nucleic acid (PNA) monomers,[80,81,82] 2′-*O*-methyl ribonucleotides,[83,84] and locked nucleic acid (LNA) monomers.[85] By minimizing background signals, the use of these modified molecular beacons ensures that the fluorescence signal reflects the process or product that is being measured.[67] Molecular beacons synthesized from unnatural nucleotides, such as the 2′-*O*-methyl ribonucleotides, are of particular importance when imaging mRNA targets in living cells; otherwise the probes (and sometimes the mRNA to which the probes bind) will be destroyed by cellular nucleases.[74,78,86]

Hundreds, and even thousands, of molecular beacons, each specific for a different target sequence, can be used simultaneously in a single assay by attaching them to predetermined locations on the surface of hybridization arrays,[87,88] or by attaching them to the surfaces of beads in "distributed arrays,"[89] in which each bead possesses molecular beacons that identify a different nucleic acid target sequence. There are two key advantages of using molecular beacons on hybridization arrays. First, because molecular beacons can be designed to be extraordinarily specific, all of the molecular beacons in the array function well under the same set of hybridization conditions. Second, because each molecular beacon contains a label pair (usually a fluorophore and a quencher), there is no need to carry out preliminary reactions to label the nucleic acid mixtures that are analyzed on the arrays. Molecular beacon arrays are therefore extraordinarily specific, "self-reporting" nucleic acid analyzers, in which the intensity of the fluorescence that develops at each position on the surface of the array (or on each bead in a distributed array) is directly proportional to the abundance of that molecular beacon's target sequence in the nucleic acid mixture being analyzed.

Molecular beacons can be attached to many different materials, including the surfaces of planar arrays, fiber-optic nucleic acid detection devices,[90] glass beads,[89,91] agarose gel membranes,[92] and bar-coded nanowires.[93] An example of a suitable linkage is the binding of molecular beacons that possess a biotin moiety to array surfaces coated with avidin or streptavidin.[80,94] As an alternative to the inclusion of a quencher in each molecular beacon, the surface to which the molecular beacons are linked can sometimes serve as the quencher. For example, gold surfaces interact with fluorophores to prevent fluorescence, but when the probe sequence in the molecular beacon binds to a target sequence, the rigidity of the resulting probe-target hybrid forces the fluorophore away from the surface, generating a fluorescence signal.[95,96] Moreover, molecular beacons have been designed in which the fluorophore and the quencher are replaced with moieties that alter an electrical signal when the binding of the probe to its target causes an electroactive reporter on one end of the molecular beacon to lift away from the surface to which the probe is attached.[97] The challenge facing developers of these devices and arrays is to use surfaces and attachment chemistries that do not negatively affect the ability of the molecular beacons to bind to their target sequences. In addition, because it is relatively expensive to prepare many different probes that possess a covalently linked fluorophore and quencher and a third moiety for binding the probes to the surface of arrays, novel designs for molecular beacons have been explored in which the fluorophore and quencher are incorporated into "universal oligonucleotides" that are then bound to easily prepared, unlabeled, target-specific oligonucleotides to generate the probes for the arrays.[98,99]

One of the most intriguing adaptations of the molecular beacon concept is the development of "aptamer beacons," which are oligonucleotide probes containing a fluorophore and a quencher that bind specifically to proteins, rather than to nucleic acid sequences,[100–102] or that bind specifically to small molecules, such as cocaine.[103] The defining feature of aptamer beacons is that they undergo a conformational reorganization when they bind to their target that changes the distance between a fluorophore and a quencher, generating a detectable signal. Some of the most innovative designs for signaling aptamers involve the use of two labeled oligonucleotides whose relationship to one another is altered by their binding to a target molecule.[104–107] Moreover, sophisticated procedures have been devised to select desired aptamers in vitro from large pools of oligonucleotides possessing random sequences.[108,109] Finally, "peptide beacons" have been developed, in which a peptide undergoes a conformational reorganization when it binds to its target protein, generating a detectable fluorescence signal.[110]

OBSERVING THE MOVEMENT OF mRNAs IN LIVE CELLS

Perhaps the most thrilling application of molecular beacons as biosensors is their use as probes to visualize mRNAs in living cells. Because molecular beacons that are not bound to mRNAs in live cells are dark and physically dispersed, their background fluorescence is low. In contrast, when molecular beacons bind to mRNA targets, they fluoresce brightly; and if the target mRNAs are localized in particular areas in the cell, those regions are highlighted when viewed with a fluorescence microscope. Initially, the molecular beacons used in live cells were synthesized from deoxyribonucleotides.[111–113] Later, a series of modifications in the design of the molecular beacons significantly lowered fluorescence background. When it was realized that molecular beacons can be digested by cellular nucleases, and that target mRNAs that are bound to molecular beacons can be digested by cellular ribonuclease H, which cleaves RNA in DNA:RNA hybrids, the molecular beacon probes were modified by synthesizing them from unnatural 2′-*O*-methyl ribonucleotides[83,84] or from PNAs,[82] which do not serve as substrates for cellular nucleases. In addition, when it was realized that cellular nucleic acid–binding proteins could open molecular beacons that are not hybridized to targets, dual-FRET probes were developed[78,79] that bind to adjacent positions on a target mRNA, generating a FRET signal that can be distinguished from the background fluorescence of each probe by itself. Finally, when it was realized that molecular beacons are rapidly sequestered in cell nuclei, the probes were tethered to the protein streptavidin, which cannot pass through pores in the nuclear membrane[86,114] or to tRNA, which is actively retained in the cytoplasm.[74]

These improvements created a biosensor toolbox that enables mRNAs in living cells to be specifically lit up in a chosen fluorescent color, to directly observe their synthesis, movement, and localization. For example, the movement and localization of *oskar* mRNA from nurse cells to the posterior pole of developing fruit fly oocytes was observed in real time,[78] demonstrating that molecular beacons enable mRNAs to be studied in much the same way that green fluorescent protein enables proteins to be studied in live cells. Using molecular beacons to light up β-actin mRNA in live cultured chicken embryo fibroblasts, which move about on glass surfaces by extending new lamellipodia (pseudopods) while withdrawing old lamellipodia, movies were taken that show β-actin mRNAs moving out of the shrinking lamellipodia and into the growing lamellipodia, where they are translated into β-actin protein, which is needed for the cell to move.[86] Finally, a modified gene was cloned into the genome of Chinese hamster ovary cells that could be experimentally induced to synthesize mRNAs containing a tandem array of ninety-six identical molecular beacon binding sites in their 3′-untranslated regions. In the presence of molecular beacons whose probe sequences are complementary to these inserted binding sites, ninety-six molecular beacons hybridize to each mRNA, creating a probe-target complex so bright that the movement of individual mRNA molecules could be followed in the living cells. Careful observations confirmed that, contrary to widely held beliefs, the mRNAs (each of which forms an individual complex with nuclear proteins), rather than exiting the nucleus through a nuclear pore near the site of their synthesis, move rapidly by Brownian motion throughout the interchromatin spaces within the nucleus, and eventually exit the nucleus through whatever nuclear pore they eventually encounter.[115] Sets of molecular beacons, each specific for different regions of the same primary gene transcript, and each labeled with a differently colored fluorophore, are now being used to assemble a detailed description of where in living cells mRNA splicing, maturation, transport, localization, and decay occur.

PRACTICAL APPLICATIONS

Molecular beacons are primarily employed as highly specific amplicon detection probes in homogeneous, real time, multiplex gene amplification assays. In addition to applications in basic research, molecular beacons are used for the detection of infectious agents in food, in donated blood, and in agricultural, veterinary, and environmental samples. Molecular beacons are also used in forensics and paternity testing – and they are even used for the detection of DNA markers added to products to prevent counterfeiting.[116] An extensive list of publications describing the many applications of molecular beacons is available at http://www.molecular-beacons.org.

By far, the most significant applications for molecular beacons occur in the field of human in vitro diagnostics. Hundreds of different PCR assays have been designed (and many have been commercialized). There are assays for the detection of specific genes[117] and specific mRNAs[118,119]; there are assays for the detection of mutations that cause genetic diseases[50,120–125]; there are assays that identify somatic mutations associated with cancer[126–128] and that provide guidance for determining prognosis and appropriate treatment[129,130]; and most importantly, there are assays that identify and quantitate an extraordinarily wide range of different infectious agents in clinical samples.[131]

PCR assays have been developed for the detection of viruses, including pathogenic retroviruses,[38,132] adenoviruses,[133–135] papillomaviruses,[136,137] cytomegaloviruses,[138] respiratory viruses,[139,140] and hepatitis viruses.[141–143] PCR assays also have been developed for the detection of pathogenic bacteria, including *Mycobacterium tuberculosis,*[53,77,144] *Salmonella,*[145] *Bordetella pertussis,*[146] *Shigella dysenteriae* and *E. coli* O157:H7,[147,148] *Clostridium difficile,*[149] *Vibrio cholerae,*[150] antibiotic-resistant *Staphylococcus aureus* strains,[151,152] bacterial bioterrorism agents (*Bacillus anthracis*, *Yersinia pestis*, *Burkholderia mallei*, and *Francisella tularensis*),[153] *Chlamydia trachomatis* and *Neisseria gonorrhoeae,*[131] and bacteria that cause pneumonia.[154–156] Lastly, PCR assays have been developed for the detection of pathogenic fungi such as *Candida dubliensis*[157] and *Aspergillus fumigatus*[158] and for the detection of pathogenic protozoa that cause malaria[159,160] and dysentery.[161]

NASBA assays are particularly amenable to the use of molecular beacons[59] because the vast majority of the amplicons produced by this isothermal exponential amplification process are single-stranded "plus" RNAs, rather than the complementary "plus and minus" DNA strands synthesized during PCR. Therefore, unlike the situation that occurs during PCR, there is no competition between the molecular beacon probes and the minus strands for binding to the target (plus) strands, and a significantly greater portion of the target strands are lit up by the molecular beacons. Assays that use LATE-PCR, in which far more plus strands than minus strands are synthesized,[65,66] have a similar advantage.

NASBA assays containing molecular beacons can distinguish single-nucleotide polymorphisms in human DNA[162] and can identify mutant mRNAs associated with cancer.[163] NASBA assays have been developed for the detection of viruses, including HIV-1,[60,61,164,165] West Nile virus and St. Louis encephalitis virus,[166] herpesvirus,[167] cytomegalovirus,[168] hepatitis viruses,[169,170] respiratory viruses,[171–173] enteroviruses,[174,175] severe acute respiratory syndrome (SARS) virus,[176] and papillomaviruses.[177] NASBA assays also have been developed for the detection of pathogenic bacteria, including bacteria that cause pneumonia,[178–180] *Vibrio cholerae,*[181] and bacteria that contaminate food.[182–185] Lastly, NASBA assays have been developed for the detection of *Plasmodium falciparum.*[186]

We take particular pride in assays that have received the approval of regulatory agencies for medical diagnostic use, including a NASBA assay used throughout the developing world for quantitating viral load in people infected with HIV-1[164,187] and a widely used PCR assay for detecting the presence of methicillin-resistant *S. aureus* in people entering hospitals,[188,189] the use of which has curtailed the spread of nosocomial infections.[190,191]

The significance of gene amplification assays that use molecular beacons is illustrated by the widespread availability of a PCR assay that detects the presence of group B *Streptococci* in samples taken from pregnant women.[192] Babies born to infected mothers can develop meningitis, which can cause blindness, deafness, and death. After performing this assay on women entering the hospital to give birth, those women infected are treated with antibiotics that cross the placenta and prevent the development of meningitis.[193] It is thus particularly gratifying to see that what began as a basic research program to explore the mechanism by which bacteriophage Qβ RNA is amplified exponentially has led (circuitously) to the invention of simple, elegant, and extraordinarily specific biosensors that enhance the use of exponential amplification assays for beneficial medical purposes.

REFERENCES

1. Haruna I, Spiegelman S (1965) The autocatalytic synthesis of a viral RNA *in vitro*. *Science* **150**: 884–886.
2. Kacian DL, Mills DR, Kramer FR, Spiegelman S (1972) A replicating RNA molecule suitable for a detailed analysis of extracellular evolution and replication. *Proceedings of the National Academy of Sciences of the United States of America* **69**: 3038–3042.
3. Nishihara T, Kramer FR (1983) Localization of the Qβ replicase recognition site in MDV-1 RNA. *Journal of Biochemistry* **93**: 669–674.
4. Miele EA, Mills DR, Kramer FR (1983) Autocatalytic replication of a recombinant RNA. *Journal of Molecular Biology* **171**: 281–295.
5. Lizardi PM, Guerra CE, Lomeli H, Tussie-Luna I, Kramer FR (1988) Exponential amplification of recombinant RNA hybridization probes. *Nature Biotechnology* **6**: 1197–1202.
6. Lomeli H, Tyagi S, Pritchard CG, Lizardi PM, Kramer FR (1989) Quantitative assays based on the use of replicatable hybridization probes. *Clinical Chemistry* **35**: 1826–1831.
7. Chu BC, Kramer FR, Orgel LE (1986) Synthesis of an amplifiable reporter RNA for bioassays. *Nucleic Acids Research* **14**: 5591–5603.
8. Kramer FR, Lizardi PM (1989) Replicatable RNA reporters. *Nature* **339**: 401–402.
9. Kramer FR, Mills DR, Cole PE, Nishihara T, Spiegelman S (1974) Evolution *in vitro*: sequence and phenotype of a mutant RNA resistant to ethidium bromide. *Journal of Molecular Biology* **89**: 719–736.

10. Pritchard CG, Stefano JE (1991) Detection of viral nucleic acids by Qβ replicase amplification. In: LM de la Maza and EM Peterson (eds), Medical Virology 10, pages 67–82. New York: Plenum Press.
11. Shah JS, Liu J, Smith J, Popoff S, Radcliffe G, O'Brien WJ, et al. (1994) Novel, ultra-sensitive, Q-beta replicase-amplified hybridization assay for detection of *Chlamydia trachomatis*. *Journal of Clinical Microbiology* **32**: 2718–2724.
12. Burg JL, Cahill PB, Kutter M, Stefano JE, Mahan DE (1995) Real-time fluorescence detection of RNA amplified by Qβ replicase. *Analytical Biochemistry* **230**: 263–272.
13. Morrissey DV, Lombardo M, Eldredge JK, Kearney KR, Groody EP, Collins ML (1989) Nucleic acid hybridization assays employing dA-tailed capture probes. I. Multiple capture methods. *Analytical Biochemistry* **181**: 345–359.
14. Tyagi S, Landegren U, Tazi M, Lizardi PM, Kramer FR (1996) Extremely sensitive, background-free gene detection using binary probes and Q-beta replicase. *Proceedings of the National Academy of Sciences of the United States of America* **93**: 5395–5400.
15. Shore D, Langowski J, Baldwin RL (1981) DNA flexibility studied by covalent closure of short fragments into circles. *Proceedings of the National Academy of Sciences of the United States of America* **78**: 4833–4837.
16. Blok HJ, Kramer FR (1997) Amplifiable hybridization probes containing a molecular switch. *Molecular and Cellular Probes* **11**: 187–194.
17. Saiki RK, Gelfand DH, Stoffel S, Scharf SJ, Higuchi R, Horn GT, et al. (1988) Primer-directed enzymatic amplification of DNA with a thermostable DNA polymerase. *Science* **239**: 487–491.
18. Abbott MA, Poiesz BJ, Byrne BC, Kwok S, Sninsky JJ, Ehrlich GD (1988) Enzymatic gene amplification: qualitative and quantitative methods for detecting proviral DNA amplified *in vitro*. *Journal of Infectious Diseases* **158**: 1158–1169.
19. Kwok S, Higuchi R (1989) Avoiding false positives with PCR. *Nature* **339**: 237–238.
20. Holland PM, Abramson RD, Watson R, Gelfand DH (1991) Detection of specific polymerase chain reaction product by utilizing the 5′–3′ exonuclease activity of *Thermus aquaticus* DNA polymerase. *Proceedings of the National Academy of Sciences of the United States of America* **88**: 7276–7280.
21. Higuchi R, Dollinger G, Walsh PS, Griffith R (1992) Simultaneous amplification and detection of specific DNA sequences. *Nature Biotechnology* **10**: 413–417.
22. Higuchi R, Fockler C, Dollinger G, Watson R (1993) Kinetic PCR analysis: real-time monitoring of DNA amplification reactions. *Nature Biotechnology* **11**: 1026–1030.
23. Erlich HA, Gelfand D, Sninsky JJ (1991) Recent advances in the polymerase chain reaction. *Science* **252**: 1643–1651.
24. Stryer L, Haugland RP (1967) Energy transfer: a spectroscopic ruler. *Proceedings of the National Academy of Sciences of the United States of America* **58**: 719–726.
25. Heller MJ, Morrison LE (1985) Chemiluminescent and fluorescent probes for DNA hybridization systems. In: DT Kingsbury and S Falkow (eds), Rapid Detection and Identification of Infectious Agents, pages 245–256. New York: Academic Press.
26. Cardullo RA, Agrawal S, Flores C, Zamecnik PC, Wolf DE (1988) Detection of nucleic acid hybridization by nonradiative fluorescence resonance energy transfer. *Proceedings of the National Academy of Sciences of the United States of America* **85**: 8790–8794.
27. Wittwer CT, Herrmann MG, Moss AA, Rasmussen RP (1997) Continuous fluorescence monitoring of rapid cycle DNA amplification. *BioTechniques* **22**: 130–138.
28. Morrison LE, Halder TC, Stols LM (1989) Solution-phase detection of polynucleotides using interacting fluorescent labels and competitive hybridization. *Analytical Biochemistry* **183**: 231–244.
29. Tyagi S, Kramer FR (1996) Molecular beacons: probes that fluoresce upon hybridization. *Nature Biotechnology* **14**: 303–308.
30. Matayoshi ED, Wang GT, Krafft GA, Erickson J (1990) Novel fluorogenic substrates for assaying retroviral proteases by resonance energy transfer. *Science* **247**: 954–958.

31. Parkhurst KM, Parkhurst LJ (1993) Kinetic studies of oligonucleotide-DNA hybridization in solution by fluorescence resonance energy transfer. Abstract W-Pos 97 presented at the 37th Annual Meeting of the Biophysical Society, Washington, DC. *Biophysical Journal* **64**: A266.
32. Parkhurst KM, Parkhurst LJ (1995) Kinetic studies by fluorescence resonance energy transfer employing a double-labeled oligonucleotide: hybridization to the oligonucleotide complement and to single-stranded DNA. *Biochemistry* **34**: 285–292.
33. Lee LG, Connell CR, Bloch W (1993) Allelic discrimination by nick-translation PCR with fluorogenic probes. *Nucleic Acids Research* **21**: 3761–3766.
34. Livak KJ, Flood SJA, Marmaro J, Giust W, Deetz K (1995) Oligonucleotides with fluorescent dyes at opposite ends provide a quenched probe system useful for detecting PCR product and nucleic acid hybridization. *PCR Methods and Applications* **4**: 357–362.
35. Heid CA, Stevens J, Livak KJ, Williams PM (1996) Real time quantitative PCR. *Genome Research* **6**: 986–994.
36. Haugland RP, Yguerabide J, Stryer L (1969) Dependence of the kinetics of singlet-singlet energy transfer on spectral overlap. *Proceedings of the National Academy of Sciences of the United States of America* **63**: 23–30.
37. Tyagi S, Bratu DP, Kramer FR (1998) Multicolor molecular beacons for allele discrimination. *Nature Biotechnology* **16**: 49–53.
38. Vet JA, Majithia AR, Marras SAE, Tyagi S, Dube S, Poiesz BJ, et al. (1999) Multiplex detection of four pathogenic retroviruses using molecular beacons. *Proceedings of the National Academy of Sciences of the United States of America* **96**: 6394–6399.
39. Marras SAE, Kramer FR, Tyagi S (1999) Multiplex detection of single-nucleotide variations using molecular beacons. *Genetic Analysis* **14**: 151–156.
40. Bernacchi S, Mély Y (2001) Exciton interaction in molecular beacons: a sensitive sensor for short range modifications of the nucleic acid structure. *Nucleic Acids Research* 29: e62.
41. Knemeyer J-P, Marmé N, Sauer M (2000) Probes for detection of specific DNA sequences at the single-molecule level. *Analytical Chemistry* **72**: 3717–3724.
42. Crockett AO, Wittwer CT (2001) Fluorescein-labeled oligonucleotides for real-time PCR: using the inherent quenching of deoxyguanosine nucleotides. *Analytical Biochemistry* **290**: 89–97.
43. Nasarabadi S, Milanovich F, Richards J, Belgrader P (1999) Simultaneous detection of TaqMan probes containing FAM and TAMRA reporter fluorophores. *BioTechniques* **27**: 1116–1118.
44. Marras SAE, Kramer FR, Tyagi S (2002) Efficiencies of fluorescence resonance energy transfer and contact-mediated quenching in oligonucleotide probes. *Nucleic Acids Research* **30**: e122.
45. Li Q, Luan G, Guo Q, Liang J (2002) A new class of homogeneous nucleic acid probes based on specific displacement hybridization. *Nucleic Acids Research* 30: e5.
46. Yi J, Zhang W, Zhang DY (2006) Molecular zipper: a fluorescent probe for real-time isothermal DNA amplification. *Nucleic Acids Research* **34**: e81.
47. Johansson MK, Fidder H, Dick D, Cook RM (2002) Intramolecular dimers: a new strategy to fluorescence quenching in dual-labeled oligonucleotide probes. *Journal of the American Chemical Society* **124**: 6950–6956.
48. Moreira RG, You Y, Behlke MA, Owczarzy R (2005) Effects of fluorescent dyes, quenchers, and dangling ends on DNA duplex stability. *Biochemical and Biophysical Research Communications* **327**: 473–484.
49. Mullah B, Livak K (1999) Efficient automated synthesis of molecular beacons. *Nucleosides, Nucleotides & Nucleic Acids* **18**: 1311–1312.
50. Giesendorf BAJ, Vet JAM, Tyagi S, Mensink EJMG, Trijbels FJM, Blom HJ (1998) Molecular beacons: a new approach for semiautomated mutation analysis. *Clinical Chemistry* **44**: 482–486.

51. Kostrikis LG, Tyagi S, Mhlanga MM, Ho DD, Kramer FR (1998) Spectral genotyping of human alleles. *Science* **279**: 1228–1229.
52. Täpp I, Malmberg L, Rennel E, Wik M, Syvänen A-C (2000) Homogeneous scoring of single-nucleotide polymorphisms: comparison of the 5′-nuclease TaqMan assay and molecular beacon probes. *BioTechniques* **28**: 732–738.
53. Piatek AS, Tyagi S, Pol AC, Telenti A, Miller LP, Kramer FR, et al. (1998) Molecular beacon sequence analysis for detecting drug resistance in *Mycobacterium tuberculosis*. *Nature Biotechnology* **16**: 359–363.
54. Mhlanga MM, Malmberg L (2001) Using molecular beacons to detect single-nucleotide polymorphisms with real-time PCR. *Methods* **25**: 463–471.
55. Bonnet G, Tyagi S, Libchaber A, Kramer FR (1999) Thermodynamic basis of the enhanced specificity of structured DNA probes. *Proceedings of the National Academy of Sciences of the United States of America* **96**: 6171–6176.
56. Roberts RW, Crothers DM (1991) Specificity and stringency in DNA triplex formation. *Proceedings of the National Academy of Sciences of the United States of America* **88**: 9397–9401.
57. Broude NE (2002) Stem-loop oligonucleotides: a robust tool for molecular biology and biotechnology. *Trends in Biotechnology* **20**: 249–256.
58. Marras SAE, Antson D-O, Tyagi S, Kramer FR (2010) Highly multiplex PCR screening assays that utilize color-coded molecular beacons for the identification of bacterial species. In preparation.
59. Leone G, van Schijndel H, van Gemen B, Kramer FR, Schoen CD (1998) Molecular beacon probes combined with amplification by NASBA enable homogeneous, real-time detection of RNA. *Nucleic Acids Research* **26**: 2150–2155.
60. de Baar MP, van Dooren MW, de Rooij E, Bakker M, van Gemen B, Goudsmit J, et al. (2001) Single rapid real-time monitored isothermal RNA amplification assay for quantification of human immunodeficiency virus type 1 isolates from groups M, N, and O. *Journal of Clinical Microbiology* **39**: 1378–1384.
61. de Baar MP, Timmermans EC, Bakker M, de Rooij E, van Gemen B, Goudsmit J (2001) One-tube real-time isothermal amplification assay to identify and distinguish human immunodeficiency virus type 1 subtypes A, B, C and circulating recombinant forms AE and AG. *Journal of Clinical Microbiology* **39**: 1895–1902.
62. Nilsson M, Gullberg M, Dahl F, Szuhai K, Raap AK (2002) Real-time monitoring of rolling-circle amplification using a modified molecular beacon design. *Nucleic Acids Research* **30**: e66.
63. Alsmadi OA, Bornarth CJ, Song W, Wisniewski M, Du J, Brockman JP, et al. (2003) High accuracy genotyping directly from genomic DNA using a rolling circle amplification based assay. *BMC Genomics* **4**: 21.
64. Vogelstein B, Kinzler KW (1999) Digital PCR. *Proceedings of the National Academy of Sciences of the United States of America* **96**: 9236–9241.
65. Sanchez JA, Pierce KE, Rice JE, Wangh LJ (2004) Linear-After-The-Exponential (LATE)-PCR: an advanced method of asymmetric PCR and its uses in quantitative real-time analysis. *Proceedings of the National Academy of Sciences of the United States of America* **101**: 1933–1938.
66. Pierce KE, Sanchez JA, Rice JE, Wangh LJ (2005) Linear-After-The-Exponential (LATE)-PCR: primer design criteria for high yields of specific single-stranded DNA and improved real-time detection. *Proceedings of the National Academy of Sciences of the United States of America* **102**: 8609–8614.
67. Marras SAE, Gold B, Kramer FR, Smith I, Tyagi S (2004) Real-time measurement of *in vitro* transcription. *Nucleic Acids Research* **32**: e72.
68. Ririe KM, Rasmussen RP, Wittwer CT (1997) Product differentiation by analysis of DNA melting curves during the polymerase chain reaction. *Analytical Biochemistry* **245**: 154–160.

69. El-Hajj HH, Marras SAE, Tyagi S, Shashkina E, Kamboj M, Kiehn TE, et al. (2009) Use of sloppy molecular beacon probes for identification of mycobacterial species. *Journal of Clinical Microbiology* **47**: 1190–1198.
70. Nazarenko IA, Bhatnager SK, Hohman RJ (1997) A closed tube format for amplification and detection of DNA based on energy transfer. *Nucleic Acids Research* **25**: 2516–2521.
71. Nazarenko I, Lowe B, Darfler M, Ikonomi P, Schuster D, Rashtchian A (2002) Multiplex quantitative PCR using self-quenched primers labeled with a single fluorophore. *Nucleic Acids Research* **30**: e37.
72. Whitcombe D, Theaker J, Guy SP, Brown T, Little S (1999) Detection of PCR products using self-probing amplicons and fluorescence. *Nature Biotechnology* **17**: 804–807.
73. Nitin N, Santangelo PJ, Kim G, Nie S, Bao G (2004) Peptide-linked molecular beacons for efficient delivery and rapid mRNA detection in living cells. *Nucleic Acids Research* 32: e58.
74. Mhlanga MM, Vargas DY, Fung CW, Kramer FR, Tyagi S (2005) tRNA-linked molecular beacons for imaging mRNAs in the cytoplasm of living cells. *Nucleic Acids Research* **33**: 1902–1912.
75. Satterfield BC, West JAA, Caplan MR (2007) Tentacle probes: eliminating false positives without sacrificing sensitivity. *Nucleic Acids Research* 35: e76.
76. Tyagi S, Marras SAE, Kramer FR (2000) Wavelength-shifting molecular beacons. *Nature Biotechnology* **18**: 1191–1196.
77. El-Hajj HH, Marras SAE, Tyagi S, Kramer FR, Alland D (2001) Detection of rifampin resistance in *Mycobacterium tuberculosis* in a single tube with molecular beacons. *Journal of Clinical Microbiology* **39**: 4131–4237.
78. Bratu DP, Cha B-J, Mhlanga MM, Kramer FR, Tyagi S (2003) Visualizing the distribution and transport of mRNAs in living cells. *Proceeding of the National Academy of Sciences of the United States of America* **100**: 13308–13313.
79. Santangelo PJ, Nix B, Tsourkas A, Bao G (2004) Dual FRET molecular beacons for mRNA detection in living cells. *Nucleic Acids Research* **32**: e57.
80. Ortiz E, Estrada G, Lizardi PM (1998) PNA molecular beacons for rapid detection of PCR amplicons. *Molecular and Cellular Probes* **12**: 219–226.
81. Kuhn H, Demidov VV, Coull JM, Fiandaca MJ, Gildea BD, Frank-Kamenetskii MD (2002) Hybridization of DNA and PNA molecular beacons to single-stranded and double-stranded DNA targets. *Journal of the American Chemical Society* **124**: 1097–1103.
82. Xi C, Balberg M, Boppart SA, Raskin L (2003) Use of DNA and peptide nucleic acid molecular beacons for detection and quantification of rRNA in solution and in whole cells. *Applied and Environmental Microbiology* **69**: 5673–5678.
83. Molenaar C, Marras SAE, Slats JCM, Truffert J-C, Lemaître M, Raap AK, et al. (2001) Linear 2′-*O*-methyl RNA probes for the visualization of RNA in living cells. *Nucleic Acids Research* **29**: e89.
84. Tsourkas A, Behlke MA, Bao G (2002) Hybridization of 2′-*O*-methyl and 2′-deoxy molecular beacons to RNA and DNA targets. *Nucleic Acids Research* **30**: 5168–5174.
85. Wang L, Yang CJ, Medley CD, Benner SA, Tan W (2005) Locked nucleic acid molecular beacons. *Journal of the American Chemical Society* **127**: 15664–15665.
86. Tyagi S, Alsmadi O (2004) Imaging native β-actin mRNA in motile fibroblasts. *Biophysical Journal* **87**: 4253–4162.
87. Ramachandran A, Flinchbaugh J, Ayoubi P, Olah GA, Malayer JR (2004) Target discrimination by surface-immobilized molecular beacons designed to detect *Francisella tularensis*. *Biosensors & Bioelectronics* **19**: 727–736.
88. Yao G, Tan W (2004) Molecular-beacon-based array for sensitive DNA analysis. *Analytical Biochemistry* **331**: 216–223.
89. Steemers FJ, Ferguson JA, Walt DR (2000) Screening unlabeled DNA targets with randomly ordered fiber-optic gene arrays. *Nature Biotechnology* **18**: 91–94.

90. Liu X, Tan W (1999) A fiber-optic evanescent wave DNA biosensor based on novel molecular beacons. *Analytical Chemistry* **71**: 5054–5059.
91. Brown LJ, Cummins J, Hamilton A, Brown T (2000) Molecular beacons attached to glass beads fluoresce upon hybridization to target DNA. *Chemical Communications (Cambridge, England)* **2000**: 621–622.
92. Wang H, Li J, Liu H, Liu Q, Mei Q, Wang Y, et al. (2002) Label-free hybridization detection of a single nucleotide mismatch by immobilization of molecular beacons on an agarose film. *Nucleic Acids Research* **30**: e61.
93. Stoermer RL, Cederquist KB, McFarland SK, Sha MY, Penn SG, Keating CD (2006) Coupling molecular beacons to barcoded metal nanowires for multiplexed, sealed chamber DNA bioassays. *Journal of the American Chemical Society* **128**: 16892–16903.
94. Fang X, Liu X, Schuster S, Tan W (1999) Designing a novel molecular beacon for surface-immobilized DNA hybridization studies. *Journal of the American Chemical Society* **121**: 2921–2922.
95. Dubertret B, Calame M, Libchaber AJ (2001) Single-mismatch detection using gold-quenched fluorescent oligonucleotides. *Nature Biotechnology* **19**: 365–370.
96. Du H, Disney MD, Miller BL, Krauss TD (2003) Hybridization-based unquenching of DNA hairpins on Au surfaces: prototypical "molecular beacon" biosensors. *Journal of the American Chemical Society* **125**: 4012–4013.
97. Fan C, Plaxco KW, Heeger AJ (2003) Electrochemical interrogation of conformational changes as a reagentless method for the sequence-specific detection of DNA. *Proceedings of the National Academy of Sciences of the United States of America* **100**: 9134–9137.
98. Nutiu R, Li Y (2002) Tripartite molecular beacons. *Nucleic Acids Research* **30**: e94.
99. Landré JBP, Ruryk A, Schlicksbier T, Russwurm S, Deigner H-P (2005) Design and applications of a novel type of hairpin probe: addressable bipartite molecular hook (ABMH). In: Nanotech 2005, Vol. 1, pages 385–388. Cambridge, MA: Nano Science and Technology Institute.
100. Yamamoto R, Kumar PKR (2000) Molecular beacon aptamer fluoresces in the presence of Tat protein of HIV-1. *Genes to Cells* **5**: 389–396.
101. Hamaguchi N, Ellington A, Stanton (2001) Aptamer beacons for the direct detection of proteins. *Analytical Biochemistry* **294**: 126–131.
102. Li JJ, Fang X, Tan W (2002) Molecular aptamer beacons for real-time protein recognition. *Biochemical and Biophysical Research Communications* **292**: 31–40.
103. Stojanovic MN, de Prada P, Landry DW (2001) Aptamer-based folding fluorescent sensor for cocaine. *Journal of the American Chemical Society* **123**: 4928–4931.
104. Nutiu R, Li Y (2003) Structure-switching signaling aptamers. *Journal of the American Chemical Society* **125**: 4771–4778.
105. Nutiu R, Li Y (2004) Structure-switching signaling aptamers: transducing molecular recognition into fluorescence signaling. *Chemistry* **10**: 1868–1876.
106. Heyduk T, Heyduk E (2002) Molecular beacons for detecting DNA binding proteins. *Nature Biotechnology* **20**: 171–176.
107. Heyduk E, Heyduk T (2005) Nucleic acid-based fluorescence sensors for detecting proteins. *Analytical Chemistry* **77**: 1147–1156.
108. Rajendran M, Ellington AD (2003) *In vitro* selection of molecular beacons. *Nucleic Acids Research* **31**: 5700–5713.
109. Nutiu R, Li Y (2005) *In vitro* selection of structure-switching signaling aptamers. *Angewandte Chemie (International ed. in English)* **44**: 1061–1065.
110. Oh KJ, Cash KJ, Hugenberg V, Plaxco KW (2007) Peptide beacons: a new design for polypeptide-based optical biosensors. *Bioconjugate Chemistry* **18**: 607–609.
111. Matsuo T (1998) In situ visualization of messenger RNA for basic fibroblast growth factor in living cells. *Biochimica et Biophysica Acta* **1379**: 178–184.
112. Sokol DL, Zhang X, Lu P, Gerwirtz AM (1998) Real time detection of DNA·RNA hybridization in living cells. *Proceedings of the National Academy of Sciences of the United States of America* **95**: 11538–11543.

113. Perlette J, Tan W (2001) Real-time monitoring of intracellular mRNA hybridization inside single living cells. *Analytical Chemistry* **73**: 5544–5550.
114. Tsuji A, Koshimoto H, Sato Y, Hirano M, Sei-Iida Y, Kondo S, et al. (2000) Direct observation of specific messenger RNA in a single living cell under a fluorescence microscope. *Biophysical Journal* **78**: 3260–3274.
115. Vargas DY, Raj A, Marras SAE, Kramer FR, Tyagi S (2005) Mechanism of mRNA transport in the nucleus. *Proceedings of the National Academy of Sciences of the United States of America* **102**: 17008–17013.
116. Wolfrum C, Josten A (2005) Oligonucleotides as coding molecules in an anti-counterfeiting system. *Nucleosides, Nucleotides & Nucleic Acids* **24**: 1069–1074.
117. Pierce KE, Rice JE, Sanchez JA, Brenner C, Wangh LJ (2000) Real-time PCR using molecular beacons for accurate detection of the Y chromosome in single human blastomeres. *Molecular Human Reproduction* **6**: 1155–1164.
118. Dracheva S, Marras SAE, Elhakem SL, Kramer FR, Davies KL, Haroutunian V (2001) *N*-methyl-D-aspartic acid receptor expression in the dorsolateral prefrontal cortex of elderly patients with schizophrenia. *American Journal of Psychiatry* **158**: 1400–1410.
119. Lai J-P, Douglas SD, Shaheen F, Pleasure DE, Ho W-Z (2002) Quantification of substance P mRNA in human immune cells by real-time reverse transcriptase PCR assay. *Clinical and Diagnostic Laboratory Immunology* **9**: 138–143.
120. Szuhai K, Van Den Ouweland JM, Dirks RW, Lemaître M, Truffert J-C, Janssen GM, et al. (2001) Simultaneous A8344G heteroplasmy and mitochondrial DNA copy number quantification in myoclonus epilepsy and ragged-red fibers (MERRF) syndrome by a multiplex molecular beacon based real-time fluorescence PCR. *Nucleic Acids Research* **29**: e13.
121. Smit ML, Giesendorf BAJ, Vet JAM, Trijbels FJM, Blom HJ (2001) Semiautomated DNA mutation analysis using a robotic workstation and molecular beacons. *Clinical Chemistry* **47**: 739–744.
122. Frei K, Szuhai K, Lucas T, Weipoltshammer K, Schöfer C, Ramsebner R, et al. (2002) Connexin 26 mutations in cases of sensorineural deafness in eastern Austria. *European Journal of Human Genetics* **10**: 427–432.
123. Rice JE, Sanchez JA, Pierce KE, Wangh LJ (2002) Real-time PCR with molecular beacons provides a highly accurate assay for detection of Tay-Sachs alleles in single cells. *Prenatal Diagnosis* **22**: 1130–1134.
124. Pierce KE, Rice JE, Sanchez JA, Wangh LJ (2003) Detection of cystic fibrosis alleles from single cells using molecular beacons and a novel method of asymmetric real-time PCR. *Molecular Human Reproduction* **9**: 815–820.
125. Orrù G, Faa G, Pillai S, Pilloni L, Montaldo C, Pusceddu G, et al. (2005) Rapid PCR real-time genotyping of M-Malton α_1-antitrypsin deficiency alleles by molecular beacons. *Diagnostic Molecular Pathology* **14**: 237–242.
126. Shih I-M, Zhou W, Goodman SN, Lengauer C, Kinzler KW, Vogelstein B (2001) Evidence that genetic instability occurs at an early stage of colorectal tumorigenesis. *Cancer Research* **61**: 818–822.
127. Martínez-López J, Lahuerta JJ, Salama P, Ayala R, Bautista JM (2004) The use of fluorescent molecular beacons in real time PCR of IgH gene rearrangements for quantitative evaluation of multiple myeloma. *Clinical and Laboratory Haematology* **26**: 31–35.
128. Yang L, Cao Z, Lin Y, Wood WC, Staley CA (2005) Molecular beacon imaging of tumor marker gene expression in pancreatic cancer cells. *Cancer Biology & Therapy* **4**: 561–570.
129. Span PN, Manders P, Heuvel JJTM, Thomas CMG, Bosch RR, Beex LVAM, et al. (2003) Molecular beacon reverse transcription-PCR of human chorionic gonadotropin-β-3, -5, and -8 mRNAs has prognostic value in breast cancer. *Clinical Chemistry* **49**: 1074–1080.
130. Zhao J, He D, He H, Li L, Zhang L-L, Wang X-Y (2007) Primary application study in early diagnosis of bladder cancer by survivin molecular beacons. *Urology* **70**: 60–64.

131. Abravaya K, Huff J, Marshall R, Merchant B, Mullen C, Schneider G, et al. (2003) Molecular beacons as diagnostic tools: technology and applications. *Clinical Chemistry and Laboratory Medicine* **41**: 468–474.
132. Lewin SR, Vesanen M, Kostrikis L, Hurley A, Duran M, Zhang L, et al. (1999) Use of real-time PCR and molecular beacons to detect virus replication in human immunodeficiency virus type 1-infected individuals on prolonged effective antiretroviral therapy. *Journal of Virology* **73**: 6099–6103.
133. Poddar SK (1999) Detection of adenovirus using PCR and molecular beacon. *Journal of Virological Methods* **82**: 19–26.
134. Poddar SK (2000) Symmetric vs asymmetric PCR and molecular beacon probe in the detection of a target gene of adenovirus. *Molecular and Cellular Probes* **14**: 25–32.
135. Claas ECJ, Schilham MW, de Brouwer CS, Hubacek P, Echavarria M, Lankester AC, et al. (2005) Internally controlled real-time PCR monitoring of adenovirus DNA load in serum or plasma of transplant recipients. *Journal of Clinical Microbiology* **43**: 1738–1744.
136. Szuhai K, Sandhaus E, Kolkman-Uljee SM, Lemaître M, Truffert J-C, Dirks RW, et al. (2001) A novel strategy for human papillomavirus detection and genotyping with SybrGreen and molecular beacon polymerase chain reaction. *American Journal of Pathology* **159**: 1651–1660.
137. Takács T, Jeney C, Kovács L, Mózes J, Benczik M, Sebe A (2008) Molecular beacon based real-time PCR method for detection of 15 high-risk and 5 low-risk HPV types. *Journal of Virological Methods* **149**: 153–162.
138. Yeo AC, Chan KP, Kumarasinghe G, Yap HK (2005) Rapid detection of codon 460 mutations in the UL97 gene of ganciclovir-resistant cytomegalovirus clinical isolates by real-time PCR using molecular beacons. *Molecular and Cellular Probes* **19**: 389–393.
139. Templeton KE, Scheltinga SA, Beersma MFC, Kroes ACM, Claas ECJ (2004) Rapid and sensitive method using multiplex real-time PCR for diagnosis of infections by influenza A and influenza B viruses, respiratory syncytial virus, and parainfluenza viruses 1, 2, 3, and 4. *Journal of Clinical Microbiology* **42**: 1654–1569.
140. O'Shea MK, Cane PA (2004) Development of a highly sensitive semi-quantitative real-time PCR and molecular beacon probe assay for the detection of respiratory syncytial virus. *Journal of Virological Methods* **118**: 101–110.
141. Yang J-H, Lai J-P, Douglas SD, Metzger D, Zhu X-H, Ho W-Z (2002) Real-time RT-PCR for quantitation of hepatitis C virus RNA. *Journal of Virological Methods* **102**: 119–128.
142. Sum SS-M, Wong DK-H, Yuen M-F, Yuan H-J, Yu J, Lai C-L, et al. (2004) Real-time PCR assay using molecular beacon for quantitation of hepatitis B virus DNA. *Journal of Clinical Microbiology* **42**: 3438–3440.
143. Waltz TL, Marras SAE, Rochford G, Nolan J, Lee E, Melegari M, et al. (2005) Development of a molecular-beacon assay to detect the G1896A precore mutation in hepatitis B virus-infected individuals. *Journal of Clinical Microbiology* **43**: 254–258.
144. Li Q, Liang J-X, Luan G-Y, Zhang Y, Wang K (2000) Molecular beacon-based homogeneous fluorescence PCR assay for the diagnosis of infectious diseases. *Analytical Sciences* **16**: 245–248.
145. Chen W, Martinez G, Mulchandani A (2000) Molecular beacons: a real-time polymerase chain reaction assay for detecting *Salmonella*. *Analytical Biochemistry* **280**: 166–172.
146. Poddar SK, Le CT (2001) *Bordetella pertussis* detection by spectrofluorometry using polymerase chain reaction (PCR) and a molecular beacon probe. *Molecular and Cellular Probes* **15**: 161–167.
147. Fortin NY, Mulchandani A, and Chen W (2001) Use of real-time polymerase chain reaction and molecular beacons for the detection of *Escherichia coli* O157:H7. *Analytical Biochemistry* **289**, 281–288.
148. Bélanger AD, Boissinot M, Ménard C, Picard FP, Bergeron MG (2002) Rapid detection of Shiga toxin-producing bacteria in feces by multiplex PCR with molecular beacons on the smart cycler. *Journal of Clinical Microbiology* **40**: 1436–1440.

149. Bélanger AD, Boissinot M, Clairoux N, Picard FJ, Bergeron MG (2003) Rapid detection of *Clostridium difficile* in feces by real-time PCR. *Journal of Clinical Microbiology* **41**: 730–734.
150. Gubala AJ, Proll DF (2006) Molecular-beacon multiplex real-time PCR assay for detection of *Vibrio cholerae*. *Applied and Environmental Microbiology* **72**: 6424–6428.
151. Elsayad S, Chow BL, Hamilton NL, Gregson DB, Pitout JDD, Church DL (2003) Development and validation of a molecular beacon probe-based real-time polymerase chain reaction assay for rapid detection of methicillin resistance in *Staphylococcus aureus*. *Archives of Pathology & Laboratory Medicine* **127**: 845–849.
152. Sinsimer D, Leekha S, Park S, Marras SAE, Koreen L, Willey B, et al. (2005) Use of a multiplex molecular beacon platform for rapid detection of methicillin and vancomycin resistance in *Staphylococcus aureus*. *Journal of Clinical Microbiology* **43**: 4585–4591.
153. Varma-Basil M, El-Hajj HH, Marras SAE, Hazbón MH, Mann JM, Connell ND, et al. (2004) Molecular beacons for multiplex detection of four bacterial bioterrorism agents. *Clinical Chemistry* **50**: 1060–1063.
154. Templeton KE, Scheltinga SA, Sillekens P, Crielaard JW, van Dam AP, Goossens H, et al. (2003) Development and clinical evaluation of an internally controlled, single-tube multiplex real-time PCR assay for detection of *Legionella pneumophila* and other *Legionella* species. *Journal of Clinical Microbiology* **41**: 4016–4021.
155. Morozumi M, Nakayama E, Iwata S, Aoki Y, Hasegawa K, Kobayashi R, et al. (2006) Simultaneous detection of pathogens in clinical samples from patients with community-acquired pneumonia by real-time PCR with pathogen-specific molecular beacon probes. *Journal of Clinical Microbiology* **44**: 1440–1446.
156. Gullsby K, Storm M, Bondeson K (2008) Simultaneous detection of *Chlamydophila pneumoniae* and *Mycoplasma pneumoniae* by use of molecular beacons in a duplex real-time PCR. *Journal of Clinical Microbiology* **46**: 727–731.
157. Park S, Wong M, Marras SAE, Cross EW, Kiehn TE, Chaturvedi V, et al. (2000) Rapid identification of *Candida dubliniensis* using a species-specific molecular beacon. *Journal of Clinical Microbiology* **38**: 2829–2836.
158. Balashov SV, Gardiner R, Park S, Perlin DS (2005) Rapid, high-throughput, multiplex, real-time PCR for identification of mutations in the *cyp*51A gene of *Aspergillus fumigatus* that confer resistance to itraconazole. *Journal of Clinical Microbiology* **43**: 214–222.
159. Durand R, Eslahpazire J, Jafari S, Delabre J-F, Marmorat-Khuong A, Di Piazza J-P, et al. (2000) Use of molecular beacons to detect an antifolate resistance-associated mutation in *Plasmodium falciparum*. *Antimicrobial Agents and Chemotherapy* **44**: 3461–3464.
160. Durand R, Huart V, Jafari S, Le Bras J (2002) Rapid detection of a molecular marker for chloroquine-resistant *falciparum* malaria. *Antimicrobial Agents and Chemotherapy* **46**: 2684–2686.
161. Roy S, Kabir M, Mondal D, Ali IKM, Petri WA Jr, Haque R (2005) Real-time-PCR assay for diagnosis of *Entamoeba histolytica* infection. *Journal of Clinical Microbiology* **43**: 2168–2172.
162. Berard C, Cazalis M-A, Leissner P, Mougin B (2004) DNA nucleic acid sequence-based amplification-based genotyping for polymorphism analysis. *BioTechniques* **37**: 680–686.
163. Fradet Y, Saad F, Aprikian A, Dessureault J, Elhilali M, Trudel C, et al. (2004) uPM3, a new molecular urine test for the detection of prostate cancer. *Urology* **64**: 311–316.
164. van Beuningen R, Marras SAE, Kramer FR, Oosterlaken T, Weusten JJAM, Borst G, et al. (2001) Development of a high throughput detection system for HIV-1 using real-time NASBA based on molecular beacons. In: R Raghavachari and W Tan (eds), *Genomics and Proteomics Technologies*, pages 66–72. Bellingham, WA: SPIE.
165. Weusten JJAM, Carpay WM, Oosterlaken TAM, van Zuijlen MCA, van de Wiel PA (2002) Principles of quantitation of viral loads using nucleic acid sequence-based

amplification in combination with homogeneous detection using molecular beacons. *Nucleic Acids Research* **30**: e26.

166. Lanciotti RS, Kerst AJ (2001) Nucleic acid sequence-based amplification assays for rapid detection of West Nile and St. Louis encephalitis viruses. *Journal of Clinical Microbiology* **39**: 4506–4513.
167. Polstra AM, Goudsmit J, Cornelissen M (2002) Development of real-time NASBA assays with molecular beacon detection to quantify mRNA coding for HHV-8 lytic and latent genes. *BMC Infectious Diseases* **2**: 18.
168. Greijer AE, Adriaanse HMA, Dekkers CAJ, Middeldorp JM (2002) Multiplex real-time NASBA for monitoring expression dynamics of human cytomegalovirus encoded IE1 and pp67 RNA. *Journal of Clinical Virology* **24**: 57–66.
169. Yates S, Penning M, Goudsmit J, Frantzen I, van de Weijer B, van Strijp D, et al. (2001) Quantitative detection of hepatitis B virus DNA by real-time nucleic acid sequence-based amplification with molecular beacon detection. *Journal of Clinical Microbiology* **39**: 3656–3665.
170. Abd el-Galil KH, el-Sokkary MA, Kheira SM, Salazar AM, Yates MV, Chen W, et al. (2005) Real-time nucleic acid sequence-based amplification assay for detection of hepatitis A virus. *Applied and Environmental Microbiology* **71**: 7113–7116.
171. Hibbitts S, Rahman A, John R, Westmoreland D, Fox JD (2003) Development and evaluation of NucliSens basic kit NASBA for diagnosis of parainfluenza virus infection with 'end-point' and 'real-time' detection. *Journal of Virological Methods* **108**: 145–155.
172. Moore C, Hibbitts S, Owen N, Corden SA, Harrison G, Fox JD, et al. (2004) Development and evaluation of a real-time nucleic acid sequence based amplification assay for rapid detection of influenza A. *Journal of Medical Virology* **74**: 619–628.
173. Deiman B, Schrover C, Moore C, Westmoreland D, van de Wiel P (2007) Rapid and highly sensitive qualitative real-time assay for detection of respiratory syncytial virus A and B using NASBA and molecular beacon technology. *Journal of Virological Methods* **146**: 29–35.
174. Capaul SE, Gorgievski-Hrisoho M (2005) Detection of enterovirus RNA in cerebrospinal fluid (CSF) using NucliSens EasyQ enterovirus assay. *Journal of Clinical Virology* **32**: 236–240.
175. Landry ML, Garner R, Ferguson D (2005) Real-time nucleic acid sequence-based amplification using molecular beacons for detection of enterovirus RNA in clinical specimens. *Journal of Clinical Microbiology* **43**: 3136–3139.
176. Keightley MC, Sillekens P, Schippers W, Rinaldo C, George KS (2005) Real-time NASBA detection of SARS-associated coronavirus and comparison with real-time reverse transcription-PCR. *Journal of Medical Virology* **77**: 602–608.
177. Molden T, Kraus I, Skomedal H, Nordstrøm T, Karlsen F (2007) PreTect HPV-Proofer: real-time detection and typing of E6/E7 mRNA from carcinogenic human papillomaviruses. *Journal of Virological Methods* **142**: 204–212.
178. Loens K, Ieven M, Ursi D, Beck T, Overdijk M, Sillekens P, et al. (2003) Detection of *Mycoplasma pneumoniae* by real-time nucleic acid sequence-based amplification. *Journal of Clinical Microbiology* **41**: 4448–4450.
179. Loens K, Beck T, Goossens H, Ursi D, Overdijk M, Sillekens P, et al. (2006) Development of conventional and real-time nucleic acid sequence-based amplification assays for detection of *Chlamydophila pneumoniae* in respiratory specimens. *Journal of Clinical Microbiology* **44**: 1241–1244.
180. Loens K, Beck T, Ursi D, Overdijk M, Sillekens P, Goossens H, et al. (2008) Development of real-time multiplex nucleic acid sequence-based amplification for detection of *Mycoplasma pneumoniae*, *Chlamydophila pneumoniae*, and *Legionella* spp. in respiratory specimens. *Journal of Clinical Microbiology* **46**: 185–191.
181. Fykse EM, Skogan G, Davies W, Olsen JS, Blatny JM (2007) Detection of *Vibrio cholerae* by real-time nucleic acid sequence-based amplification. *Applied and Environmental Microbiology* **73**: 1457–1466.

182. Gore HM, Wakeman CA, Hull RM, McKillip JL (2003) Real-time molecular beacon NASBA reveals *hblC* expression from *Bacillus* spp. in milk. *Biochemical and Biophysical Research Communications* **311**: 386–390.
183. Rodríguez-Lázaro D, Lloyd J, Herrewegh A, Ikonomopoulos J, D'Agostino M, Pla M, et al. (2004) A molecular beacon-based real-time NASBA assay for detection of *Mycobacterium avium* subsp. *paratuberculosis* in water and milk. *FEMS Microbiology Letters* **237**: 119–126.
184. Churruca E, Girbau C, Martínez I, Mateo E, Alonso R, Fernández-Astorga A (2007) Detection of *Campylobacter jejuni* and *Campylobacter coli* in chicken meat samples by real-time nucleic acid sequence-based amplification with molecular beacons. *International Journal of Food Microbiology* **117**: 85–90.
185. Nadal A, Coll A, Cook N, Pla M (2007) A molecular beacon-based real time NASBA assay for detection of *Listeria monocytogenes* in food products: role of target mRNA secondary structure on NASBA design. *Journal of Microbiological Methods* **68**: 623–632.
186. Schneider P, Wolters L, Schoone G, Schallig H, Sillekens P, Hermsen R, et al. (2005) Real-time nucleic acid sequence-based amplification is more convenient than real-time PCR for quantification of *Plasmodium falciparum*. *Journal of Clinical Microbiology* **43**: 402–405.
187. de Mendoza C, Koppelman M, Montès B, Ferre V, Soriano V, Cuypers H, et al. (2005) Multicenter evaluation of the NucliSens EasyQ HIV-1 v1.1 assay for the quantitative detection of HIV-1 RNA in plasma. *Journal of Virological Methods* **127**: 54–59.
188. Huletsky A, Giroux R, Rossbach V, Gagnon M, Vaillancourt M, Bernier M, et al. (2004) New real-time PCR assay for rapid detection of methicillin-resistant *Staphylococcus aureus* directly from specimens containing a mixture of *Staphylococci*. *Journal of Clinical Microbiology* **42**: 1875–1884.
189. Warren DK, Liao RS, Merz LR, Eveland M, Dunne WM Jr (2004) Detection of methicillin-resistant *Staphylococcus aureus* directly from nasal swab specimens by a real-time PCR assay. *Journal of Clinical Microbiology* **42**: 5578–5581.
190. Paule SM, Hacek DM, Kufner B, Truchon K, Thompson RB Jr, Kaul KL, et al. (2007) Performance of the BD GeneOhm methicillin-resistant *Staphylococcus aureus* test before and during high-volume clinical use. *Journal of Clinical Microbiology* **45**: 2993–2998.
191. Robicsek A, Beaumont JL, Paule SM, Hacek DM, Thompson RB Jr, Kaul KL, et al. (2008) Universal surveillance for methicillin-resistant *Staphylococcus aureus* in three affiliated hospitals. *Annals of Internal Medicine* **148**: 409–418.
192. Davies HD, Miller MA, Faro S, Gregson D, Kehl SC, Jordan JA (2004) Multicenter study of a rapid molecular-based assay for the diagnosis of group B *Streptococcus* colonization in pregnant women. *Clinical Infectious Diseases* **39**: 1129–1135.
193. Goodrich JS, Miller MB (2007) Comparison of culture and two real-time polymerase chain reaction assays to detect group B *Streptococcus* during antepartum screening. *Diagnostic Microbiology and Infectious Disease* **59**: 17–22.

4 Rapid polymerase chain reaction and melting analysis

Carl T. Wittwer, Randy P. Rasmussen, and Kirk M. Ririe

The polymerase chain reaction (PCR) is conceptually divided into three reactions, each usually assumed to occur over time at a single temperature. Such an "equilibrium paradigm" of PCR is naïve, but widely accepted. It is easy to think of three reactions (denaturation, annealing, and extension) occurring at three temperatures over three time periods in each cycle (Figure 4–1, *left*). However, this equilibrium paradigm does not fit well with physical reality. Instantaneous temperature changes do not occur; it takes time to change the sample temperature. Furthermore, individual reaction rates vary with temperature, and after primer annealing occurs, polymerase extension immediately follows. More accurate is a kinetic paradigm for PCR in which reaction rates and the temperature are always changing (Figure 4–1, *right*). Holding the temperature constant during PCR is not necessary as long as the products denature and the primers anneal. Under the kinetic paradigm of PCR, product denaturation, primer annealing, and polymerase extension may temporally overlap and their rates continuously vary with temperature. Under the equilibrium paradigm, three temperatures each

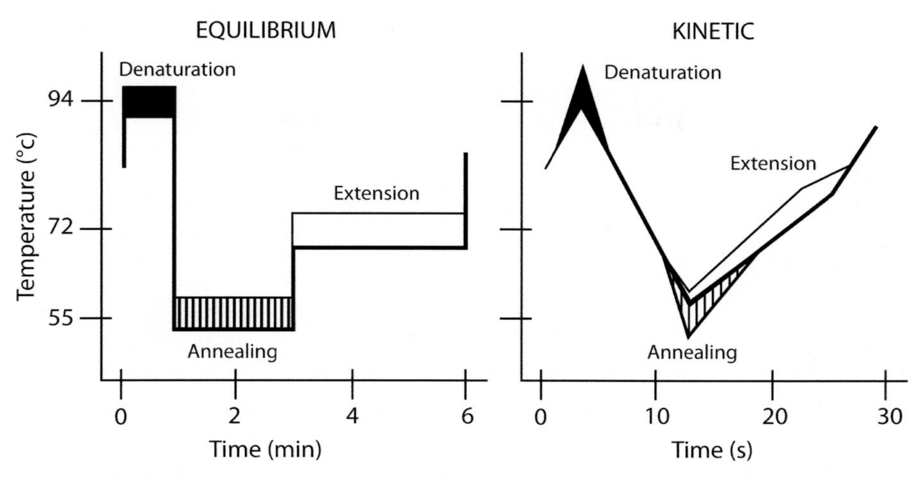

Figure 4–1. Equilibrium and kinetic paradigms of the polymerase chain reaction (PCR). Each paradigm focuses on three reactions (denaturation, annealing, and extension) during each PCR cycle. In the conventional equilibrium paradigm for PCR (*left*), each reaction occurs at a single temperature over a certain time period. Temperature transitions are not considered. In contrast, in the kinetic paradigm (*right*), the temperature is always changing. Each reaction occurs over a temperature range, rates depend on temperature, and more than one process can occur simultaneously.

held for finite time periods define a cycle, whereas the kinetic paradigm requires transition rates and target temperatures.

Paradigms are not right or wrong, but should be judged by their usefulness. The equilibrium paradigm is simple to understand and lends itself well to the engineering mindset and instrument manufacturing. The kinetic paradigm is more relevant to biochemistry, rapid PCR, and melting curve analysis.

When PCR was first popularized in the late 1980s, the process was slow. A typical protocol was 1 minute for denaturation at 94°C, 2 minutes for annealing at 55°C, and 3 minutes for extension at 72°C. When the time for transition between temperatures was included, 8-minute cycles were typical, resulting in completion of 30 cycles in 4 hours. Twenty-five percent of the cycling time was spent in temperature transitions. As cycling speeds increased, the proportion of time spent in temperature transitions also increased and the kinetic paradigm became more and more relevant. During rapid PCR, the temperature is usually changing. For rapid PCR of short products (<100 base pairs), 100% of the time is spend in temperature transition and no holding times are necessary. For rapid PCR of longer products, a temperature hold at an optimal extension temperature may be included.

Another nice fit for the kinetic paradigm is DNA melting. DNA melts as the temperature increases, and the melting transitions of both PCR products and internal probes can be monitored. Just as "old" (slow) PCR was viewed as an equilibrium process, "old" (dot blot) hybridizations were performed at a single temperature. Much more powerful is dynamic monitoring of the entire melting curve as the temperature changes. Hybridization can be monitored during PCR or after completion of temperature cycling. Although continuous monitoring during PCR is more powerful, end-point melting is simpler and has become more popular. A relatively new development is high-resolution melting analysis for mutation scanning and genotyping.

Both rapid PCR and melting analysis, enabled by kinetic considerations of PCR, are detailed in this chapter.

RAPID PCR

The term "rapid PCR" is both relative and vague. A 1-hour PCR is rapid compared to 2 hours, but long compared to 15 minutes. Furthermore, PCR protocols can be made shorter if you start with higher template concentrations or by using fewer cycles. More specific is the time required for each cycle. Rapid-cycle PCR was defined in 1994 as 30 cycles completed in 10 to 30 minutes[1] so that the cycles are 20 to 60 seconds each. This time is the actual time of each cycle, and it is longer than the sum of the times often programmed for denaturation, annealing, and extension. Initial work in the early 1990s established the feasibility of rapid cycling using capillary tubes and hot air temperature control. Over the years, systems have become faster, and the kinetic requirements of denaturation, annealing, and extension have become clearer.

Capillary tubes and hot air

Before commercial instruments were available for PCR, we started work on a simple system to automate the repetitive task of temperature cycling. A heating element and fan from a hair dryer, a thermocouple, and PCR samples in capillary tubes were enclosed in a chamber.[2] The fan created a rapid flow of heated air past the thermocouple and capillaries. By matching the thermal response of the thermocouple to the sample, the temperature of the thermocouple closely tracked the temperature of the samples, even during temperature changes. Although air has a low heat capacity, rapidly moving air against the large surface area exposed by the capillaries was adequate to cycle the sample between denaturation, annealing, and extension temperatures. Off-the-shelf or home brew electronic controllers monitored the temperature, adjusted the power to the heating element, and provided the required timing and number of cycles. For cooling, the controller activated a solenoid that opened a portal to outside air, introducing cooling air to the otherwise closed chamber.

Minimizing the time for PCR

Although our initial objective was not speed, we soon realized that with the air/capillary system, temperatures could be rapidly changed. Using a low thermal mass chamber, circulating air, and samples in glass capillaries, PCR products greater than 500 base pairs were visualized on ethidium bromide–stained gels after only 10 minutes of PCR (30 cycles of 20 seconds each).[3] Product yield increased by increasing the extension time or the concentration of polymerase. Such rapid protocols used momentary or "0"-second holds at the denaturation and annealing temperatures. That is, the temperature time profile shows temperature spikes for denaturation and annealing, without holding the top and bottom temperatures. Apparently, denaturation and annealing can occur quickly.

Optimal temperatures and times for rapid-cycle PCR

Rapid and accurate control of temperature allowed analytical study of the required temperatures and times for PCR. Because optimal temperatures and times may be target specific, we chose a representative single copy gene from human genomic DNA (β-globin) and amplified a 536-base pair fragment. Optimal temperatures and times were determined by varying only one parameter at a time and viewing the products on agarose gels.[4] Denaturation temperatures between 91°C and 97°C were equally effective, as were denaturation times from less than 1 second to 16 seconds. Denaturation times longer than 16 seconds decreased product yield. Specific products in good yield were obtained with annealing temperatures of 50°C to 60°C as long as the time for primer annealing was limited. That is, the best specificity was obtained by rapid cooling from denaturation to annealing and an annealing time of less than 1 second. Yield was best at extension temperatures of 75°C to 79°C, and increased with extension time up to approximately 40 seconds. Conventional and rapid-cycle profiles with their reaction products separated on a gel are shown in Figure 4–2.

Conclusions from this early work were as follows: (1) denaturation of PCR products is rapid with no need to hold the denaturation temperature, (2) annealing of primers can occur quickly and annealing temperature holds are not necessary, and (3) the required extension time depends on PCR product length and polymerase concentration. We also found rapid PCR not only faster, but better in terms of specificity and yield[4,5] as long as the temperature was controlled precisely. PCR speed was not limited by biochemistry, but by instrumentation that did not control the sample temperature closely or rapidly.

Commercial PCR instrumentation

Most laboratory PCR instruments perform poorly with momentary denaturation and annealing times. This is reflected by their aversion towards 0-second holding periods. Time delays from thermal transfer through the walls of conical tubes, low surface area-to-volume ratios, and heating of large samples forces most instruments to rely on extended times at denaturation and annealing to assure that the sample reaches the desired temperatures. Because the temperature-versus-time course is indefinite, reproducibility is limited.[6] Nearly all instruments show marked temperature variance during temperature transitions.[7,8] Undershoot and/or overshoot of temperature is a chronic problem that is seldom solved by attempted software prediction that depends on sample volume and thermal properties of the instrument that may change with age.

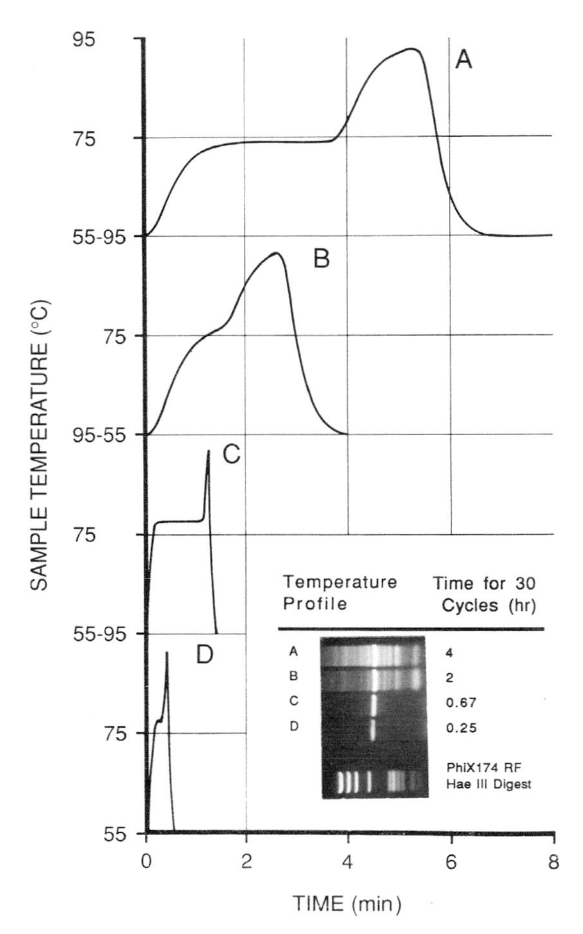

Figure 4–2. Effect of polymerase chain reaction (PCR) cycling speed on product specificity and yield. A 536-base pair β-globin fragment was amplified from human genomic deoxyribonucleic acid (DNA) by 30 temperature cycles between 55°C and 93°C. Different temperature profiles were obtained on a conventional heating block instrument **(A, B)** and on a rapid-cycling system **(C, D)** with the sample temperature monitored with a miniature thermocouple. Agarose gel electrophoresis of the PCR products was performed to monitor specificity and yield. As the time for PCR decreases from 4 hours to 15 minutes, specificity increases. [Reprinted by permission of the publisher. From Wittwer CT, Garling DJ (1991) Rapid cycle DNA amplification: time and temperature optimization. *BioTechniques* 10: 76–83. © 1991 Eaton Publishing.]

Over time, conventional heating block instruments have become faster, with incremental improvements in "thin wall" tubes, more conductive heat distribution between samples, low heat capacity blocks, and other "fast" modifications. Nevertheless, it is unusual for these systems to cycle rapidly enough to complete a cycle in less than 60 seconds. A few heating block systems can achieve cycles of less than 60 seconds, usually restricted to two-temperature cycling between a limited range of temperatures. By flattening the sample container, rapid cycling can be achieved by resistive heating and air cooling[9] or by moving the sample in a flexible tube between heating zones kept at a constant temperature.

Commercial versions of the air/capillary system for PCR have been available since 1991[1] and for real-time PCR since 1996.[10,11] Rapid cycling capabilities of other instruments are often compared against the air/capillary standard that first demonstrated 20- to 60-second cycles. Oddly enough, there has been a trend to run the air/capillary systems slower over the years, perhaps reflecting discomfort with 0-second denaturation and annealing times by many users. For example, heat-activated enzymes require longer denaturation periods, often doubling run times even when "fast" activation enzymes are used. Another compromise away from rapid cycling is the use of plastic capillaries. These capillaries are not thermally matched to the instrument, so 20-second holds at denaturation and annealing are required to reach the target temperatures.[12]

Microsystem PCR

Substantial progress with rapid-cycle PCR has occurred in microsystems, where small volumes are naturally processed.[13,14] However, even with high surface area-to-volume sample chambers, cycles may be long if the heating element has a high thermal mass and is external to the chamber.[15] With thin-film resistive heaters and temperature sensors close to the samples, rapid cycling can be achieved.[16,17]

Although low-thermal-mass systems are usually cooled by passive thermal diffusion and/or by forced air, several interesting heating methods have been developed. Infrared radiation can be used for heating[18] with calibrated infrared pyrometry for temperature monitoring.[19] Alternatively, thin metal films on glass capillaries can serve as both a resistive heating element and a temperature sensor for rapid cycling.[20] Finally, direct Joule heating and temperature monitoring of the PCR solution by electrolytic resistance are possible and have been implemented in capillaries.[21] All of the above methods transfer heat to and from fixed samples.

Instead of heat transfer to and from stationary samples, the samples can be moved through fixed temperature zones. Microfluidic methods have become popular with the PCR fluid passing within channels through different segments kept at denaturation, annealing, and extension temperatures. Continuous-flow PCR has been demonstrated within serpentine channels that pass back and forth through three temperature zones[22] and within loops of increasing or decreasing radius that pass through three temperature sectors.[23] A variant with a serpentine layout uses stationary thermal gradients instead of isothermal zones to more closely fit the kinetic paradigm of PCR.[24] To limit the length of the microchannel necessary for PCR, some systems shuttle samples back and forth between temperature zones by bidirectional pressure-driven flow,[25] pneumatics,[26] or electrokinetic forces.[27] Instead of linear shuttling of samples, a single circular channel can be used with sample movement driven as a magnetic ferrofluid[28] or by convection.[29] One potential advantage of microsystem PCR, including continuous flow methods, is cycling speed.

Although some microsystems still require cycles of greater than 60 seconds, many operate in the 20- to 60-second cycle range of rapid-cycle PCR.[13,30] Minimum cycle times ranging from 16 to 37 seconds have been reported for infrared heating.[18,19] Metal-coated capillaries have achieved 40-second PCR cycles,[20] whereas direct electrolytic heating has amplified with 21-second cycles.[20]

Table 4–1. Optimal rates and target temperatures during rapid-cycle PCR under the kinetic paradigm

PCR step	Approach rate (°C/sec)	Target temperature (°C)
Denaturation	10–30	Product $T_m + 3$
Annealing	10–30	Primer $T_m - 5$
Extension	1–10 (usually 2–5)	65–80 (usually 70–74)

Minimum cycle times reported for closed loop convective PCR range from 24 to 42 seconds.[29,31] Several groups have focused on reducing PCR cycle times to less than 20 seconds, faster than the original definition of rapid-cycle PCR that was first demonstrated in 1990. Thin-film resistive heating of stationary samples has reduced cycle times down to 17 seconds for 25-μl samples[32] and 8.5 seconds for 100-nl samples.[17] Continuous-flow systems have achieved 12- to 14-second cycles with thermal gradient PCR[24] and sample shuttling,[26] whereas a ferrofluid loop claims successful PCR with 9-second cycles[28] Finally, the fastest reported cycle times are 6.9 seconds[22] and 5.2 seconds[23] for various size PCR products in continuous-flow systems.

Although engineers may be motivated by speed, less attention has been paid to the usefulness and quality of the resulting PCR. As a general rule, as cycles become shorter and shorter, claims for successful PCR correlate with lower complexity targets (bacteria, phage, or even PCR products) used at higher starting concentrations. Two-step PCR and a small range between denaturation and annealing/extension temperatures simplify cycling requirements. With short PCR cycle times, amplification efficiency and yield are poor compared to control reactions.[22,23] Exaggerated heating and cooling rates (up to 175°C/sec) are reported based on modeling and measurements without PCR samples present.[17] A typical engineering report focuses extensively on the design and modeling of the thermal cycling device with a final brief PCR demonstration using a high concentration of a low complexity target. There is promise for PCR with 5- to 10-second cycles (30 cycles in 2.5 to 5.0 minutes), but general use will depend on a system robust enough to amplify complex DNA (human genomic DNA) at low copy number with good PCR efficiency.

Ideal temperature profile for rapid-cycle PCR

Instrument limitations aside, what is an ideal temperature profile for rapid-cycle PCR? How can the cycle time be shortened without sacrificing PCR efficiency? Optimizing PCR parameters to fit the target is quite different than starting with universal cycling conditions (e.g., 60°C for 60 seconds and 94°C for 15 seconds) and forcing the design to fit the protocol. Optimal protocols depend on the melting temperature (T_m) and concentration of the primers, the length and T_m of the product, and the activity and stability of the polymerase at different temperatures. In turn, these factors depend on the buffer, ionic strength, Mg^{2+} concentration, and presence of additives. Table 4–1 summarizes an ideal kinetic protocol for rapid-cycle PCR, followed by specific considerations for denaturation, annealing, and extension.

Denaturation

Inadequate denaturation is a common reason for PCR failure. The goal is complete denaturation in each cycle, providing quantitative template availability for primer annealing. Initial denaturation of template before PCR, particularly genomic DNA, usually requires more severe conditions than does denaturation of the product during PCR. The original optimization of rapid-cycle PCR[4] was performed after boiling the template – a good way to assure initial denaturation of genomic DNA. Incomplete initial denaturation can occur with high-T_m targets, particularly those with flanking regions of high stability.[33] If minor temperature differences between samples are present during initial incomplete denaturation, quantitative PCR for genomic insertions or deletions can be compromised.[33–35] The solution is to ensure complete initial denaturation. If prior boiling or restriction digestion[33] is not desired, and higher denaturation temperatures compromise the polymerase, adjuvants that lower product T_m (DMSO, betaine) can be used.

The approach rate to denaturation can be as fast as possible, and is listed in Table 4–1 as 10 to 30°C/sec. At these rates, only approximately 1 second is required to reach denaturation during PCR. Although faster rates could be used, the risk of overshooting the target temperature with polymerase inactivation or boiling the solution increases.

Momentary (0 second) denaturation at 2 to 3°C above the T_m of the product assures complete denaturation. If the product melts in multiple domains, the target denaturation temperature should be 2 to 3°C above the highest melting domain. As long as the sample reaches this temperature, denaturation is fast, even for long products. Using capillaries and water baths,[36] complete denaturation of PCR products larger than 20 kB occurred in less than 1 second (data not shown). Product T_ms and melting domains are best determined experimentally with DNA dyes and high-resolution melting.[37] Although T_m estimates can be obtained by software predictions,[38] their accuracy is limited. Furthermore, observed T_ms strongly depend on local reaction conditions, such as salt concentrations, and the presence of any dyes and adjuvants. It is better to observe rather than predict, especially when the observation can be made easily under matched conditions.

Although 94°C is often used as a default target temperature for denaturation, it is seldom optimal. PCR products melt over a 40°C range depending on guanine–cytosine (GC) content and length.[39] Low denaturation target temperatures have both a speed and specificity advantage when the PCR product melts low enough that they can be used. The lower the denaturation temperature, the faster PCR can be performed. Added specificity arises from eliminating all potential products with higher denaturation temperatures. Because they are not denatured, they cannot be amplified. To amplify high denaturation products, the target temperature may need to be increased above 94°C. However, most heat-stable polymerases start to denature above 97°C, and the PCR solution may boil between 95 and 100°C depending on the altitude, so there is not much room to increase the temperature. Lowering the monovalent salt and Mg^{2+} concentration lowers product T_m. Similarly, incorporating 2′-deoxyuridine 5′-triphosphate (dUTP) and/or 7-deaza-2′-deoxyguanosine 5′-triphosphate (7-deaza-dGTP) also lowers product T_m, but may decrease polymerase extension rates. Most proprietary PCR "enhancers" are simple organics that lower product T_m, enabling denaturation (and amplification) of high-T_m products. Most popular among these are dimethyl sulfoxide (DMSO), betaine, glycerol, ethylene glycol, and formamide. In addition to lowering T_m, some of these additives also raise the boiling point of the PCR mixture (useful at high altitudes). As the concentration of enhancer increases, product T_ms decrease and polymerase inhibition increases.

Annealing

Incomplete and/or misdirected primer annealing can result in poor PCR. Low efficiency results if all template sites are not primed. If priming occurs at undesired sites, alternative products may be produced. The goal is complete primer annealing to only the desired sites during each cycle, providing quantitative primed templates for polymerase extension.

The approach rate to the annealing target temperature should be as fast as possible. The 10 to 30°C/sec range given in Table 4–1 requires only a few seconds, yet is not so fast that the risk of overshoot is too great. Besides speed, the main reason for rapid cooling is to minimize full-length product hybridization. To the extent that duplex product forms during cooling, PCR efficiency is reduced because primers cannot anneal to duplex product. Although duplex product formation is limited early in PCR, as the product concentration increases, more and more duplex forms during cooling. Continuous monitoring with SYBR® Green I suggests that such product annealing is a major cause of the PCR plateau.[40]

Table 4–1 indicates momentary (0 second) primer annealing at 5°C below the lowest T_m of the two primers. Complete primer annealing is assured if the primer concentration is high enough. Annealing rates are directly proportional to the concentration of the primer. Typical primer concentrations for rapid-cycle PCR range from 0.5 μM[10] to 5 μM.[41] These concentrations are higher than those used in conventional PCR where long annealing times are used. Limiting the primer concentration may improve specificity in conventional PCR, but with rapid cycling, specificity is obtained by limiting the annealing time. For reasons analogous to those for product T_ms, primer T_ms are best obtained experimentally rather than by prediction. Primer T_ms can be measured by melting analysis using saturating DNA dyes and oligonucleotides under the same buffer conditions used for amplification. The primer is combined with its complementary target (including a 5′ extension as a dangling end) and melting analysis performed.

Extension

Complete extension of the primed template during each cycle is necessary for optimal PCR efficiency. In contrast to denaturation and annealing where approach rates of 10 to 30°C are optimal, the heating rate toward the extension temperature should be limited to 1 to 10°C. After primer annealing, polymerase extension is relatively slow because of the low temperature. The polymerase needs some time to extend the primer so that the growing duplex is stable enough for the higher, optimal extension temperature. However, if the temperature exceeds the stability of the growing duplex, extension will terminate. Depending on the sequence and annealing temperature, rates of 1 to 10°C/sec may be appropriate, although usually 2 to 5°C/sec is used.

Taq polymerase extension rates increase with temperature, reaching a maximum of approximately 100 nucleotides/sec at 75 to 80°C. For a 536-base pair β-globin product, 76°C was found optimal in rapid-cycle PCR.[4] If low stability domains (low GC regions) are present within the PCR product, the extension target temperature must be reduced (≤65°C) to prevent dissociation.[42] The extension temperatures most commonly used in PCR (70 to 74°C) are a compromise between higher extension rates at higher temperatures and greater stability of the extending product at lower temperatures.

Specificity against long PCR products can be obtained if the approach to extension is rapid and a 0-second extension time is used. Holding the extension target temperature is seldom necessary for PCR products of 100 or fewer base pairs. For efficient amplification of longer products, a hold at the extension temperature may be necessary. The time required at extension depends on the polymerase extension rate at the temperature selected. For Taq polymerase at 70 to 74°C, a 3-second hold for every additional 100 base pairs has been recommended.[42] Faster polymerases recently have been introduced with commercial claims that they can reduce overall PCR times, suggesting that they may be able to eliminate or shorten extension holding times for longer products.

MELTING ANALYSIS

Thermal melting of DNA is traditionally monitored by ultraviolet absorbance. For high-quality melting curves, microgram amounts of DNA and rates of 0.1 to 1.0°C/min are required. In contrast to absorbance, fluorescent analysis of DNA melting is more sensitive and only nanogram amounts are needed, conveniently provided by PCR. Methods that monitor DNA melting by fluorescence became popular with the advent of real-time PCR[43] and were introduced in 1997 with the LightCycler®.[10,39,40] Smaller sample volumes in capillaries allowed better temperature control, enabling much faster melting rates of 0.05 to 0.3°C/sec.

Melting curves of PCR product duplexes or probes hybridized to one PCR strand can be observed. In both cases, the melting curves are usually obtained without

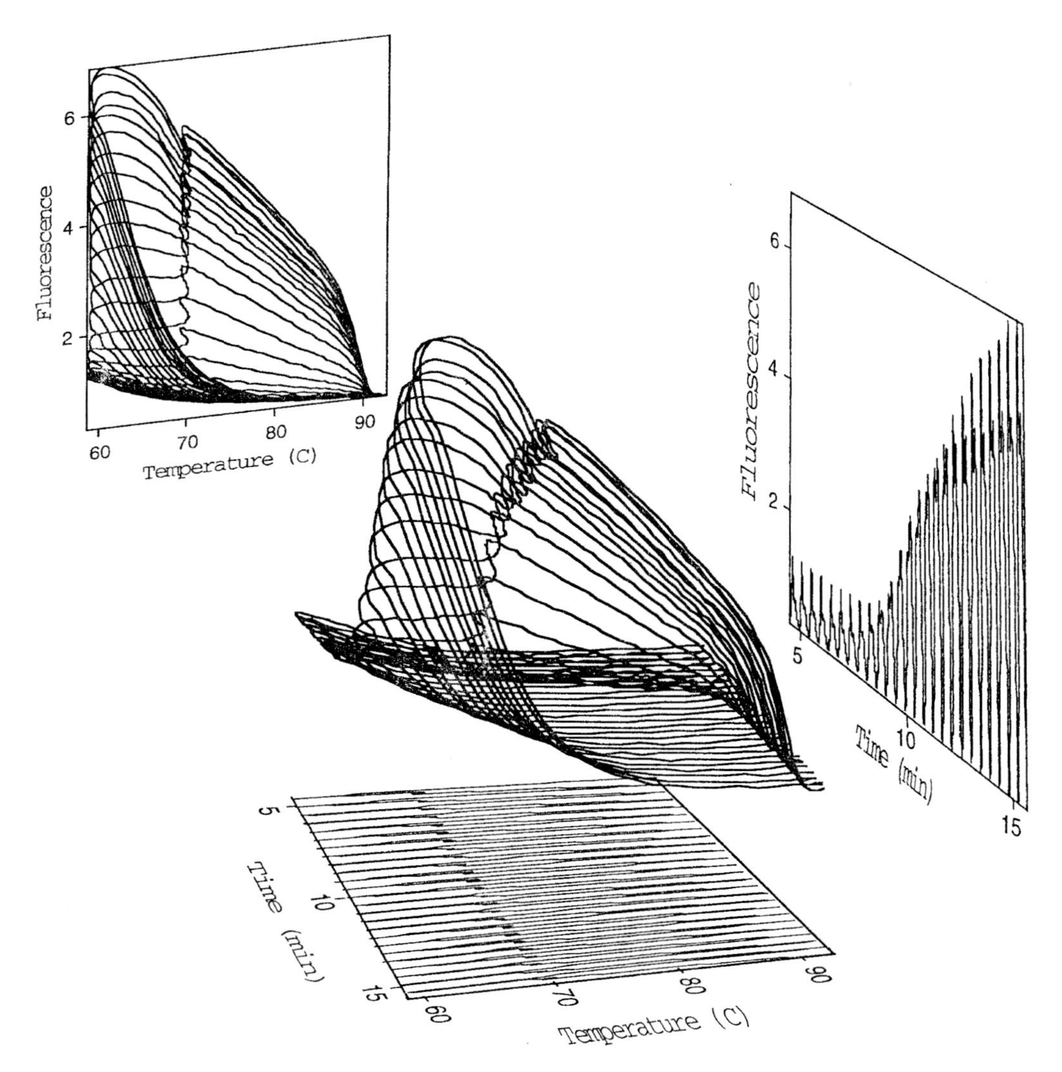

Figure 4–3. Continuous monitoring of rapid-cycle polymerase chain reaction (PCR). A 110-base pair β-globin fragment was amplified from human genomic deoxyribonucleic acid (DNA) by cycling between 60°C and 90°C over 15 minutes in the presence of SYBR® Green I. Center: Three-dimensional plot of temperature, time, and fluorescence. Bottom: Temperature profile over time. Right: Fluorescence profile over time. Top left: Fluorescence-versus-temperature plot that continuously monitors hybridization.

any processing; all dyes or probes are typically added before the start of PCR. Probe melting curves are often used for genotyping. PCR product melting curves are commonly used to verify the product amplified and to scan the product for heteroduplexes.

PCR product melting

Figure 4–3 shows continuous monitoring of PCR using the dye SYBR® Green I, a dye that specifically fluoresces in the presence of double-stranded DNA. If time, temperature, and fluorescence are plotted during rapid-cycle PCR, a complex three-dimensional spiral results. Two-dimensional plots of only two parameters are easier to visualize, and there are three possible combinations of the three parameters taken two at a time. The time course of rapid-cycle PCR is nicely displayed using the parameters temperature and time. Plotting fluorescence versus time was first demonstrated using ethidium bromide with the introduction of real-time PCR.[44] Finally, hybridization during PCR is best displayed on fluorescence-versus-temperature plots.[10,40] As the temperature is decreased each cycle, the fluorescence rises, resulting from both increased quantum efficiency at lower temperatures and hybridization of duplex PCR products. During each extension cycle, fluorescence also increases as more double-stranded DNA is synthesized. As the sample is heated toward denaturation, a rapid drop in fluorescence indicates PCR product melting. Although hybridization can be measured continuously during amplification, melting curves are usually obtained after PCR is complete and therefore do not require real-time PCR.

The melting temperature of a PCR product depends on its GC content, length, and sequence, and is usually monitored with the asymmetric cyanine dyes SYBR® Green I or LCGreen® Plus. SYBR® Green I was first used in real-time PCR for quantification in 1997.[10,40] Because probes were not required, costs for quantitative real-time PCR were greatly reduced.[45] Soon after, SYBR® Green I was used for melting curve analysis of PCR products after amplification.[39] Different PCR products had different melting temperatures, allowing a simple, closed-tube method of analysis. In contrast to gel analysis, where products are identified by size, melting analysis categorizes PCR products by melting temperature. It is now common practice to use melting curve analysis at the end of PCR to confirm amplification of the intended products. Furthermore, real-time quantitative PCR with SYBR® Green I can be improved by collecting fluorescence at each cycle just below the melting transition of the intended product to exclude any contribution from low-melting side products.[45]

High-resolution melting analysis

The T_m of a PCR product is a convenient metric, but it is only one point on the melting curve. Much more information is contained in the complete melting curve than just the T_m. High-resolution fluorescence methods to precisely follow the entire melting transition were introduced in 2002.[46] Melting analysis resolution was dramatically improved by increased temperature and fluorescence precision, accuracy, and data density, while eliminating smoothing procedures. Most conventional real-time PCR instruments have low resolution and do not perform well by comparison. Detailed technical evaluations of sixteen different melting instruments have been reported,[7,8] and additional comments on instrumentation can be found in a recent review.[47]

High-resolution fluorescent melting analysis was first performed with fluorescently labeled primers that generated a 113-base pair amplicon of β-globin bracketing the hemoglobin S, C, and E single base variants. All homozygotes (AA, SS,

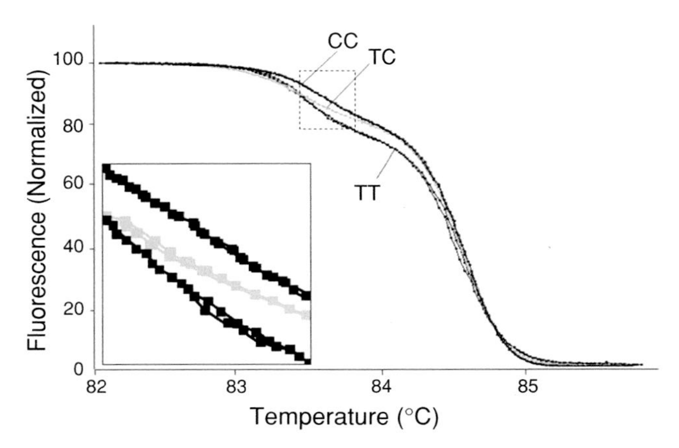

Figure 4–4. Genotyping by amplicon melting. A single base variant within a 544-base pair fragment of human *HTR2A* is genotyped after rapid-cycle polymerase chain reaction (PCR) by high-resolution melting analysis. The three genotypes, wild-type homozygote (TT), mutant homozygote (CC), and heterozygote (TC), are clearly distinguished in the low-temperature melting domain. Inset: Magnification of the data in the box. [Adapted by permission of the publisher from Wittwer CT, Reed GH, Gundry CN, et al. (2003) High resolution genotyping by amplicon melting analysis using LCGreen. *Clinical Chemistry* 49: 853–860. AACC.]

CC, and EE) and heterozygotes (AS, AC, AE, SC) could be distinguished from each other. The melting curves of the heterozygotes were different from homozygotes because multiple duplexes (homoduplexes and heteroduplexes) of different stability were formed from the heterozygotes resulting in a change in shape of the melting transition. Homozygotes were distinguished from each other by T_m, and all heterozygotes differed in shape from each other. Genotyping by amplicon melting became more difficult as the amplicon size increased. Similarly, differences between alleles decreased as the distance from the labeled primer increased, and the labeled primer had to be in the same melting domain as the sequence variant. This problem was solved in 2003 with the introduction of saturation dyes.[37]

Although SYBR® Green I can detect homozygous differences, it usually fails to detect heterozygotes[37,48] With SYBR® Green I, it is difficult to saturate the PCR product with dye because only limited concentrations can be used before it inhibits PCR. Much better results are possible with a new generation of "saturation" dyes, specifically developed for high-resolution melting. The first of these dyes to be introduced under the trade name "LCGreen®" is compatible with PCR over a wide range of concentrations. Single base variants and small insertions or deletions are easily detected and genotyped with these dyes.

With saturation dyes, the PCR product is labeled along its entire length, so that all melting domains are detected as shown in Figure 4–4. All genotypes of a C/T single base variant in a two-domain melting curve of a 544-base pair PCR product are shown. The difference between genotypes is revealed in the lower temperature domain, whereas the higher melting domain remains constant. Genotyping of most variants is possible by high-resolution melting of PCR products. However, robust, specific PCR is critical. Use of a gradient thermal cycler followed by melting analysis and/or gel electrophoresis is a good method for optimization of conditions.

With high-resolution melting, heterozygotes are easy to distinguish from homozygotes by a change in curve shape. However, it is more difficult to differentiate between different homozygotes or between different heterozygotes. Most homozygotes can be distinguished from each other by small T_m differences using short amplicons.[49] Approximately 84% of human single base changes result in an A:T to G:C interchange with a T_m difference between alternative homozygotes of approximately 1°C. In the remaining 16%, the base pair is inverted or neutral (A:T to T:A or G:C to C:G) and the T_m difference is smaller. In approximately 4% of single base changes, nearest neighbor symmetry predicts no difference in T_m. In such cases, quantitative heteroduplex analysis by mixing of samples may be necessary for complete genotyping.[50] Similar to homozygotes, different heterozygotes usually can be distinguished from each other by differences in curve shape.[51,52] For example, in a study of *CFTR* variants, 93% of heterozygotes in the same amplicon were distinguishable, but 7% were not.[53]

High-resolution PCR product melting has been used to genotype many human (diploid) and microbial (monoploid) variants. A recent review provides a comprehensive compilation.[47] Multiplexing of two[54,55] and even four[56] short PCR products in the same reaction can provide genotyping of multiple products. Synthetic oligonucleotides can be included as internal temperature controls and may improve the resolution between homozygotes.[55–57]

In some cases, specific genotyping may be less relevant than determining whether DNA sequences from two different sources are the same. For example, in tissue transplantation, genotype–phenotype correlation, and forensics, establishing the sequence identity of highly polymorphic regions is more important than knowing what the sequence actually is. Human leukocyte antigen (HLA) sequence identity (matching) by high-resolution melting of the highly polymorphic HLA-A locus has been demonstrated in all seven cases of shared alleles among two individuals.[58] HLA genotype identity is suggested when two individuals have the same melting curves and is confirmed by comparing the melting curve of a 1:1 mixture with the individual melting curves. If the samples are not identical, different heteroduplexes are formed that change the shape of the melting curves.

Heterozygote scanning

Many methods are available to screen for differences between the two copies of DNA within an individual. They include single-strand conformational polymorphism (SSCP) analysis, denaturing gradient gel electrophoresis (DGGE), denaturing high-pressure liquid chromatography (DHPLC), and temperature gradient capillary electrophoresis (TGCE). Sequencing not only screens for differences, but provides a complete genotype except in cases where haplotyping is ambiguous. However, all of these methods require separation of the sample on a gel or other matrix, some after additional processing and/or enzymatic reactions.

The simplicity of heterozygote scanning by melting analysis without any processing in a closed-tube system is attractive by comparison.

Heterozygote scanning by high-resolution melting depends on the presence of heteroduplexes that alter the melting curve shape.[37,46] Single base changes, insertions, and deletions are all detected, as long as the PCR primers bracket the variation. This limitation is similar to sequencing: Heterozygous deletions of entire genes and exons will be missed by both methods. The sensitivity and specificity of scanning for heterozygous single base changes have been systematically studied using a set of engineered plasmids.[59] All single base combinations in PCR products from 50 base pairs to 1 kb in a background of 40%, 50%, or 60% GC content were studied. Sensitivity and specificity were 100% for PCR products less than 400 base pairs. Sensitivity dropped to 96.1% and specificity to 99.4% in PCR products between 400 and 1,000 base pairs. Scanning accuracy was not affected by the position of the variation within the PCR product. Subsequent studies on many different genomic targets confirm the high accuracy of heterozygote scanning by high-resolution melting.[47] For example, in a blinded study scanning all 27 exons of *CFTR* in 20 samples enriched for disease-causing variants, 87 heterozygous variants (36 unique) were correctly identified for a sensitivity of 100%. The sensitivity and specificity of high-resolution melting appears better than those of DHPLC in some studies,[60] whereas in others false negatives have been reported using variants initially detected by DHPLC.[61] High-resolution melting methods have been compared to other techniques in recent reviews.[47,62–64] It is likely that results obtained depend on the care and experience of the user, as well as the specific instrument, reagents, and software used for melting analysis.

With sensitivities approaching 100%, can high-resolution melting replace sequencing for whole gene analysis where the vast majority of PCR products covering exons and flanking splice sites are normal? Conservative wisdom suggests that any sample found negative by scanning should be fully sequenced to detect missed heterozygous or homozygous variants.[61] Time will tell whether high-resolution melting can detect variants at sensitivities high enough to displace sequencing. Although most heterozygote scanning methods do not detect homozygous variants, high-resolution melting analysis is an exception. In a blinded study of *CFTR* scanning,[53] 75% of homozygotes were detected, and up to 96% should be detectable in short amplicons.[49] Of course, mixing of homozygous variants with wild-type DNA can also be used to convert them to heterozygotes that are easier to detect.

Even if all variants are detected, benign variants still need to be distinguished from disease-causing variants. Sequencing can accomplish this, but if benign variants are common, many PCR products will require sequencing. Alternatively, common benign variants can be identified by comparing them to standards. Identical amplicon melting curves are strong evidence of sequence identity. Identity can be confirmed by mixing the unknown with the standard and remelting, by specific genotyping, or by sequencing. Small amplicon melting or unlabeled probes (see next section) are attractive for genotyping common variants because they use the same methods (instruments and saturation dyes) as

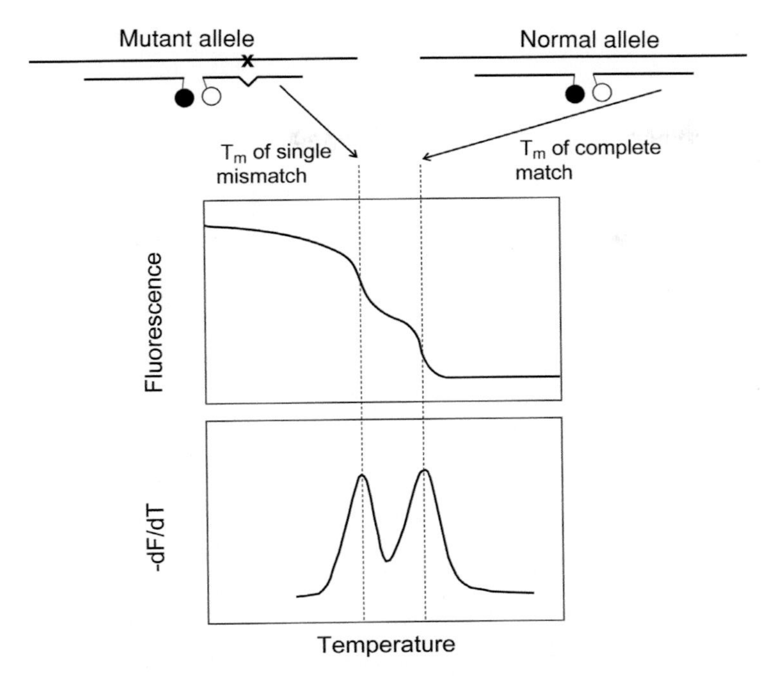

Figure 4–5. Melting curve genotyping with probes. A heterozygous sample is polymerase chain reaction (PCR) amplified and melted in the presence of classical dual hybridization probes. Two temperature transitions result, one from the mutant allele that is mismatched with the probe and dissociates at a lower temperature, and one from the normal allele that is completely matched with the probe and dissociates at a higher temperature. The lower derivative plot shows the melting temperatures of both the mutant-probe and the normal-probe duplexes as peaks.

heteroduplex scanning. In one example that scanned 24 exons,[52] benign polymorphisms were present in 96% of normal samples, greatly reducing the positive predictive value of heteroduplex scanning for mutation detection. When common polymorphisms were identified by amplicon melting, the positive predictive value for mutation detection increased to 100%. Melting curves of the same genotype were mathematically clustered together, eliminating the guesswork of genotype assignment. In the large majority of cases, common polymorphisms were eliminated by amplicon melting alone. Results were confirmed by unlabeled probe melting analysis, although secondary genotyping or sequencing appeared unnecessary.

Genotyping by probe melting

Genotyping by probe melting in solution was initially performed with a labeled primer and a labeled probe,[65] although adjacent single-labeled probes subsequently became more popular.[66] The method is inherently more powerful than allele-specific techniques because many different alleles are distinguished. Hybridization is monitored over a range of temperatures, rather than at only a single temperature, producing a "dynamic dot blot" (Figure 4–5). Depending on the sequence under the probe, different alleles result in different probe melting

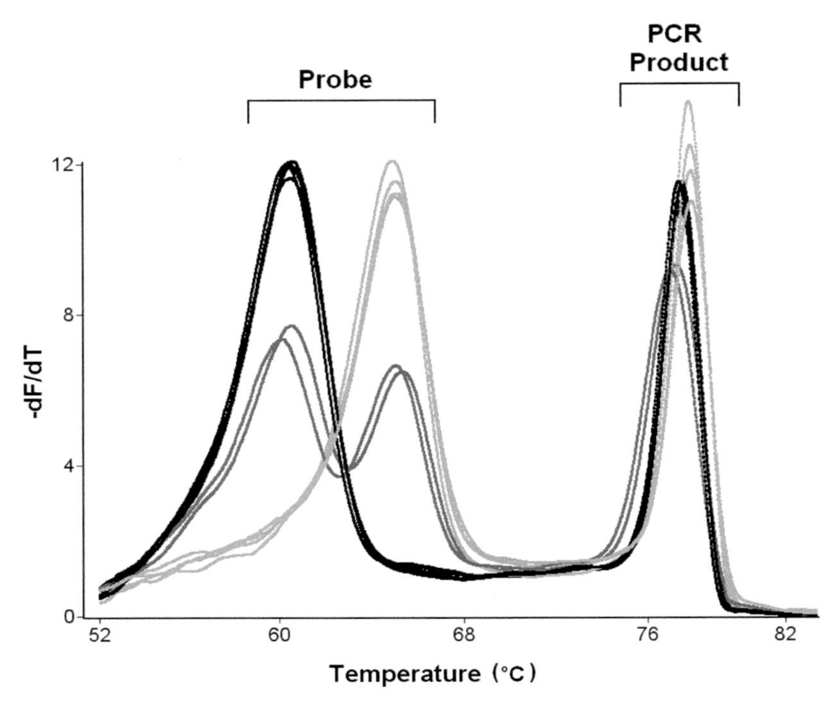

Figure 4–6. Unlabeled probe genotyping of an A/C single base variant. An 86-base pair polymerase chain reaction (PCR) product was amplified by asymmetric rapid-cycle PCR in the presence of a 25-base pair unlabeled probe and LCGreen® Plus. Melting data over both the probe region and the amplicon region were obtained at 0.3°C/sec and displayed as a derivative plot. The probe region clearly allows genotyping. Less obvious is the fact that the same genotype can be discerned from the PCR product melting region. Unlabeled probe melting analysis can provide simultaneous heterozygote scanning and genotyping.[74]

temperatures. Heterozygous PCR products are easily distinguished from homozygous samples by a double peak on derivative melting curve plots. The first U.S. Food and Drug Administration (FDA)-approved genetic assays in the United States (*F5* and *F2* single base variants) used this method. Both fluorescence color and T_m can be exploited for multiplexing[67]; for example, in genotyping HbC, HbS, and HbE of human β-globin.[68]

Subsequently, genotyping by probe melting was simplified to reduce the number of labeled probes required. First, by relying on nucleobase quenching,[69] only one labeled probe was needed.[70] Later, all labeled probes were eliminated by using a saturating DNA dye to monitor melting of an unlabeled probe.[48] To prevent polymerase extension from the probe, the 3′ end is terminated with a phosphate or other blocker.[71] Unequal primer concentrations are used to generate one strand in excess that hybridizes to the complementary unlabeled probe. Both probe and amplicon duplexes are saturated with dye, giving melting regions for both the probe and the amplicon (Figure 4–6). The probes are usually present during PCR, but can be added after amplification is complete without breaking the closed-tube environment.[72]

Genotyping by probe melting can discriminate multiple variants under the probe. High-resolution melting improves the quality of the probe melting curves and allows even more variants to be distinguished from each other. Probes can be designed to mask certain variants or segments by incorporating deletions, mismatches, or universal bases.[73] Multiple unlabeled probes can interrogate different regions in the same PCR product. For example, two unlabeled probes strategically positioned within exon 10 of the cystic fibrosis gene were used to genotype six different variants.[74] Genotyping with unlabeled probes has been recently reviewed.[47,75] Because no covalent labels or separations are required, it is the simplest, least expensive probe-based genotyping method available. However, software for adequate elimination of background fluorescence is necessary.[75]

CONCLUSION

The utility of any DNA analysis method depends on how fast it can be performed, how much information is obtained, and how much trouble it is to do. Compared to conventional cloning techniques, PCR is fast and simple. Rapid-cycle PCR focuses on continued reduction of the time required. Real-time PCR increases the information content by acquiring data during each cycle. Melting analysis can be performed during or after PCR and monitors DNA hybridization continuously as the temperature is increased. High-resolution melting provides simple solutions for genotyping, sequence matching, and heterozygote scanning. PCR, real-time PCR, and melting analysis will be with us for a long time because they are rapid, inexpensive, and rich in information content. Continuous monitoring of PCR during all stages may yet provide additional information without increasing the time or expense required.

REFERENCES

1. Wittwer CT, Reed GB, Ririe KM (1994) Rapid cycle DNA amplification. In: K Mullis, F Ferre, and R Gibbs (eds), The Polymerase Chain Reaction, pages 174–181. Deerfield Beach, FL: Springer-Verlag.
2. Wittwer CT, Fillmore GC, Hillyard DR (1989) Automated polymerase chain reaction in capillary tubes with hot air. *Nucleic Acids Research* **17**: 4353–4357.
3. Wittwer CT, Fillmore GC, Garling DJ (1990) Minimizing the time required for DNA amplification by efficient heat transfer to small samples. *Analytical Biochemistry* **186**: 328–331.
4. Wittwer CT, Garling DJ (1991) Rapid cycle DNA amplification: time and temperature optimization. *BioTechniques* **10**: 76–83.
5. Wittwer CT, Marshall BC, Reed GH, Cherry JL (1993) Rapid cycle allele-specific amplification: studies with the cystic fibrosis delta F508 locus. *Clinical Chemistry* **39**: 804–809.
6. Schoder D, Schmalwieser A, Schauberger G, Hoorfar J, Kuhn M, Wagner M (2005) Novel approach for assessing performance of PCR cyclers used for diagnostic testing. *Journal of Clinical Microbiology* **43**: 2724–2728.
7. Herrmann MG, Durtschi JD, Wittwer CT, Voelkerding KV (2007) Expanded instrument comparison of amplicon DNA melting analysis for mutation scanning and genotyping. *Clinical Chemistry* **53**: 1544–1548.

8. Herrmann MG, Durtschi JD, Bromley LK, Wittwer CT, Voelkerding KV (2006) Amplicon DNA melting analysis for mutation scanning and genotyping: cross-platform comparison of instruments and dyes. *Clinical Chemistry* **52**: 494–503.
9. Raja S, El-Hefnawy T, Kelly LA, Chestney ML, Luketich JD, Godfrey TE (2002) Temperature-controlled primer limit for multiplexing of rapid, quantitative reverse transcription-PCR assays: application to intraoperative cancer diagnostics. *Clinical Chemistry* **48**: 1329–1337.
10. Wittwer CT, Ririe KM, Andrew RV, David DA, Gundry RA, Balis UJ (1997) The LightCycler: a microvolume multisample fluorimeter with rapid temperature control. *BioTechniques* **22**: 176–181.
11. Wittwer CT, Ririe KM, Rasmussen RP (1998) Fluorescence monitoring of rapid cycle PCR for quantification. In: F Ferre (ed), Gene Quantification, pages 129–144. New York: Birkhauser.
12. Elenitoba-Johnson O, David D, Crews N, Wittwer CT (2008) Plastic vs glass capillaries for rapid-cycle PCR. *BioTechniques* **44**: 487–488, 490, 492.
13. Roper MG, Easley CJ, Landers JP (2005) Advances in polymerase chain reaction on microfluidic chips. *Analytical Chemistry* **77**: 3887–3893.
14. Zhang C, Xing D (2007) Miniaturized PCR chips for nucleic acid amplification and analysis: latest advances and future trends. *Nucleic Acids Research* **35**: 4223–4237.
15. Cheng J, Shoffner MA, Hvichia GE, Kricka LJ, Wilding P (1996) Chip PCR. II. Investigation of different PCR amplification systems in microfabricated silicon-glass chips. *Nucleic Acids Research* **24**: 380–385.
16. Woolley AT, Hadley D, Landre P, deMello AJ, Mathies RA, Northrup MA (1996) Functional integration of PCR amplification and capillary electrophoresis in a microfabricated DNA analysis device. *Analytical Chemistry* **68**: 4081–4086.
17. Neuzil P, Zhang C, Pipper J, Oh S, Zhuo L (2006) Ultra fast miniaturized real-time PCR: 40 cycles in less than six minutes. *Nucleic Acids Research* **34**: e77.
18. Oda RP, Strausbauch MA, Huhmer AF, Borson N, Jurrens SR, Craighead J, et al. (1998) Infrared-mediated thermocycling for ultrafast polymerase chain reaction amplification of DNA. *Analytical Chemistry* **70**: 4361–4368.
19. Roper MG, Easley CJ, Legendre LA, Humphrey JA, Landers JP (2007) Infrared temperature control system for a completely noncontact polymerase chain reaction in microfluidic chips. *Analytical Chemistry* **79**: 1294–1300.
20. Friedman NA, Meldrum DR (1998) Capillary tube resistive thermal cycling. *Analytical Chemistry* **70**: 2997–3002.
21. Heap DM, Herrmann MG, Wittwer CT (2000) PCR amplification using electrolytic resistance for heating and temperature monitoring. *BioTechniques* **29**: 1006–1012.
22. Kopp MU, Mello AJ, Manz A (1998) Chemical amplification: continuous-flow PCR on a chip. *Science* **280**: 1046–1048.
23. Hashimoto M, Chen PC, Mitchell MW, Nikitopoulos DE, Soper SA, Murphy MC (2004) Rapid PCR in a continuous flow device. *Lab on a Chip* **4**: 638–645.
24. Crews N, Wittwer C, Gale B (2008) Continuous-flow thermal gradient PCR. *Biomedical Microdevices* **10**: 187–195.
25. Chiou JT, Matsudaira PT, Ehrlich DJ (2002) Thirty-cycle temperature optimization of a closed-cycle capillary PCR machine. *BioTechniques* **33**: 557–558, 560, 562.
26. Frey O, Bonneick S, Hierlemann A, Lichtenberg J (2007) Autonomous microfluidic multi-channel chip for real-time PCR with integrated liquid handling. *Biomedical Microdevices* **9**: 711–718.
27. Chen J, Wabuyele M, Chen H, Patterson D, Hupert M, Shadpour H, et al. (2005) Electrokinetically synchronized polymerase chain reaction microchip fabricated in polycarbonate. *Analytical Chemistry* **77**: 658–666.
28. Sun Y, Kwok YC, Nguyen NT (2007) A circular ferrofluid driven microchip for rapid polymerase chain reaction. *Lab on a Chip* **7**: 1012–1017.
29. Agrawal N, Hassan YA, Ugaz VM (2007) A pocket-sized convective PCR thermocycler. *Angewandte Chemie (International ed. in English)* **46**: 4316–4319.

30. Zhang C, Xu J, Ma W, Zheng W (2006) PCR microfluidic devices for DNA amplification. *Biotechnology Advances* **24**: 243–284.
31. Wheeler EK, Benett W, Stratton P, Richards J, Chen A, Christian A, et al. (2004) Convectively driven polymerase chain reaction thermal cycler. *Analytical Chemistry* **76**: 4011–4016.
32. Belgrader P, Benett W, Hadley D, Long G, Mariella R Jr, Milanovich F, et al. (1998) Rapid pathogen detection using a microchip PCR array instrument. *Clinical Chemistry* **44**: 2191–2194.
33. Wilhelm J, Hahn M, Pingoud A (2000) Influence of DNA target melting behavior on real-time PCR quantification. *Clinical Chemistry* **46**: 1738–1743.
34. Zuna J, Muzikova K, Madzo J, Krejci O, Trka J (2002) Temperature non-homogeneity in rapid airflow-based cycler significantly affects real-time PCR. *BioTechniques* **33**: 508, 510, 512.
35. von Kanel T, Adolf F, Schneider M, Sanz J, Gallati S (2007) Sample number and denaturation time are crucial for the accuracy of capillary-based LightCyclers. *Clinical Chemistry* **53**: 1392–1394.
36. Wittwer CT, Herrmann MG (1999) Rapid thermal cycling and PCR kinetics. In: M Innis, D Gelfand, and J Sninsky (eds), PCR Methods Manual, pages 211–229. San Diego: Academic Press.
37. Wittwer CT, Reed GH, Gundry CN, Vandersteen JG, Pryor RJ (2003) High-resolution genotyping by amplicon melting analysis using LCGreen. *Clinical Chemistry* **49**: 853–860.
38. von Ahsen N, Wittwer CT, Schutz E (2001) Oligonucleotide melting temperatures under PCR conditions: nearest-neighbor corrections for Mg(2+), deoxynucleotide triphosphate, and dimethyl sulfoxide concentrations with comparison to alternative empirical formulas. *Clinical Chemistry* **47**: 1956–1961.
39. Ririe KM, Rasmussen RP, Wittwer CT (1997) Product differentiation by analysis of DNA melting curves during the polymerase chain reaction. *Analytical Biochemistry* **245**: 154–160.
40. Wittwer CT, Herrmann MG, Moss AA, Rasmussen RP (1997) Continuous fluorescence monitoring of rapid cycle DNA amplification. *BioTechniques* **22**: 130–131, 134–138.
41. Weis JH, Tan SS, Martin BK, Wittwer CT (1992) Detection of rare mRNAs via quantitative RT-PCR. *Trends in Genetics* **8**: 263–264.
42. Brown RA, Lay MJ, Wittwer CT (1998) Rapid cycle amplification for construction of competitive templates. In: RM Horton and RC Tait (eds), Genetic Engineering with PCR, pages 57–70. Norfolk: Horizon Scientific Press.
43. Wittwer CT, Kusukawa N (2004) Real-time PCR. In: DH Persing, FC Tenover, J Versalovic, YW Tang, ER Unger, DA Relman, et al. (eds), Diagnostic Molecular Microbiology: Principles and Applications, pages 71–84. Washington, DC: ASM Press.
44. Higuchi R, Dollinger G, Walsh PS, Griffith R (1992) Simultaneous amplification and detection of specific DNA sequences. *Biotechnology (NY)* **10**: 413–417.
45. Morrison TB, Weis JJ, Wittwer CT (1998) Quantification of low-copy transcripts by continuous SYBR Green I monitoring during amplification. *BioTechniques* **24**: 954–958, 960, 962.
46. Gundry CN, Vandersteen JG, Reed GH, Pryor RJ, Chen J, Wittwer CT (2003) Amplicon melting analysis with labeled primers: a closed-tube method for differentiating homozygotes and heterozygotes. *Clinical Chemistry* **49**: 396–406.
47. Reed GH, Kent JO, Wittwer CT (2007) High-resolution DNA melting analysis for simple and efficient molecular diagnostics. *Pharmacogenomics* **8**: 597–608.
48. Zhou L, Myers AN, Vandersteen JG, Wang L, Wittwer CT (2004) Closed-tube genotyping with unlabeled oligonucleotide probes and a saturating DNA dye. *Clinical Chemistry* **50**: 1328–1335.
49. Liew M, Pryor R, Palais R, Meadows C, Erali M, Lyon E, et al. (2004) Genotyping of single-nucleotide polymorphisms by high-resolution melting of small amplicons. *Clinical Chemistry* **50**: 1156–1164.

50. Palais RA, Liew MA, Wittwer CT (2005) Quantitative heteroduplex analysis for single nucleotide polymorphism genotyping. *Analytical Biochemistry* **346**: 167–175.
51. Graham R, Liew M, Meadows C, Lyon E, Wittwer CT (2005) Distinguishing different DNA heterozygotes by high-resolution melting. *Clinical Chemistry* **51**: 1295–1298.
52. Vandersteen JG, Bayrak-Toydemir P, Palais RA, Wittwer CT (2007) Identifying common genetic variants by high-resolution melting. *Clinical Chemistry* 53: 1191–1198.
53. Montgomery J, Wittwer CT, Kent JO, Zhou L (2007) Scanning the cystic fibrosis transmembrane conductance regulator gene using high-resolution DNA melting analysis. *Clinical Chemistry* **53**: 1891–1898.
54. Liew M, Nelson L, Margraf R, Mitchell S, Erali M, Mao R, et al. (2006) Genotyping of human platelet antigens 1 to 6 and 15 by high-resolution amplicon melting and conventional hybridization probes. *The Journal of Molecular Diagnostics* **8**: 97–104.
55. Seipp MT, Durtschi JD, Liew MA, Williams J, Damjanovich K, Pont-Kingdon G, et al. (2007) Unlabeled oligonucleotides as internal temperature controls for genotyping by amplicon melting. *The Journal of Molecular Diagnostics* **9**: 284–289.
56. Seipp MT, Pattison D, Durtschi JD, Jama M, Voelkerding KV, Wittwer CT (2008) Quadruplex genotyping of F5, F2, and MTHFR variants in a single closed tube by high-resolution amplicon melting. *Clinical Chemistry* **54**: 108–115.
57. Liew M, Seipp M, Durtschi J, Margraf RL, Dames S, Erali M, et al. (2007) Closed-tube SNP genotyping without labeled probes/a comparison between unlabeled probe and amplicon melting. *American Journal of Clinical Pathology* **127**: 1–8.
58. Zhou L, Vandersteen J, Wang L, Fuller T, Taylor M, Palais B, et al. (2004) High-resolution DNA melting curve analysis to establish HLA genotypic identity. *Tissue Antigens* **64**: 156–164.
59. Reed GH, Wittwer CT (2004) Sensitivity and specificity of single-nucleotide polymorphism scanning by high-resolution melting analysis. *Clinical Chemistry* **50**: 1748–1754.
60. Chou LS, Lyon E, Wittwer CT (2005) A comparison of high-resolution melting analysis with denaturing high-performance liquid chromatography for mutation scanning: cystic fibrosis transmembrane conductance regulator gene as a model. *American Journal of Clinical Pathology* **124**: 330–338.
61. Laurie AD, Smith MP, George PM (2007) Detection of Factor VIII gene mutations by high-resolution melting analysis. *Clinical Chemistry* **53**: 2211–2214.
62. Dujols VE, Kusukawa N, McKinney JT, Dobrowolski SF, Wittwer CT (2006) High-resolution melting analysis for scanning and genotyping. In: MT Dorak (ed), Real-Time PCR, pages 157–171. New York: Garland Science.
63. Gingeras TR, Higuchi R, Kricka LJ, Lo YM, Wittwer CT (2005) Fifty years of molecular (DNA/RNA) diagnostics. *Clinical Chemistry* **51**: 661–671.
64. Wittwer CT, Kusukawa N (2005) Nucleic acid techniques. In: C Burtis, E Ashwood, and D Bruns (eds), Tietz Textbook of Clinical Chemistry and Molecular Diagnostics. Fourth edition, pages 1407–1449. New York: Elsevier.
65. Lay MJ, Wittwer CT (1997) Real-time fluorescence genotyping of factor V Leiden during rapid-cycle PCR. *Clinical Chemistry* **43**: 2262–2267.
66. Bernard PS, Ajioka RS, Kushner JP, Wittwer CT (1998) Homogeneous multiplex genotyping of hemochromatosis mutations with fluorescent hybridization probes. *American Journal of Pathology* **153**: 1055–1061.
67. Wittwer CT, Herrmann MG, Gundry CN, Elenitoba-Johnson KS (2001) Real-time multiplex PCR assays. *Methods* **25**: 430–442.
68. Herrmann MG, Dobrowolski SF, Wittwer CT (2000) Rapid beta-globin genotyping by multiplexing probe melting temperature and color. *Clinical Chemistry* **46**: 425–428.
69. von Ahsen N (2003) Labeled primers for mutation scanning: making diagnostic use of the nucleobase quenching effect. *Clinical Chemistry* **49**: 355–356.
70. Crockett AO, Wittwer CT (2001) Fluorescein-labeled oligonucleotides for real-time PCR: using the inherent quenching of deoxyguanosine nucleotides. *Analytical Biochemistry* **290**: 89–97.

71. Dames SA, Margraf RL, Pattison D, Wittwer CT, Voelkerding KV (2007) Characterization of aberrant melting peaks in unlabeled probe assays. *The Journal of Molecular Diagnostics* **9**: 290–296.
72. Poulson MD, Wittwer CT (2007) Closed-tube genotyping of apolipoprotein E by isolated-probe PCR with multiple unlabeled probes and high-resolution DNA melting analysis. *BioTechniques* **43**: 87–91.
73. Margraf RL, Mao R, Wittwer CT (2006) Masking selected sequence variation by incorporating mismatches into melting analysis probes. *Human Mutation* **27**: 269–278.
74. Zhou L, Wang L, Palais R, Pryor R, Wittwer CT (2005) High-resolution DNA melting analysis for simultaneous mutation scanning and genotyping in solution. *Clinical Chemistry* **51**: 1770–1777.
75. Erali M, Palais B, Wittwer C (2008) SNP genotyping by unlabeled probe melting analysis. In: O Seitz and A Marx (eds), Molecular Beacons – Signaling Nucleic Acid Probes, Methods and Protocols, pp. 199–206 (Methods in Molecular Biology Series, Vol. 429). Totowa, NJ: Humana Press.

5 Polymerase chain reaction and fluorescence chemistries: deoxyribonucleic acid incarnate

Ben Sowers

ILLUMINATING THE UNSEEN

The polymerase chain reaction (PCR) is a revolutionary piece of chemistry that has upended the science of biology by allowing manipulation of individual molecules of deoxyribonucleic acid (DNA). With a broad reach into drug development, forensics, and the sequencing of the human genome, PCR has already touched persons who may never pick up a pipette. Relying upon awkward language like oligonucleotide and amplicon makes explanation of its mechanism a difficult lesson, but replicating any sequence of DNA is of fundamental importance to modern biotechnology.

And replicate it does. Not only does PCR succeed in finding the needle in the haystack, it proceeds to make an entire haystack out of needles! This unrelenting amplification of the DNA molecule evokes images of the Sorcerer's Apprentice, where the utility of a single broom is recognized and duplicated until it takes on a life of its own.[1]

In certain laboratories, PCR can also wear out its welcome when the sorcery becomes hard to control. Replicating the very sequence that one intends to detect is both a blessing and a curse because residual molecules from previous experiments make it hard to start again with a clean slate. They can splash onto the laboratory bench, cling to gloves, and even launch into the air to leisurely float down onto an inconvenient location (inconvenient for the researcher, at least).

Without broomlike proportions, how does one know that these molecules are even there – let alone multiplying? How does the scientist peer into the broth of the test tube and gauge whether anything is happening? When concentrated,

DNA has the consistency of mucus, but for the most part these molecules drift about within transparent solutions, remaining hidden to the naked eye.

Wavelengths beyond the perception of our eyesight can reveal nucleic acid. The molecule absorbs ultraviolet (UV) light around 260 nanometers, permitting measurement according to Beer's Law: "The deeper the glass, the darker the brew, the less of the incident light that gets through" – anon. DNA is certainly not the only thing that absorbs this energetic light: Other biological materials like proteins absorb here as does the glass of the test tube, obscuring our view.

Even with illumination by UV light and the assistance of a spectrometer to discriminate its presence, this method has a limitation in detecting DNA: An astronomical number of molecules barely can be discerned; even more are needed to quantify with confidence. Rather than grasping at shadows, this molecule deserves the limelight, so methods were devised for DNA to emit a signal rather than absorb one.

ISOTOPES AND FLUOROPHORES

Radioactive isotopes were among the first tools used to signal the presence of a DNA molecule and were quickly embraced for genetic analysis. Atomic beacons like the ^{32}P isotope could be embedded into specific sequences, or probes, and their location traced by the trail of radiation. DNA has an unprecedented ability to recognize and bind its complementary sequence even if countless others entangle its search. A short strand of DNA that has been radiolabeled can be used to highlight a particular gene from within a background of other nucleic acids.[2] Autoradiograms record a ghostly photograph, with bands of DNA emerging from the abyss.

Radioisotopes have been integral to many developments in DNA technology. Sanger's method of sequencing DNA was originally demonstrated using radiolabeled fragments and, through optimization, became the dominant sequencing method for thirty years.[3–5] Radioisotopes also have allowed researchers to peer into the structure of genes and expose important correlations between genetic variation and inherited disease. For example, a telltale pattern of DNA bands could be correlated with individuals afflicted by sickle-cell anemia.[6] Despite these successes, the obvious hazards of handling and disposing of radioactive material have kept these isotopes from becoming embraced more fully. Methods that emit luminescence rather than radiation are a sensible alternative.

Originally developed to treat parasites in livestock, ethidium bromide is the quintessential substance used to expose DNA. More properly known as 2,7-diamino, 9-phenylphenanthridinium 10-ethyl bromide, this molecule with the lengthy name is actually quite tiny. In fact, it is small enough to intercalate into the DNA itself, finding comfort between the consecutive bases that dictate the sequence of the strand.[7]

In the late 1960s, Jean-Bernard Le Pecq and others made several important discoveries upon combining nucleic acid with ethidium: When UV light irradiates the DNA molecule, the DNA in turn passes on this energy onto its stowaway.[8–10] The small molecule becomes elevated to an excited state, with all of its excitement released as fluorescence. To the naked eye it appears as a cool blue glow, like moonlight upon water. Ethidium is only weakly fluorescent without the protection of the DNA molecule but intensifies greatly when enveloped by the nucleotide bases. DNA that is heated does not provide this same shelter from the surrounding solution because the partnered strands of the double helix separate from one another.

Ethidium bromide is certainly not the only molecule to glow upon contact with DNA. Fluorophores with myriad properties were developed over the years and given fanciful names like Acridine orange, TOTO, and SYBR Green®.[11] Emitting fluorescence in a painter's palette of colors, these dyes bind to nucleic acid through distinct methods, and occasionally fire off a brighter signal to allow for the detection of even fewer molecules.[12] Yet none of them have quite risen to the prominence of ethidium bromide as the hallowed tool of molecular biology, ubiquitous across university campuses and the occasional appearance in a school classroom.

Ethidium sparked a tiny flare in the core of the DNA molecule but kindled a fire under molecular biology research. Le Pecq quickly catalogued a range of uses for this versatile molecule and even predicted with tremendous foresight that it could be "applied to the study of DNA polymerase in action."[13] It was only a matter of time before researchers thought to stain gels with ethidium, revealing glowing bands of DNA pulled through the porous matrix of agarose or polyacrylamide.[14–16] The earliest triumphs at cloning genes such as insulin, transplanting them from nature and into the test tube, were achieved by interpreting ethidium-stained gels.[17,18] Considering the limited arsenal of techniques before PCR, these researchers were clever to pick the correct DNA fragments to manipulate. With ethidium providing the illumination, whole genes could be collected now from far-flung species like jellyfish or fireflies and dropped into the familiar environment of the *Escherichia coli* bacterium where they were manipulated with ease. Recombinant DNA technology had dawned.

DNA BY DESIGN

Extracting DNA from a specialized tissue was a struggle, but remained the only way to isolate a rarely expressed gene. Resolving that particular sequence from the other transcripts cluttering the background required meticulous purification, along with a little bit of luck. With the invention of PCR in 1983, a tremendous burden was lifted off of genetic research, and DNA sequences began replicating on the benchtop of every biology laboratory.

PCR amplifies trace molecules to make their presence much more obvious, and the drive to boost detection sensitivity was suddenly relaxed. Why improve visualization when the specimen can be brought within reach? All of these pre-PCR detection systems were hardly obviated – if anything, quite the contrary. The fragments of DNA enriched by the PCR process still need to be identified.

Visualizing these fragments has become imperative, and fluorescent dyes, indispensable.

PCR was founded upon a major advancement in organic chemistry: the ability to manufacture short pieces of DNA, or oligonucleotides (oligos), entirely synthetically. The building blocks, or phosphoramidites, used by the chemist to sequentially add each base to the growing strand differ from those DNA building blocks found in the cell.[19] Despite this deviation from nature, the final piece of synthesized DNA behaves indistinguishably from its natural counterpart: It is embraced by enzymes, incorporated into genomes, and transcribed into ribonucleic acid (RNA).

DNA synthesizers now automate nucleotide chemistry, allowing the researcher to summon arbitrary sequences at will, but cannot outrival nature in one principal aspect: length. The repeated cycle of chemical reactions does not proceed with perfect efficiency and is quickly exhausted after 100 bases, whereas the longest chromosome in the human genome is approximately 250 million bases long! Enzymatic methods like the PCR are capable of replicating thousands of bases at a time, but only according to a preexisting template delineated by the synthetic oligos. Thus, whole genes are amplified through a poetic partnership of synthetic and enzymatic means. Techniques that fuse biology with industry continue to push the envelope of human capability.

The modular nature of DNA synthesis has allowed for enhancements never before conceived by nature. Besides the nucleotide bases, other unusual compounds can be introduced into the growing strand of this biomolecule, creating a hybrid with artificial capabilities. "Labels" have been introduced with a variety of features: They can attach to proteins, strengthen the binding of the oligo, resist cellular degradation, and glow in a variety of colors.[20,21] By tethering fluorophores directly onto the oligo, it is the DNA itself that is emitting light rather than a separate molecule with only a transient association. Labeling well-defined oligos in this manner achieves the same sequence fidelity as with radioactivity but without such toxicity, and with a much easier means of detection. Ethidium bromide remains indispensable for its illumination of all the nucleic acid within a particular sample, but a fluorescence-labeled oligo can focus this illumination exclusively onto that sequence of interest.

Prior to DNA synthesis, a limited collection of labels could be introduced through the use of enzymes but with less flexibility over the site of modification. Even when oligo manufacturing became routine, fluorophores were attached by hand after the synthesis using ester coupling, which is not amenable to automation.[22–24] Formulating fluorophores into phosphoramidites, the same format as the nucleotide building blocks, has allowed their incorporation at precise positions in an automated manner upon an oligo synthesizer (Figure 5–1).[21] Simply by inserting reagents into a computer-controlled instrument,fluorescence-labeled oligos are rapidly produced and purified, available for amplification within a matter of days (Figure 5–2).

All of these striking modifications create unlimited possibilities for design. Vivid fluorophores emit light from the UV to the infrared (Figure 5–3). Chains of carbon atoms can disrupt the sugar phosphate backbone of the DNA. Even

Figure 5–1. A fluorophore is slowly forged under the supervision of the dye chemist (Matt Lyttle, PhD). *See Color Plates.*

branching points can be introduced so that a single DNA strand forks into two. Fanciful designs are obtained by stringing together multiple modifications. All of these capabilities permit some degree of control over PCR's biggest pitfall.

SUCCESS BREEDS FAILURE

The ability to replicate a molecule billions of times over is both a fantastic capability and a tragic flaw. If even a handful of the replicated molecules escape into the laboratory environment, they can become impossible to avoid amplifying in the future. Yet detection systems involving chromatography, electrophoresis, and radioisotopes necessarily expose the contents of the PCR. The amplified fragments are pulled across gels, immobilized onto sheets of nitrocellulose, or fixed onto glass and plastic surfaces. All of this exposure increases the chances of laboratory contamination, and in this game of astronomical numbers the odds are very much in favor of the house.

Carry-over contamination is only a minor nuisance when the goal is a robust yield of DNA to manipulate further, perhaps to splice into a plasmid, transform a bacterium, and ultimately express the gene product, but the PCR process also has incredible utility as a tool of medicine and forensics, with DNA sequences proclaiming the diagnosis. In this context of molecular diagnostics, contamination becomes much more than an inconvenience because the intent is to confirm the presence or absence of a particular gene, or even quantify its prevalence.

PCR can reveal genetic predispositions toward cancers, measure the progress of infections such as human immunodeficiency virus (HIV), and screen for biosecurity threats such as anthrax. Increasingly, the culprits responsible for viral

Figure 5–2. David Seebach, oligo technician, operating the SuperSAM^{TM} high-throughput DNA synthesizer. *See Color Plates.*

infections are unmasked only through DNA amplification and analysis.[25] Within these applications, false positives can have disastrous consequences. Detection systems encapsulated entirely within a closed tube would avoid this risk because the amplified molecules cannot escape into the environment. Such a homogeneous detection format would allow PCR to be deployed outside of the molecular biology laboratory into environments that are not necessarily equipped with electrophoresis machines, UV-visible spectroscopy cameras, and radioactive waste disposal.[5]

Exactly this sort of system is achieved by spiking ethidium bromide into a transparent PCR tube before rather than after amplification.[26] Fluorescence

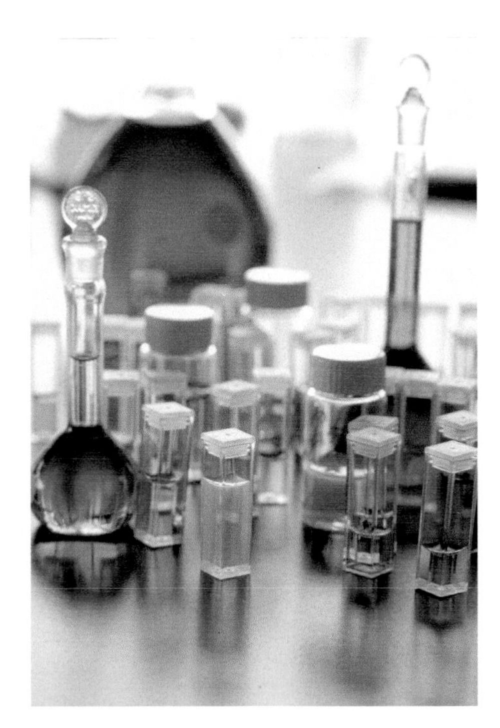

Figure 5–3. Flasks and cuvettes display an assortment of fluorophores and other dye labels intended to modify oligos. *See Color Plates.*

emitting from the tube during thermal cycling confirms the presence of a particular target sequence, which is now replicating. By simply combining fluorescent dyes into the reaction tube, PCR has been brought to the threshold of numerous diagnostic applications – but the system is not yet foolproof.

When given the opportunity, the PCR also can replicate undesirable sequences. In fact, in the absence of the true target, the transient association of primers with one another is sometimes sufficient to trigger amplification: "Idle hands do the devil's work and the same is true with primers" – Greg Shipley, PhD. Binding dyes fluoresce in response to any and all duplex DNA, and cannot provide complete sequence identification without further scrutinizing the contents.[27,28]

FLUOROGENIC PROBES

To avoid misinterpreting the fluorescent signal, oligo conjugates can be used in place of the DNA binding dyes. With fluorophores attached to either end of an oligo, sophisticated probes are engineered to fluoresce only upon disrupting a molecule's geometry.[29–32] The signal in such probes is carefully regulated by a molecular interaction called Förster Resonance Energy Transfer (FRET), causing the signal from one modification to be immediately captured by the other, rather than released as fluorescence. Alternately, the labels may be physically binding one another to extinguish the signal through static quenching.[33] Within either of these arrangements, the signal from the first cannot escape the grasp of the second until they become separated.[5]

The double helix formed upon the binding of a probe to its target is quite rigid, extending the reach of an oligo from its flexible unbound conformation, which more closely resembles a wet noodle.[34] Probes that bind to their target with a sufficiently strong grasp also can become severed by the polymerase during PCR, permanently liberating the two labels from one another.[35] Both of these mechanisms can disrupt the geometry of the probe to produce luminescence from within a closed tube, even visible to the naked eye.

The properties of these probes were further refined by the introduction of dark quenchers: molecules that have no intrinsic fluorescence of their own, but that can absorb the signal of adjacent fluorophores. These dark quenchers can be used to cloak multiple fluorophores, each capable of emitting a distinct color of light.[36] By introducing several fluorescence-quenched probes within the same reaction tube, it becomes possible to amplify multiple targets simultaneously but detect them independently.[37,38]

TRANSFORMING TOMORROW

Fluorophore–DNA conjugates have revolutionized our ability to detect sequences amplified during the PCR. These exquisite molecules have emerged only through incremental advancements in DNA chemistry spanning many decades of research. Oligo manufacturing is now automated using solid-phase DNA synthesizers. Phosphoramidite chemistry permits precise modification of the DNA strand. Striking fluorophores and dark quenchers build into a single molecule both the mechanism of quenching and signal release.

Fluorescence-labeled DNA is thus an artifice, born on the benchtop of a laboratory rather than in a primordial tide pool. Yet, this synthetic partnership is showing no sign of unraveling any time soon. In fact, methods of organic chemistry and biological wizardry have become even more entangled with each passing decade, producing new technologies that further augment our genetic acuity. With all of our technological sophistication, it is important to not lose sight of the forest for the trees. The culmination of this technology is a remarkably elegant system: a beacon of light emitting from a transparent plastic tube, and illuminating our path toward a new world of molecular diagnostics.

As humanity marches into an increasingly global society, the need to rapidly diagnose will only intensify. Repeated outbreaks of infectious agents (severe acute respiratory syndrome [SARS], avian influenza virus, drug-resistant tuberculosis) have highlighted the urgency of an early diagnosis and the consequences of delay. Many persons involved in health care argue that the spiraling costs of modern medicine would be best controlled by changing from a system based upon treatment to one based upon diagnosis.[39]

It is easy to envision compact instruments that identify pathogens in a matter of hours rather than weeks and distinguish the strains causing infection. Cancers will be distinguished to guide patients to the most appropriate treatment. Outside of medicine, agricultural crops can be more carefully bred according to their desirable genetic signatures. Food processes should be continuously screened for the emergence of spoilage organisms. In fact, these new procedures already have been implemented and are now becoming widespread; they all rely on fluorescence-labeled DNA to produce an answer. With each tiny flicker of light glancing across subtle optical sensors, our understanding of biology is becoming further refined and the very fabric of society is changing.

With an already impressive track record advancing biological research, PCR is on the verge of transforming science and technology once again. Even so, it is important to anticipate these new developments for both their positive and negative impact. Do we personally seek out our genetic predispositions toward disease? How do we regard DNA variations that intimately define our individuality? Society will slowly absorb all of this information and we, as individuals, will struggle to react to new questions. All the while, the clockwork repetition of DNA synthesis will hum in the background – in the nuclei of our cells and in the microfluidic chambers of machines.

REFERENCES

1. Dukas P (1897) The sorcerer's apprentice (L'apprenti sorcier). New York: E. F. Kalmus.
2. Southern EM (1975) Detection of specific sequences among DNA fragments separated by gel electrophoresis. *Journal of Molecular Biology* **98**(3): 503–517.
3. Sanger F, Nicklen S, Coulson AR (1977) DNA sequencing with chain-terminating inhibitors. *Proceedings of the National Academy of Sciences of the United States of America* **74**(12): 5463–5467.
4. Ansorge W, Sproat B, Stegemann J, Schwager C, Zenke M (1987) Automated DNA sequencing: ultrasensitive detection of fluorescent bands during electrophoresis. *Nucleic Acids Research* **15**(11): 4593–4602.
5. Didenko VV (2001) DNA probes using fluorescence resonance energy transfer (FRET): designs and applications. *BioTechniques* **31**(5): 1106–1116, 1118, 1120–1101.
6. Kan YW, Dozy AM (1978) Polymorphism of DNA sequence adjacent to human beta-globin structural gene: relationship to sickle mutation. *Proceedings of the National Academy of Sciences of the United States of America* **75**(11): 5631–5635.
7. Morgan AR, Lee JS, Pulleyblank DE, Murray NL, Evans DH (1979) Review: ethidium fluorescence assays. Part 1. Physicochemical studies. *Nucleic Acids Research* **7**(3): 547–569.
8. Waring MJ (1965) Complex formation between ethidium bromide and nucleic acids. *Journal of Molecular Biology* **13**(1): 269–282.
9. Le Pecq JB, Paoletti C (1967) A fluorescent complex between ethidium bromide and nucleic acids. Physical-chemical characterization. *Journal of Molecular Biology* **27**(1): 87–106.
10. Paoletti C, Le Pecq JB, Lehman IR (1971) The use of ethidium bromide-circular DNA complexes for the fluorometric analysis of breakage and joining of DNA. *Journal of Molecular Biology* **55**(1): 75–100.
11. Rye HS, Yue S, Wemmer DE, Quesada MA, Haugland RP, Mathies RA, et al. (1992) Stable fluorescent complexes of double-stranded DNA with bis-intercalating asymmetric cyanine dyes: properties and applications. *Nucleic Acids Research* **20**(11):2803–2812.
12. Mullis KB, Ferrâe F, Gibbs R (1994) The Polymerase Chain Reaction. Boston: Birkhäuser.
13. Le Pecq JB (1971) Use of ethidium bromide for separation and determination of nucleic acids of various conformational forms and measurement of their associated enzymes. *Methods of Biochemical Analysis* **20**: 41–86.
14. Aaij C, Borst P (1972) The gel electrophoresis of DNA. *Biochimica et Biophysica Acta* **269**(2): 192–200.
15. Sharp PA, Sugden B, Sambrook J (1973) Detection of two restriction endonuclease activities in *Haemophilus parainfluenzae* using analytical agarose–ethidium bromide electrophoresis. *Biochemistry* **12**(16): 3055–3063.
16. Borst P (2005) Ethidium DNA agarose gel electrophoresis: how it started. *IUBMB Life* **57**(11): 745–747.
17. Cohen SN, Chang AC, Boyer HW, Helling RB (1973) Construction of biologically functional bacterial plasmids in vitro. *Proceedings of the National Academy of Sciences of the United States of America* **70**(11): 3240–3244.
18. Ullrich A, Shine J, Chirgwin J, Pictet R, Tischer E, Rutter WJ, et al. (1977) Rat insulin genes: construction of plasmids containing the coding sequences. *Science (New York)* **196**(4296): 1313–1319.

19. Caruthers MH, Beaucage SL, Becker C, Efcavitch JW, Fisher EF, Galluppi G, et al. (1983) Deoxyoligonucleotide synthesis via the phosphoramidite method. *Gene Amplification and Analysis* **3**: 1–26.
20. Li P, Medon PP, Skingle DC, Lanser JA, Symons RH (1987) Enzyme-linked synthetic oligonucleotide probes: non-radioactive detection of enterotoxigenic *Escherichia coli* in faecal specimens. *Nucleic Acids Research* **15**(13): 5275–5287.
21. Goodchild J (1990) Conjugates of oligonucleotides and modified oligonucleotides: a review of their synthesis and properties. *Bioconjugate Chemistry* **1**(3): 165–187.
22. Eckstein F (1991) Oligonucleotides and Analogues: A Practical Approach. Oxford: IRL Press at Oxford University Press.
23. Hochstrasser RA, Chen SM, Millar DP (1992) Distance distribution in a dye-linked oligonucleotide determined by time-resolved fluorescence energy transfer. *Biophysical Chemistry* **45**(2): 133–141.
24. Theisen P, McCollum C, Andrus A (1992) Fluorescent dye phosphoramidite labelling of oligonucleotides. *Nucleic Acids Symposium Series* (27): 99–100.
25. Scott JD, Gretch DR (2007) Molecular diagnostics of hepatitis C virus infection: a systematic review. *JAMA* **297**(7): 724–732.
26. Higuchi R, Dollinger G, Walsh PS, Griffith R (1992) Simultaneous amplification and detection of specific DNA sequences. *Bio/technology* (Nature Publishing Company) **10**(4): 413–417.
27. Wittwer CT, Herrmann MG, Moss AA, Rasmussen RP (1997) Continuous fluorescence monitoring of rapid cycle DNA amplification. *BioTechniques* **22**(1): 130–131, 134–138.
28. Giglio S, Monis PT, Saint CP (2003) Demonstration of preferential binding of SYBR Green I to specific DNA fragments in real-time multiplex PCR. *Nucleic Acids Research* **31**(22): e136.
29. Morrison LE, Halder TC, Stols LM (1989) Solution-phase detection of polynucleotides using interacting fluorescent labels and competitive hybridization. *Analytical Biochemistry* **183**(2): 231–244.
30. Livak KJ, Flood SJ, Marmaro J, Giusti W, Deetz K (1995) Oligonucleotides with fluorescent dyes at opposite ends provide a quenched probe system useful for detecting PCR product and nucleic acid hybridization. *PCR Methods and Applications* **4**(6): 357–362.
31. Parkhurst KM, Parkhurst LJ (1995) Kinetic studies by fluorescence resonance energy transfer employing a double-labeled oligonucleotide: hybridization to the oligonucleotide complement and to single-stranded DNA. *Biochemistry* **34**(1): 285–292.
32. Tyagi S, Kramer FR (1996) Molecular beacons: probes that fluoresce upon hybridization. *Nature Biotechnology* **14**(3): 303–308.
33. Johansson MK, Fidder H, Dick D, Cook RM (2002) Intramolecular dimers: a new strategy to fluorescence quenching in dual-labeled oligonucleotide probes. *Journal of the American Chemical Society* **124**(24): 6950–6956.
34. Parkhurst KM, Parkhurst LJ (1995) Donor-acceptor distance distributions in a double-labeled fluorescent oligonucleotide both as a single strand and in duplexes. *Biochemistry* **34**(1): 293–300.
35. Holland PM, Abramson RD, Watson R, Gelfand DH (1991) Detection of specific polymerase chain reaction product by utilizing the 5′–3′ exonuclease activity of *Thermus aquaticus* DNA polymerase. *Proceedings of the National Academy of Sciences of the United States of America* **88**(16): 7276–7280.
36. Lee LG, Livak KJ, Mullah B, Graham RJ, Vinayak RS, Woudenberg TM (1999) Seven-color, homogeneous detection of six PCR products. *BioTechniques* **27**(2): 342–349.
37. Rudert WA, Braun ER, Faas SJ, Menon R, Jaquins-Gerstl A, Trucco M (1997) Double-labeled fluorescent probes for 5′ nuclease assays: purification and performance evaluation. *BioTechniques* **22**(6): 1140–1145.
38. Tyagi S, Bratu DP, Kramer FR (1998) Multicolor molecular beacons for allele discrimination. *Nature Biotechnology* **16**(1): 49–53.
39. Rieff D (2005) Illness as more than metaphor. *The New York Times*.

6 Analysis of microribonucleic acid expression by quantitative real-time polymerase chain reaction

Vladimir Benes, Jens Stolte, David Ibberson, Mirco Castoldi, and Martina Muckenthaler

Alteration of microribonucleic acid (miRNA) expression in a disease compared to a healthy state and/or correlation of miRNA expression with clinical parameters (like disease progression or therapy response) may indicate that miRNAs can serve as clinically relevant biomarkers.[1–3] An important first step for further functional characterization is the information about differential miRNA expression in cellular processes such as differentiation,[4,5] proliferation, or apoptosis[6] that may determine which disease-causing genes are specifically regulated by miRNAs or, vice versa, which genes regulate miRNA expression. Whatever question you would like to address, the precise information about the level of miRNA expression in a specific cell type or tissue is often considered an important first step. A range of methods can be used for the isolation and profiling of miRNAs. Two recent reviews on microRNA[7] and quantitative polymerase chain reaction (PCR)[8] in *European Pharmaceutical Review* addressed both topics individually in great detail but not their combination. This chapter aims to provide an insight into the application of quantitative real-time PCR (qPCR) to assay microRNA expression.

miRNA EXPRESSION PROFILING USING qPCR

miRNA-specific qPCR assays are frequently used to confirm data obtained from microarray experiments but can be used independently of course. On first glance, the major advantages of this technology over microarrays are (1) the speed of the assay, (2) the increased sensitivity with which miRNAs expressed even at low levels can be measured, (3) the extended dynamic range compared to microarray analysis, and (4) the low requirement for starting material (10 ng per reaction). If, however, the aim is to perform genome-wide miRNA profiling of all 750 human miRNAs currently annotated in miRBase (http://www.mirbase.org/index.shtml, release 14; September 2009) by qPCR, together with technical replicates and non-template controls (NTCs), six 384-well plates will be required for each condition analyzed. This scale will bring the amount of total RNA required to more than 10 μg and the overall cost and time required for miRNA profiling by qPCR well above those required for microarray analysis.

Generally, miRNAs represent challenging molecules to assay. The challenges of miRNA analysis reside in (1) the small size of the mature miRNAs (18 to 24 nt), (2) absence of a common anchor sequence, such as a poly(A) tail or a cap, (3) highly heterogenous sequence composition of individual miRNAs resulting in a relatively large interval of melting temperatures (T_m) of nucleic acid duplexes (45°C to 74°C), (4) the presence of miRNA families within which individual members may differ by just one base (let-7 family, e.g.) but each family member may have a different level of expression, and (5) the total number of miRNAs that is still growing.

The absence of a common anchor sequence makes it necessary for complementary DNA (cDNA) synthesis that either an miRNA-specific reverse transcription (RT) primer must be designed for each miRNA analyzed or RNA molecules must be enzymatically modified (polyadenylated) to prime RT of miRNAs with oligo(dT) primer. A large T_m interval can be narrowed and normalized by use of locked nucleic acids (LNA),[9,10] for example. Currently, several different approaches to determine expression levels of mature miRNAs by qPCR analysis are described in the literature, from which the ones discussed now are employed most often.

DETECTION OF miRNA PRECURSORS BY qPCR

Schmittgen and colleagues[11] have developed a technique for detection of miRNA precursors (pre-miRNA) by qPCR. Based upon an assumption that pre-miRNAs and matured miRNAs are present in a one-to-one ratio, it was thought that this technique could ultimately substitute for the detection of mature miRNAs. It is now recognized, however, that miRNA biogenesis is a complex and tightly regulated process with the potential of each of its steps being individually regulated, resulting in different ratios of expression levels between pre-miRNA and the mature miRNA. It is important to note that this ratio may be affected in pathophysiological states.[12] The underlying experimental principle is similar to that for standard qPCR analysis of mRNAs, with the exception that a primer specific for pre-miRNA is used to prime RT. Certainly, standard protocols for priming of the first-strand cDNA synthesis (by random oligonucleotides) can also be used for this step. Extensive removal of the genomic DNA and the enrichment for the low-molecular-weight RNA fraction[11] are prerequisites for this method because the amplification primers will also recognize genomic DNA contaminants and miRNA primary transcript.

qPCR DETECTION OF MATURE miRNAs VIA SPECIFIC RT PRIMERS

Despite the short length of mature miRNAs, specific complementary RT primers with an adaptor 5′ part can be directly annealed to the miRNA-specific sequence to prime the RT step. The resulting cDNA is then used as a substrate for the qPCR reaction with one miRNA-specific primer and a second, universal primer, the annealing site of which is included within the adaptor part of the RT primer. SYBR® Green incorporated into the amplification products during qPCR enables detection. Approximately 50 ng of total RNA as starting material is required for each qPCR. The often-observed disadvantages of this method are (1) a lack of discrimination among mature, precursor, and primary miRNA and (2) the absence of a multiplexed RT step. This system is also available commercially and can be purchased from Applied Biosystems (AB)/Ambion (mirVana™) but the content of the set is not kept up to date.[9,13]

qPCR DETECTION OF MATURE miRNAs VIA A SYNTHETIC POLY(A) TAIL–MEDIATED RT

In this alternative approach *Escherichia coli* poly(A) polymerase is employed to synthesize a nontemplated homopolymeric polyadenosine tail at the 3′ end of each RNA molecule including miRNAs.[14] Then, RT is primed by a primer consisting of two parts: oligo(dT), usually anchored, at its 3′ part and a specific adaptor universal primer binding site at its 5′ part. The resulting cDNA is then used as a template for qPCR analysis using an miRNA-specific primer and the universal primer matching the adaptor sequence of the RT primer.

One of the advantages offered by this method is that the introduction of a sequence common to all miRNAs [the poly(A) tail] allows RT of all miRNAs present in the sample and thus facilitates their analysis. However, assays for some miRNAs suffer from lack of specificity requiring optimization of the miRNA-qPCR primer, which due to the constraints imposed by the length of mature miRNA and their base composition can be difficult. Incorporation of LNA-modified nucleotide(s) into the primer increases its T_m and thus better matches the requirements of the assay, particularly the annealing temperature of both primers, and runs the assay under more stringent conditions. SYBR® Green is usually used for detection of resulting amplicons, but it will work also with dual-labeled probes, in principle. However, their design is not a trivial task due to the reasons mentioned.

Another advantage is the "openness" of this system; for example, assays for newly discovered miRNAs can be developed and optimized quickly and in a straightforward manner. This system is also commercially available from Invitrogen (NCode), Qiagen (miScript), and Agilent/Stratagene.

qPCR DETECTION OF MATURE miRNAs VIA RT WITH STEM–LOOP PRIMERS

A third approach uses special stem–loop structured primers,[15] the 3′ end parts of which are complementary to and perfectly match a couple (~6) of bases at the 3′ end of a particular miRNA. The stem–loop primer also encompasses an additional sequence with a universal primer-binding site (complete sequence information of the stem–loop primer is available in supplementary data to the original article). Due to their structure and sequence, these primers can prime RT reactions only from mature miRNAs. The resulting cDNAs are amplified during qPCR with an miRNA-specific and a universal primer. In this setup, another level of specificity is implemented by addition to the qPCR mixture of the individual miRNA-specific dual-labeled, hydrolysis TaqMan® probe that is also used for detection of the amplification product. This high specificity is due to hybridization of the probe to the central region of the amplified PCR product,[16] in this case to a particular miRNA. The assay uses the 5′-nuclease activity of the DNA Taq polymerase to hydrolyze the hybridization probe, which carries both a fluorescent reporter and a quencher in adjacent positions, bound to its target amplicon. The emission of fluorescence released upon separation of reporter and its quencher is proportional to the amount of PCR products generated, which will allow accurate quantitation of assayed cDNA. The issue of the miRNA length constraint is dealt with by using so-called 3′-minor groove binders (MGBs).[17] This method is highly specific and sensitive; only approximately 10 ng of total RNA is required as starting material. Data obtained highly correlate with data obtained from microarray analysis. However, a certain disadvantage of this concept can be occasional difficulty for some miRNAs assays to bring all three components (miRNA-specific forward qPCR primer, TaqMan® probe, and universal reverse qPCR primer) into harmony in the limited space provided by mature miRNA features so that they fit into the relatively narrow window of required reaction conditions. In addition, TaqMan® probes must meet certain specifications to function properly. Therefore, their design may not be easy due to the miRNA-space constraints; their optimization may be more demanding than SYBR® Green-based assays. Also any change of the cognate sequence of mature miRNA can lead to lowering of the corresponding assay's performance, as it was reported recently.[18] The qPCR system marketed by AB is based upon this principle.

Recently, an application of RT using a stem–loop primer in combination with SYBR® Green detection in the real-time PCR step (thus alleviating the necessity for rather expensive TaqMan® probes) was published.[19]

qPCR DETECTION OF MATURE miRNAs USING MULTIPLEXING OF STEM–LOOP PRIMERS FOR RT

Genome-wide analysis of miRNA expression using qPCR is both time and reagent consuming. Tang and coworkers[20,21] described an approach to reduce both the handling time of the samples and the amount of material required for genome-wide analysis of miRNAs through the introduction of a multiplexed RT step. For this purpose they modified the approach described by Chen and colleagues[15] to apply the multiplex RT–PCR principle to the quantification of miRNAs. In a first step, stem–loop RT primers from their mixture hybridize (anneal) to their corresponding mature miRNAs to enable a multiplexed RT step. Then, the resulting cDNA is preamplified in the presence of low amounts of qPCR primers (pre-PCR; this step is also carried out in a single tube). The pre-PCR amplification product is then diluted, and a fraction of the dilution is used for qPCR in 96- or 384-well plates using qPCR primers and TaqMan® probes (AB). AB provides a set of eight different stem–loop primer pools, and recent release of TaqMan® assays in their low density arrays format can certainly facilitate setup of RT reactions for miRNAs represented in the particular pool and following qPCR assays on the array.

Duncan and colleagues[22] describe an interesting concept incorporating a miRNA-derived cDNA-templated ligation with T4 DNA ligase into the qPCR assay workflow. This should – together with an miRNA-specific RT primer and with the individual miRNA-specific dual-labeled, hydrolysis TaqMan® probe – provide the assay specificity. RT is primed by a short miRNA-specific primer without any adaptor. Two long oligonucleotide adaptors, each with miRNA specific and universal (either forward or reverse qPCR primer) parts, are annealed to a particular miRNA-derived cDNA with a gap in between. This gap is closed by T4 ligase, which tolerates and discriminates mismatches. The quantity of ligated amplicon is determined by a standard qPCR assay with a probe and universal primer pair. A disadvantage of this approach can be that, to start with, each assay requires a set of four miRNA-specific oligonucleotides, including a dual-labeled MGB probe, for it is not likely that it works with SYBR® Green without additional optimization. Also, assay optimization can be demanding because it is not certain that, on the first attempt, each step will perform quantitatively, which is an essential prerequisite for successful qPCR assay.

miRNA qPCR ASSAYS ALSO REQUIRE CONTROLS

As with any other assay, analysis of miRNA expression by qPCR also requires controls that detect quantification errors due to variation in the amount of starting material, sample harvesting, RNA preparation, and quality as well as of efficiency of enzymatic steps. The current consensus is that, for qPCR, normalization to endogenous control (reference) genes is possibly the appropriate method to correct for variation and efficiency biases. Requirements of these endogenous controls for miRNA qPCR analysis are identical to those of controls used in mRNA profiling: Their gene expression should be stable, levels should be similar to those of the targets, and they should be detectable in as many sample types as possible. It is essential that the controls share similar properties such as stability and size with the assayed sample; in addition, they must be adaptable to miRNA assay design, which is considerably different from the designs of other qPCR assays. Abundant and stable expression of small noncoding nucleolar RNAs that are not related to miRNAs makes them suitable candidates. miRNA microarray data can be a good source to identify miRNAs that can be used for that purpose. However, it is essential to validate the control candidates, because there is no universal control for all experimental conditions. Vendors specialized in products for miRNA qPCR assays usually provide some assays that can be used as controls, but they can be difficult to obtain if the model organism of choice is not human, mouse, or rat.

CONCLUSION

miRNA expression profiling using qPCR is undoubtedly a powerful approach exploiting all qPCR features: high specificity and sensitivity, wide dynamic range, speed, straightforward setup, and scalability. The results are generally robust and reliable, although discrimination of expression levels of individual members of miRNA families is not always easy to achieve. However, in the design of the assay and analysis of its results it is important to pay attention to the still dynamic status of many mature miRNA and from them derived iso-miR sequences (for example, according to the miRBase Release 10, sequences of not less than 25% of previously annotated miRNAs were changed to some extent compared with the previous release).

Another point to consider is the sample itself. Regardless of the type of qPCR assay used to determine miRNA expression levels, or any miRNA profiling assay generally, it is critical to realize that some methods applied to purify total RNA, in particular those employing column filtration, can considerably influence the resulting miRNA profile. One should always verify that the sample is not depleted of its miRNAs. It is advisable not to use purification columns for isolation of total RNA whenever analysis of its microRNA is anticipated but rather to use methods based on extraction of RNA by acid phenol in combination with guanidinium–thiocyanate and chloroform[23] (also known as "Tri-reagents" and obtainable from several vendors under various brand names). The sample source is also important. Generally, formalin-fixed paraffin-embedded (FFPE) specimens are considered a treasure trove of invaluable clinical information but are also difficult for purification of total RNA of acceptable quality suitable for profiling approaches. Although a recent report[24] on analysis of samples isolated from fresh frozen and FFPE specimens describes acceptable correlation between results of miRNA profiling by microarray and qPCR from both specimens' types, another article published earlier this year,[25] the authors of which analyzed the stability of selected miRNAs and their corresponding cDNA after RT, advises caution. It has been observed that degradation of total RNA affects expression profiles of miRNAs.[26]

REFERENCES

1. Zhang B, Pan X, Cobb GP, Anderson TA (2007) MicroRNAs as oncogenes and tumor suppressors. *Developmental Biology* **302**(1): 1–12.
2. Mattie MD, Benz CC, Bowers J, Sensinger K, Wong L, Scott GK, et al. (2006) Optimized high-throughput microRNA expression profiling provides novel biomarker assessment of clinical prostate and breast cancer biopsies. *Molecular Cancer* **5**: 24.
3. Lee EJ, Gusev Y, Jiang J, Nuovo GJ, Lerner MR, Frankel WL, et al. (2007) Expression profiling identifies microRNA signature in pancreatic cancer. *International Journal of Cancer* **120**(5): 1046–1054.
4. Song L, Tuan RS (2006) MicroRNAs and cell differentiation in mammalian development. *Birth Defects Research. Part C, Embryo Today: Reviews* **78**(2): 140–149.
5. Chen CZ, Li L, Lodish HF, Bartel DP (2004) MicroRNAs modulate hematopoietic lineage differentiation. *Science* **303**(5654): 83–86.
6. Hwang HW, Mendell JT (2006) MicroRNAs in cell proliferation, cell death, and tumorigenesis. *British Journal of Cancer* **94**(6): 776–780.
7. Clarke N, Edbrooke M (2007) How will microRNAs affect the drug discovery landscape? *European Pharmaceutical Review* **5**(4): 24–31.
8. Kubista M, Sjögreen, B, Forootan, A, Sindelka R, Jonák, J Andrade, JM (2007) Real-time PCR gene expression profiling, *European Pharmaceutical Review* **5**(1): 56–60.
9. Kauppinen S, Vester B, Wengel J (2006) Locked nucleic acids: high affinity targeting of complementary RNA for RNomics. *Handbook of Experimental Pharmacology* **173**: 405–422.
10. Raymond CK, Roberts BS, Garrett-Engele P, Lim LP, Johnson JM (2005) Simple, quantitative primer-extension PCR assay for monitoring of microRNAs and short-interfering RNAs. *RNA* **11**: 1737–1744.
11. Schmittgen TD, Jiang J, Liu Q, Yang L (2004) A high-throughput method to monitor the expression of microRNA precursors. *Nucleic Acids Research* **32**(4): e43.
12. Jiang J, Lee EJ, Gusev Y, Schmittgen TD (2005) Real-time expression profiling of microRNA precursors in human cancer cell lines. *Nucleic Acids Research* **33**(17): 5394–5403.
13. Wang X, Wang X (2006) Systematic identification of microRNA functions by combining target prediction and expression profiling. *Nucleic Acids Research* **34**(5): 1646–1652.
14. Shi R, Chiang VL (2005) Facile means for quantifying microRNA expression by real-time PCR. *BioTechniques* **39**(4): 519–525.
15. Chen C, Ridzon DA, Broomer AJ, Zhou Z, Lee DH, Nguyen JT, et al. (2005) Real-time quantification of microRNAs by stem-loop RT-PCR. *Nucleic Acids Research* **33**(20): e179.
16. Kuimelis RG, Livak KJ, Mullah B, Andrus A (1997) Structural analogues of TaqMan probes for real-time quantitative PCR. *Nucleic Acids Symposium Series* (**37**): 255–256.
17. Kutyavin IV, Afonina IA, Mills A, Gorn VV, Lukhtanov EA, Belousov ES, et al. (2000) 3′-minor groove binder-DNA probes increase sequence specificity at PCR extension temperatures. *Nucleic Acids Research* **28**(2): 655–661.
18. Wu H, Neilson JR, Kumar P, Manocha M, Shankar P, Sharp P, et al. (2007) miRNA profiling of naïve, effector and memory CD8 T cells. *PLoS ONE* **1**(10): e1020.
19. Varkonyi-Gasic E, Wu R, Wood M, Walton EF, Hellens RP (2007) Protocol: a highly sensitive RT-PCR method for detection and quantification of microRNAs. *Plant Methods* **3**(12). doi: 10.1186/1746-4811-3-12

20. Tang F, Hajkova P, Barton SC, Lao K, Surani MA (2006) MicroRNA expression profiling of single whole embryonic stem cells. *Nucleic Acids Research* **34**(2): e9.
21. Tang F, Hajkova P, Barton SC, O'Carroll D, Lee C, Lao K, et al. (2006) 220-plex microRNA expression profile of a single cell. *Nature Protocols* **1**(3): 1154–1159.
22. Duncan DD, Eshoo M, Esau C, Freier SM, Lollo BA (2006) Absolute quantitation of microRNAs with PCR based assay, *Analytical Biochemistry* **359**(2): 268–270.
23. Chomczynski P, Sacchi N (2006) The single-step method of RNA isolation by acid guanidinium thiocyanate-phenol-chloroform extraction: twenty something years on. *Nature Protocols* **1**(2): 581–585.
24. Xi Y, Nakajima G, Gavin A, Morris CG, Kudo K, Hayashi K, et al. (2007) Systematic analysis of microRNA expression of RNA extracted from fresh frozen and formalin-fixed paraffin-embedded samples. *RNA* **13**(10): 1–7.
25. Bravo V, Rosero S, Ricordi C, Pastori RL (2007) Instability of miRNA and cDNA derivatives in RNA preparations *Biochemical and Biophysical Research Communications* **353**: 1052–1055.
26. Ibberson D, Benes V, Muckenthaler MU, Castoldi M (2009) RNA degradation compromises the reliability of microRNA expression profiling. *BMC Biotechnology* accepted.

7 Miniaturized polymerase chain reaction for quantitative clinical diagnostics

Melissa Mariani, Lin Chen, and Philip J. Day

CURRENT QUANTITATIVE POLYMERASE CHAIN REACTION USES IN QUANTITATIVE CLINICAL DIAGNOSTICS

Clinical diagnostics is progressively embracing the incorporation of translational medicine, as the diagnostic industry starts to lean toward personalized treatment. Direct benefits to clinical diagnostics will be achieved as a result of technical advancements providing unambiguous quantitative analysis of the transcriptome,[1] although the quantitative clinical diagnostic sector is a relatively small part of the overall clinical diagnostic market. Interests in deoxyribonucleic acid (DNA) and messenger ribonucleic acid (mRNA) detection and quantification have improved current knowledge of cell functions, including cell regulation, growth, expression markers, and transcription.[2] The polymerase chain reaction (PCR) is one such research technique in mainstream clinical diagnostics that can provide quantitative analysis and assist in closing the "bench to bedside" gap found with translational medicine.[3]

The reverse transcription–PCR (RT–PCR) is at present the most sensitive technique for mRNA detection, although quantitative RT–PCR (qRT–PCR) increasingly provides robust PCR product measurements from each cycle. Additionally, qPCR is commonly applied to clinical diagnostics, making the technique an industry standard for RNA product detection and quantification.[4–6] An increase in qPCR applications, combined with the growing importance of clinical diagnostics, has permitted both areas to develop in parallel. Reflecting these advancements, the value of qPCR in quantitative clinical diagnostics has increased largely through the stochastic integration of fluorochrome that can be directly related to qualitative measurements.[7] Qualitative end-point PCR measurements benefit from many of the detection strategies used in qPCR, but obviously lack quantitative application.

Multiple unparalleled benefits have contributed to the extensive use of qPCR, including reliable quantification from its heightened sensitivity, assay flexibility, high throughput, and the ability to screen various cell sample sizes.[8–10] Consequently, these advantages illustrate qPCR's relevance in quantitative medical diagnostics, with applications including viral and pathogen detection, disease-specific marker detection, and disease-associated detection along with recognizing genetic responses to therapy in hematological malignancies and disease-specific transcriptome product detection and assessment.[8,9,11–15]

A further, critical benefit to qPCR integration in potential quantitative clinical diagnostic assays includes the significant decline in result variability and related increase in confidence with respect to interpreting the bioassay.[16] Resulting from modern technological advances and progress with miniaturization innovation, the prospect of miniaturizing qPCR onto a microfluidic chip has become highly possible. Overall, this development can provide more accurate gene-based analyses, reducing the risk of contamination and human error, saving time, and increasing test reproducibility and precision.[17]

Applications

qPCR can be applied to a variety of areas in medical diagnostics. Viral RNA pathogen detection is one primary use due to the abundance of RNA viruses. Despite high mutation rates in viral RNA, qRT–PCR has been illustrated to produce meaningful quantitative data, greatly benefiting studies on viral agents through the characterization of infectious disease processes, along with recognizing links between patient symptoms and identified viral sequences.[9,18] Aside from these benefits, qRT–PCR analysis of viral RNA still possesses shortcomings, as depicted from the limited reagent kits specifically formulated for viral use and technique standardization.[19]

Negative-strand RNA viral detection, including viruses producing measles and mumps, can be assessed by qRT–PCR. Other negative-strand RNA viruses include influenza types A and B, parainfluenza virus, and respiratory syncytial viruses. All of these viruses contribute to respiratory infection development in patients of all ages, creating a need for a sensitive diagnostic test capable of viral identification, and qRT–PCR has advantages that include the ability to test multiple samples while screening for up to two viruses per assay.[20]

Retroviruses and disease-specific markers are also frequently explored with qRT–PCR. Examples of disease involvement include the commonly known retrovirus HIV (human immunodeficiency virus) and various forms of cancer, including breast and colorectal. Specific benefits of qRT–PCR application to HIV detection have been clearly displayed, particularly when compared with previous methods of plasma detection tests that provide a sensitivity of 50 copies/mL or greater.[21,22] Often following treatment, patient viremia levels are detected at 50 copies/mL or lower, making detection difficult with previous techniques. Viral resistance to prolonged treatment may also occur, commonly resulting in mounting viral copy levels. By applying qRT–PCR to HIV RNA detection, potentially as little as 1 copy/mL of viral RNA may be detected and quantified, enabling monitoring of patient response to treatment.[23] Moreover, HIV detection via qPCR can be simultaneous to the detection of hepatitis B and C viruses, allowing three viral screenings to be carried out concurrently.[24]

qRT–PCR application to quantitative clinical diagnostics has also been demonstrated by Hochhaus and colleagues and others to detect RNA break points in peripheral blood specimens and disease-specific translocations, producing readily evaluated results.[25–29] qPCR can be employed to quantitatively verify levels of disease-associated nucleic acids present among individual patients. Perhaps among the most celebrated applications of qRT–PCR include investigating minimal residual disease (MRD), specifically among follicular lymphoma and acute lymphoblastic leukemia.[30,31] However, translating numerical data into a form usable in clinical practice for the provision of more precise and specific treatment remains a major caveat. This arena is where miniaturization may offer a bridge: the linking of sample preparation to downstream analysis. Connecting qPCR data to assess a patient's ability to combat disease or understand his or her level of infection, and to determine the amount of treatment necessary, is

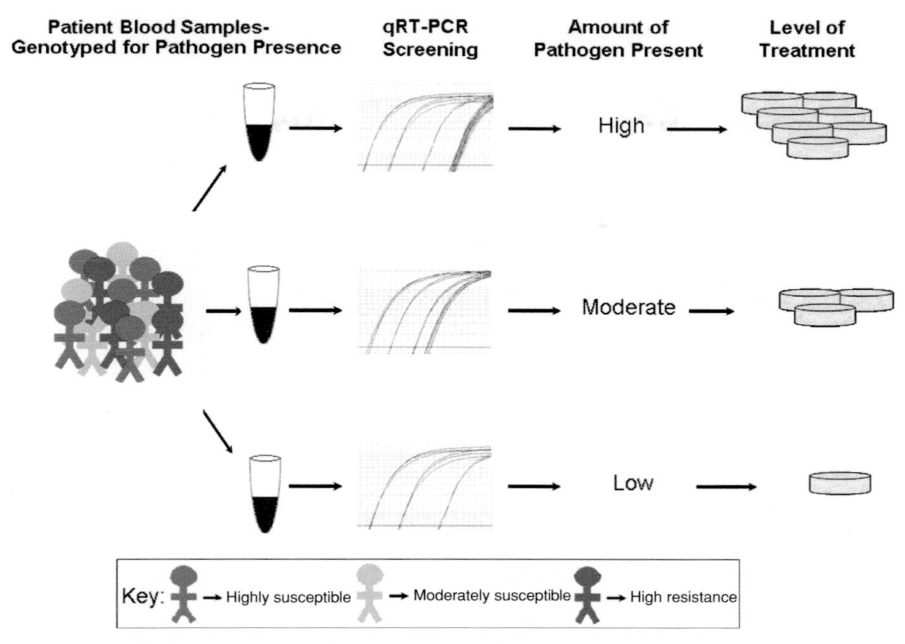

Figure 7–1. Course of clinical treatment against an infectious agent is determined by both patient genotype and quantitative polymerase chain reaction (qPCR) of the pathogenic agent. qRT–PCR, quantitative reverse transcriptase–PCR. *See Color Plates.*

within the scope of translational medicine. In this context, miniaturized qPCR may be mandatory to achieve the vision of translational medicine. Figure 7–1 illustrates the idea of qualitative patient genotyping in relation to the ability to combat a quantitatively determined infectious disease.

Levels of pathogenesis can be assessed using qPCR. From this knowledge, the appropriate intensity of treatment can be prescribed. Individuals possessing high infectivity levels can then receive elevated treatment dosages whereas individuals possessing moderate or low levels of infections receive drug dosages specific to their needs.

Combining specific PCR techniques (like multiplexing or primer nesting) creates unparalleled specificity and assay sensitivity. In particular, the ability to quantify multiple mRNA targets in small clinical samples by including primer nesting can supply trustworthy quantitative results, as exhibited in studies of salivary RNA in oral cancer patients.[32]

Quality control

To ensure assay success and reliability in medical diagnostics, several parameters must be explored. These include optimal and reproducible target sample preparation and appropriate primer selection due to their influence on assay sensitivity and specificity.[33,34] Repeated hardware calibrations (using standard reagents) are also necessary from threshold cycle (C_q) susceptibility to fluctuations resulting from frequent machine use with operator-mediated parameter changes, and varying reagent formulation.[5] Assay reliability is also central for qPCR standardization between laboratories and includes criteria like replicate continuity during and between runs. Repeatability across experiments is determined by concordance between plated replicates among different runs and within the same laboratory.[7,35,36] Recently, guidelines have been proposed for achieving standardization by suggestions for minimum information for publication of quantitative real-time PCR experiments, MIQE (see http://gene-quantification.com/miqe.html).

The quality control of qPCR's technical shortcomings also remains to be addressed. Technical validation, assay optimization, and result consistency capabilities are all vital for technique dependability in molecular medicine.[16,36,37] Specifically, addressing sources of error in machinery, pipettes, or other hardware can improve reproducibility.[38] PCR testing in duplicate or triplicate can identify problematic areas, in addition to permitting the use of statistical analysis to infer significance. Moreover, reproducibility is improved following the implementation of the Laboratory Information Management System (LIMS) and standardization across laboratories. For these criteria to be upheld, a series of standard operating procedures (SOPs) have been developed, including the use of appropriate control samples and result-reporting guidelines.[39,40]

Finally, it is also important to prevent false-positive results, which can be avoided by the inclusion of appropriate negative controls or reference genes, in addition to no template control (NTC) samples. The insertion of these controls into qPCR assays permits easier determination of error sources including inhibitory effects from poor sample purification.[41] All quality control parameters and SOPs can be incorporated into the nominal operation of a miniaturized device for qPCR, much along the lines of a mandatory requirement to use a marker ladder for assessment of DNA molecular weight on an Agilent 2100 BioAnalyzer™ chip.

CURRENT LIMITATIONS OF qPCR IN CLINICAL DIAGNOSTICS

Aside from the many potential advantages of qPCR use in quantitative clinical diagnostics, significant technical and conceptual limitations remain, preventing full integration into medical diagnostics.[35,42–44] These limitations heavily reside with assay preparation, labeling chemistries, and their resulting fluorescence detection. In addition, enzymology (complementary DNA [cDNA] synthesis, RT, PCR) allowance for statistical normalization and analysis of transcript or gene variability currently limits analyte measurement.

cDNA synthesis

The production of cDNA is a requisite step in transcript assay development. It is also a significant source of error, often leading to result ambiguity and compromising reproducibility. cDNA can be generated by multiple methods, including specific sequence or random priming. However, certain components found in blood and tissue (especially mammalian heme-related compounds) are known to inhibit RT.[45] RNA degradation is a major contributing factor that reduces predictability of cDNA synthesis. Following cell lysis and extraction, analyte nucleic acid is released from the cell and thus increases the likelihood of degradation. To guarantee accurate sample measurements, sample RNA must be at the highest purity, both DNA and nuclease free, devoid of copurified potential inhibitors of RT or PCR, and the presence of dead cells as DNA serves as usable template is important, as extracted DNA from dead cells can produce false-positive results.[46,47] However, RT has been repeatedly blamed for causing inaccurate results.[48]

Single-step cDNA synthesis is an alternative method of RT and can be carried out prior to a PCR in a closed-tube system. Advantages of this protocol include a lower risk of contamination, reduced sample degradation, less reagent, and increasing overall assay sensitivity (tenfold) resulting from use of greater amounts of cDNA compared to two-step cDNA and qRT–PCR.[49] However, disadvantages associated with single-step cDNA synthesis relate to a reduced number of assays following higher use of cDNA.

Likewise, as two-step cDNA synthesis can assess multiple transcripts, reference transcripts can be employed to ensure high-quality cDNA. Prior to assay setup, target sample integrity by gel or capillary electrophoresis, or via NanoDrop, is requisite. In contrast, this process is costly, takes time to perform, and requires training. In addition, high probe and primer binding specificities can prevent nonspecific binding, false-positive fluorescence, and primer–dimer formation. Appropriate template concentrations are also imperative for accurate PCR product measurements and for making certain that low copy analyte templates are not masked.[16,50,51]

PCR

Following target sample amplification during qPCR, quantification of resulting products is imperative. To obtain such quantitative results, the use of sensitive fluorescence detection and measurement is imperative. Multiple methods for including fluorescence in a qPCR assay exist, and their limitations are described later in this chapter.

PCR with fluorescent resonance energy transfer probes

Specific reporter dyes (like fluorescent resonance energy transfer [FRET] dyes) are attached to the probe and work by transferring incident energy from the donor fluorophore to the quencher fluorophore. The resulting fluorescence is monitored in terms of either a gain or interruption of FRET signal. Probes (such as conformational, hydrolysis, and hybridization) all employ a similar method for fluorescing through the use of fluorophores and quenchers, but vary in method that quantitatively relates hybridization and enzymology to fluorescence. All of these probes provide nearly identical sensitivity but differ in individual disadvantages.[52] Specifically, conformational probes must possess the exact sequence to the target amplicon binding site for the probe to hybridize to analyte sequences, denaturing from its hairpin formation and producing fluorescence. For hybridization probes, two probes are needed to hybridize in close proximity on the amplicon for fluorescence to occur.[10] What is more, in both of these methods the fluorescence is reversible due to the probes not being hydrolyzed. The final probe category is the ablative hydrolysis probes. This category has reduced assay versatility given the requirement of modifying the polymerization phase temperature to promote probe binding throughout primer extension, combined with forcing the polymerase to function at suboptimal temperatures[10] for both polymerization and exonuclease digestion.

PCR with intercalating dyes

Nonspecific detection involves the addition of intercalating dyes in free solution, including SYBR® Green, to the master mixes for sample quantification. Once added, the dye preferentially binds to double-stranded target DNA as it is generated during the PCR, with amounts of bound dye increasing accordingly during each elongation stage. Advantages include the ability to incorporate dyes into optimization and other areas of protocol with little trouble, while being significantly more affordable.[53]

Despite these benefits, it is not uncommon to detect fluorescence in NTC wells because of the dye's ability to bind indiscriminately to any double-stranded DNA, primer–dimers, or even to single-stranded DNA.[54] Furthermore, multiplexing is not readily accomplished, and the formation of a melt curve is requisite to confirm result accuracy, absence of amplification artifacts like primer–dimers, and proper product amplification. However, melt curves can also prove difficult data to analyze. Because the probability of multiple dye molecules' binding to single dye molecules is high, melt curves can lead to higher signals – particularly when larger products are amplified[8] – which may detract from quantitative meaning. Should amplification rates vary during the reaction, the quantification of dye will also provide false results. Furthermore, the monitoring of fluorescence is an indirect means of charting PCR progression, and hence, any deviation from a linear relationship between dye signal and PCR process will derive an imprecise assessment of initial amount of template.

Normalization

The field of qPCR is slowly moving away from ubiquitously employing glyceraldehyde 3-phosphate dehydrogenase (GAPDH) or β-actin (ACTB) as the choice for reference gene or transcript in qPCR-related assays; energetic activity and structural compositions across cell types and within cell populations have not been shown to be consistent. The identification of reference genes from gene-array screens[55] or literature searches has revealed candidate reference transcripts, and these have been applied in bioanalytical software. These methods compare (a) the difference between amounts of reference transcript and those of test transcripts from a nominated reference sample to (b) the same difference observed in all other tested samples. The units are complex, but the rank order of expression across a cohort of samples is a useful reliable outcome from such analyses. It is better perhaps, and certainly in the context of measurement units, to establish assays that are quantitative and optimized around calibration curves drawn from dilutions of vectoral cloned or synthetic targets.[55–57] This approach obviates the requirement to elect a reference sample, and units are now plausible as numbers of test transcript copies per single copy of a reference transcript. If taken further, and a single cell is used as a progenitor for qPCR assays, the units take a superior biologically relevant definition as numbers of copies per cell. To understand the meaning of the data will require populations of cells to be analyzed, and the high throughput analysis of single cells can be enabled by miniaturization.

CAN CURRENT TECHNOLOGY ACHIEVE MINIATURIZED QUANTITATIVE CLINICAL DIAGNOSTICS?

Associated advantages, weaknesses, and applications of PCR are widely known because of its high popularity. As a result, there is little doubt that PCR will become an omnipresent technique in the area of clinical diagnostics. Major problems restraining the immediate unilateral uptake of qPCR relate to its packaging and related processes. Many recent developments in microfluidics have contributed to the progression and adaptation of various techniques into the medical diagnostics sector, particularly for point-of-care applications.[58–60] By exploiting centrifugal and capillary force interactions, entire processes, including pre- and post-preparation stages, can be integrated on a microfluidic chip.[61,62] The reoptimization of analytical techniques related to sample preparation, treatment, and analysis for application on miniaturized microfluidic platforms is a major area of growth and investment. The notion that shrinking assay platforms enable precise manipulation of known numbers of molecules or cells, enhancing quantitative measurements, has a growing following. Along with these benefits, miniaturization also reduces the opportunity for human error and risk of contamination, providing significantly more reliable sample analysis.

As a result, qPCR adaptation to a microfluidic platform will increase dependability, reproducibility, and technique motility while also being capable of using reduced sample sizes and consuming less reagent.[5,61,62] From the many recent technological innovations in microfluidics, the adaptation of pre- and post-PCR setup with qPCR into a microfluidic platform would be ideal and is a realistic notion for the imminent future.[61] Huge achievements can be rightly claimed through the miniaturization of enzyme-linked immunosorbent assay (ELISA) and various routinely available microarray platforms. The production capability for manufacturing microfluidic devices is rapidly maturing, but the two limitations preventing the uptake of miniaturized qPCR relate to the integration of the different components of a qPCR assay (from raw sample to data) and to the proper control of surface derivatization and functionality of the miniaturized devices.

Table 7–1. Conventional PCR vs. miniaturized PCR

PCR parameters	Conventional PCR 60–120 min.	Miniaturized PCR 1–120 min.
Thermal cycle time	1–2 h	mins–2 h
Required sample volume	≥3 μL	pL–μL
Integrated sample preparation	Difficult with automation	Highly possible
Integrated PCR product detection	Yes	Yes
Sample throughput	High	Very high
Limit of detection	Single molecule	Single molecule
Instrument cost	10–50 k US $	30–40 k US $

BENEFITS OF MINIATURIZATION

Despite its significant impact in fields like clinical diagnostics and forensics, because of intrinsic technical and methodological drawbacks, qRT–PCR has yet to reach its potential. Fortunately, the concept of micro total analysis system (μ–TAS), also known as "lab on a chip," has been applied extensively to the PCR process more than any other molecular biology technique. We now discuss the advantages attributed to miniaturized PCR.

Following its application to μ–TAS, miniaturized PCR technologies have facilitated the development of DNA amplification through increased ramping rates and reduced sample consumption.[63–69] This nucleic acid amplification results from the lower thermal capacity and higher heat transfer rate between the sample and thermal components, permitting decreased sample size volumes, quicker cycling, and the potential for high integration in a single device.[63–70] Additionally, rapid thermocycling results in more specific PCR product formation.[71] Comparisons between conventional PCR and miniaturized PCR are listed in Table 7–1.

It is remarkable that there are not major differences in the criteria in Table 7–1 between conventional and miniaturized PCR. Instrument costs are comparable, potential limits of detection are the same at a single molecule, sample volume and cycle times are similar, and detection is often integrated for qPCR. The big difference, however, relates to the on-chip: higher throughput, feasibility for population studies (of molecules, cells, and tissues), and increased prospects for reproducibility and integrated sample preparation.

The integration of bioassays on-chip is a major challenge and, as noted earlier, requires complete reworking of classical sample preparation, certain aspects related to qRT–PCR protocols, optical engineering to measure fluorescence, and overall coordination of the process using passive or active motive forces.

Materials and fabrication of miniaturized PCR devices

Silicon/glass-based

Early in technique development, many PCR microchambers were constructed from silicon following adaptation from the microelectromechanical systems

(MEMS) industry. In addition, this silicon material provides superior thermal conductivity allowing for a quick ramping rate.[72–74] Although silicon-based microdevices for miniaturized PCR initially dominated, these also proved to be problematic. First, bare silicon will inhibit the PCR, ultimately hindering target amplification. Second, the nontransparent properties of silicon substrates further limit the applications of real-time optical detection. Glass has since become an alternative substrate material for miniaturized PCR, originating in the late 1990s, primarily due to its vast benefits including well-defined surface chemistries and superior optical transparency.[75–77]

The fabrication of a silicon/glass-based microdevice typically involves a series of micromachining processes such as photolithography, film deposition, and wet etching. However, because of high fabrication cost, PCR microdevices made from silicon or glass are not conducive to disposable applications.[72–77]

Polymer-based electrophoresis microchips

Currently, polymer-based electrophoresis microchips for DNA analysis have been widely developed, primarily as a result of their lower costs, wide range of tailor-made material properties, and relatively simple fabrication procedures. Polymeric substrates facilitate the mass production of disposable microdevices, critical to successful commercialization. Many efforts also have been focused on polymer-based miniaturized PCR platform development. Polymers including poly-dimethylsiloxane (PDMS), polycarbonate, poly(methylmethacrylate) (PMMA),[78] polyimide, SU-8, and poly(cyclic olefin) have been used for miniaturized PCR.

Manufacturing polymer-based PCR microdevices can be divided into two methods: direct fabrication and replication.[79] Direct fabrication often uses mechanical machining or laser ablation techniques to form the chamber shape or connecting channels in polymeric substrates. Each part must be fabricated separately, making the assembly process uneconomical for mass production. The replication method includes injection molding, imprinting, and casting. Ultimately, a microstructure is produced, possessing channels and chambers both faithfully and repeatedly through the aid of high-quality templates. After substrate fabrication, the microdevices are formed by irreversible or reversible bonding of the base layer to other planar polymer layers or glass slides.

Types of miniaturized PCR

Chamber-based stationary PCR

Chamber-based stationary PCR works in a similar manner to conventional PCR, where the PCR sample mixture is kept stationary and the temperature of the reaction chamber is cycled between different temperatures.[72–74,76,77] Compared to the total sample volumes of conventional PCRs (between 3 μL and 0.2 mL), the volume of stationary chambers for miniaturized PCR is predominately less than 20 μL and can potentially even decrease to nanoliter order. The reaction chamber also can serve as a sample reservoir following microchip capillary electrophoresis analysis, permitting further applications of chamber stationary PCR. See Figure 7–2A.

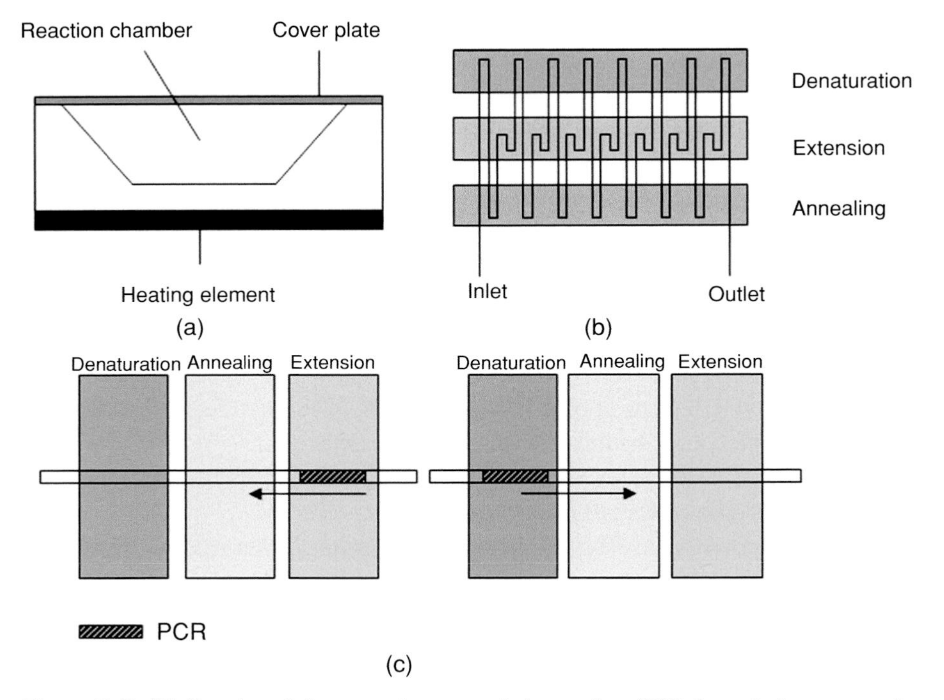

Figure 7–2. (A) Chamber stationary polymerase chain reaction (PCR), in a similar manner to conventional PCR with relatively small volume. **(B)** Continuous-flow PCR: Samples are thermocycled in a spatial way instead of a temporal way. **(C)** Shunting PCR, a combination of the former two methods with flexibility and high speed.

Continuous-flow PCR

Chamber-based stationary PCR lacks the flexibility to modify reaction rate, as it depends entirely on the thermal capacity of the substrate materials. Applying a time–space conversion concept (Figure 7–2B), miniaturized PCR has been realized through continuous flow. As opposed to the sample remaining stationary inside a chamber, movement of the PCR solution permits repeated, continuous flow through the different temperature zones required for PCR amplification.[75,80] Several attractive and interesting features are present with this technique, including (1) rapid heat transfer and fast thermal cycling speed, allowing for total run completion in a matter of minutes; (2) low possibility of cross-contamination; (3) high potential of further application by incorporating other functionalities; and (4) facilitation of liquid sample transport.

Shunting PCR

Shunting PCR combines the advantages of both stationary and continuous-flow PCR into one method. Fast DNA amplification has been achieved through sample shunting back and forth along a microchannel between different temperature zones, allowing the sample to dwell at temperatures for known residence times (Figure 7–2C). This method possesses elevated ramping rates for heating and cooling, which, inherited from continuous-flow PCR, provide increased flexibility for optimization from stationary PCR and enable an infinite number of thermal cycles.

CURRENT DEMONSTRATION OF INTEGRATED μ–TAS FOR qPCR

For the adaptation of μ–TAS PCR into quantitative clinical diagnostics, two important issues warrant addressing. First, the system requires small amounts of original sample, which can be further decreased in complexity by dilution. This small sampling requirement fits well with reduced volumes required by invasive testing and given that clinical samples of nonfluid origin (such as tissue biopsies) are often in limited supply. Second, by multiparallelization, the system is capable of high-throughput processing, which accommodates the ability to analyze a sufficient amount of clinical sample to yield a meaningful result, plus inclusion of a series of quality control standards to enable absolute or relative quantification.

Thus far, most reported miniaturized devices for PCR are stand-alone structures replacing only the role of conventional PCR thermocyclers, with the amplified product being transferred to another analysis platform for the acquisition of qualitative and quantitative information. However, this is often time consuming, is unlikely to lead to real-world applications, and possesses a high level of risk for sample contamination.

One of the most important trends within μ–TAS has been the integration of multiple functional units. Presently, the integration of miniaturized PCR is under rapid development and is coupled with online detection for real-time monitoring or post-PCR units, like capillary gel electrophoresis (CGE) and DNA microarrays, on a single microdevice. The various approaches to integrated gene analysis that encompass PCR are relatively finite and illustrated in Figure 7–2.

qRT–PCR

Currently, qRT–PCR is the most effective sensitive method to acquire quantitative information from nucleic acids and can be readily adapted to miniaturized stationary PCR. Real-time detection is one of the most important directions for the future development of miniaturized PCR. Despite this promise, only a few miniaturized real-time PCR systems have been reported, due to the difficulty of coupling detection systems.[64,66,70]

Northrup and colleagues and others initially developed a miniaturized analytical thermal cycling instrument, which was a scaled-down version of a conventional RT–PCR instrument, for detection of single-base differences in viral and human DNA.[81,82] Later, Cady and colleagues developed integrated miniaturized real-time detection equipped with microprocessor, pumps, thermocycler, and light-emitting diode (LED)–based fluorescence excitation/detection.[83] Monolithic DNA purification and RT–PCR enable quick detection of *Listeria monocytogenes* cells (10^4 to 10^7) within 45 minutes. Xiang and colleagues[84] reported real-time detection of a 150-bp DNA segment of *Escherichia coli* stx1 on a well-based PDMS microchip. Single- and three-well RT–PCRs were tested with different initial concentrations of DNA template and both were found to amplify the 150-bp DNA segment of *E. coli* stx1. Hu and colleagues[85] also performed this RT–PCR inside a PDMS-based microchannel using Joule heating effects, avoiding the necessity to incorporate a thermal cycler. Additionally, Nakayama and colleagues[86] devised a real-time online PCR microfluidic device for continuous flow through the use of laser beam scanning within the annealing area.

Post-PCR

PCR integration with post-PCR product analysis is the most developed area of PCR integrated within a single microfluidic device. Most likely, the association of PCR and PCR product analysis is attributed to the availability of highly characterized detection methods such as CGE, DNA microarray, and immunoassay, which have previously been adapted and applied to chip format. Normal CGE microfluidic devices can be changed fairly simplistically into a PCR–CGE monolithic platform, dually serving as a sample reservoir and PCR amplification chamber. After on-chip PCR, the amplified products can be directly injected into the CGE separation channel for detection. However, thus far most post-PCR analysis provides only qualitative information by defining the size or sequence of amplified products. To obtain quantitative information for unknown samples, calibration curves of known samples are always needed, increasing assay complexity and time.

Currently, several practical clinical applications of PCR–CGE systems are known. Glass-based PCR–CGE microchips can be applied to determine severe acute respiratory syndrome (SARS)–coronavirus specimens from SARS patients, and exhibit high potential for fast clinical diagnoses.[87] PDMS–glass hybrid PCR–CGE microchips also have been used to assess the risk of BK virus–associated nephropathy in renal transplant recipients, implying a wider clinical application of microchip-based systems.[88] More recently, a four-lane fully integrated PCR–CGE array microdevice was developed to amplify femtogram amounts of DNA in 380-nL volumes, followed by direct CGE separation of PCR amplicons, all in less than 30 minutes.[89]

CONSTRAINTS ASSOCIATED WITH μ–TAS FOR qPCR

The integration of PCR with technologies permitting the detection of PCR products, including online fluorescence detection and online CGE separation in a single microdevice, is a widely investigated field in miniaturized PCR research. Despite many attempts, technical constraints are still present, requiring further modifications of μ–TAS for qPCR to be achieved. This section describes some of the current prevalent limitations.

Biocompatibility of microdevice materials

Miniaturized PCR is currently undergoing rapid development but still retains one critical problem: biocompatibility of microdevice materials. Microstructures offer a significantly increased surface-to-volume ratio. The literature predictably stresses the modification of surfaces applied to miniaturized PCR through the implementation of current available materials for microdevice (silicon, glass, and several polymers).[64,66,67,70] Such surface chemistry alternation processes occur prior to or during the PCR process, increasing both the cost and time for miniaturized PCR. Furthermore, the consistency and stability of surface chemistry can affect the quantitative information derived from miniaturized PCR. Therefore, new PCR-friendly materials need to be perfected, because they are required to satisfy the demands of microfabrication. Alternatively, the development of a surface modification method that can provide reproducible performance and/or long-term stability would be of benefit. A dynamic chip coating is favorable for surface passivation of single-use devices because of their ease of use. This procedure is carried out by adding passivation agents (including bovine serum albumin, polyethylene glycol, and polyvinylpyrrolidone) into PCR solutions, essentially reducing undesired adsorption of enzymes and DNA.[75,80,90,91] However, if the device is intended for repeated use, chemical modification prior to PCR is the preferred method. Additionally, silanization is a well-established method that introduces aprotic organic groups onto the microdevice surface to enhance PCR compatibility.[75,80,91]

Detection

As the reaction volume needed for miniaturized PCR can decrease to the nanoliter order, highly sensitive detection is requisite for capturing quantitative data. Commonly adapted instruments for current real-time or post-PCR analysis are based on fluorescence detection; typically an external energy source such as a mercury, tungsten, or xenon lamp or laser is needed for the provision of high intensity and stable excitation energy. Unfortunately, most currently available sources are bulky benchtop instruments, severely inhibiting the portability of miniaturized PCR devices, thus losing a key attribute of miniaturization.

Currently, few reports concerning miniaturized detection systems are present. LEDs are one good option due to their multiple advantages, including low cost, high efficiency, small size, and considerable durability, and have been applied to miniaturized excitation sources for RT–PCR detection.[83] An alternative to portable detection instruments is electrochemical detection for miniaturized PCR. The small size of electrodes and no need for an external optical source make electrochemical detection a further option.[92,93]

μ–TAS shows a great capability for assembling different functional units for genetic analysis of various samples or diverse purposes. It possesses a sample-in/answer-out capability that shows a promising future for integrated genetic analysis point-of-care devices.

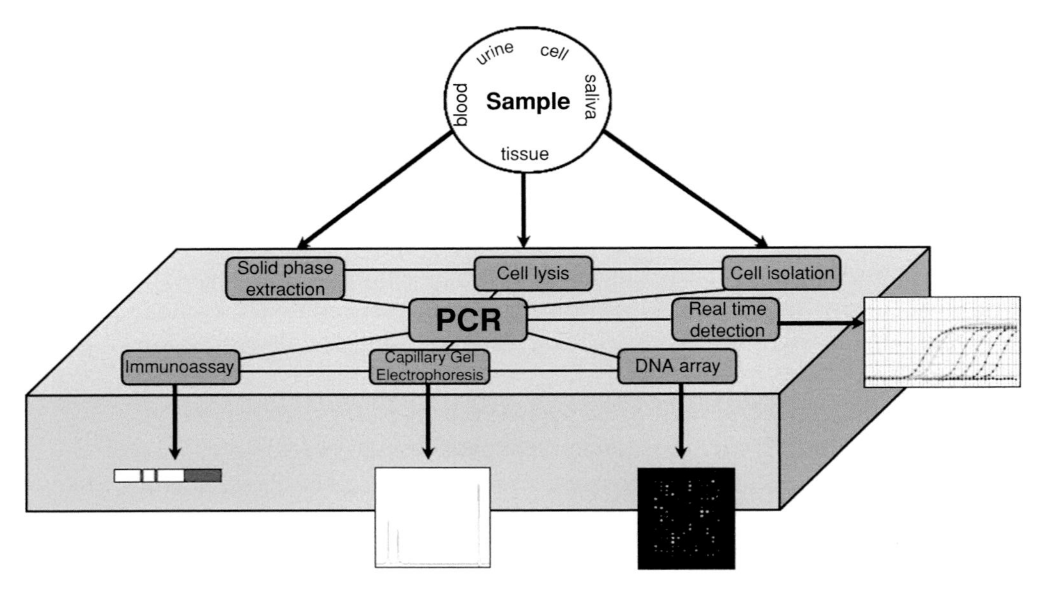

Figure 7–3. Micro total analysis system platform integrating pre-polymerase chain reaction (PCR), PCR, and post-PCR stages onto one microfluidic chip. DNA, deoxyribonucleic acid. *See Color Plates.*

APPLICATIONS OF QUANTITATIVE CLINICAL DIAGNOSTICS

Medical diagnostics is a sector with continued growing importance. Techniques capable of quantitative assessment of nucleic acid levels also have gained acclaim in research. Through the use of end-point PCR and qPCR, combined with current developments of miniaturized qPCR, applications in clinical diagnostics are vast (Figure 7–3). As a result, the possibility exists (among many other applications) to quantify levels of regulation and expression resistance markers resulting from drug treatments present within tumor cells.[2]

qPCR is of specific importance for μ–TAS integration due to its abundant current use in molecular medicine, as exhibited by numerous studies in areas including MRD, cancer presence, treatment success, and infection.[14,94–97]

Ease of qPCR use also dramatically increases with integration onto a microfluidic chip. With the improved portability of miniaturized PCR, assay preparation and analysis can be carried out in a multitude of locations and provide results rapidly. An example of how miniaturized qPCR could greatly assist diagnostics is assessing the presence of circulating tumor cells (e.g., colorectal and melanoma cells) when a microfluidics chip could be brought directly to the patient for gene expression, clonal gene signature, or fusion gene analysis, exemplifying the "bench to bedside" approach.[3,58–60,98] Moreover, the ability to conduct qPCR gene presence assessment can facilitate rapid and easy screening of target genes like breast cancer oncogenes.[99] Patient response to chemotherapy also can be measured, in addition to detecting bacterial or viral presence and quantifying specific transcription factors, like c-myc or ERBB2 for breast cancer.[12,98,100] Likewise, μ–TAS PCR permits reduced sample volumes for genetic disease markers.[101]

PERSONALIZED MEDICATION

The concept of personalized medication has been continually increasing in popularity as a result of abundant technical advancements enabling reliable quantitative clinical diagnosis. Tailoring treatment to the specific needs of each patient has become a prevailing concept in medicine, due to the recent capability of assessing gene expression, mutations, and the effects of treatment.[18,23] Personalized medicine aims to enable a move from classical medicine to the transport of quantitative measurements, such as qPCR measurements, directly to the patient, permitting rapid, personalized diagnosis and ultimately determining treatment and administration of (preventative) therapy. In addition to the aforementioned numerous benefits of miniaturization, high-throughput analyses can be maintained without the need for expensive automation, upholding assay sensitivity and specificity.[102] However, despite the advantages of personalized medicine, the reality of its becoming a widespread, readily available option in common medical practice still warrants additional field trials and technical fine-tuning. Miniaturizing PCR to assist continued development of μ–TAS increases the probability of making accessible widely available personalized medication.

CONCLUSIONS AND PERSPECTIVES

Applications of qPCR have been widely adapted to include areas such as identifying genetic markers to quantifying disease-associated gene abundance amid various sample types and quantities. With the development and increasing popularity of molecular medicine, the use of qPCR has grown in importance and evolved into a critical technique for RNA and DNA quantification. Clinical diagnostics may in the future use more quantitative diagnostic testing and rely less on qualitative tests. However, the necessity to develop μ–TAS for qPCR is not a commonly held view. Clinical diagnostics already offers medicine excellent testing for current treatment strategies. The highly characterized bioassays developed by the pharmaceutical sector are linked to effective therapy and have been accomplished using relatively standard assay volumes and laboratory equipment. The promise of miniaturization with techniques such as qPCR is that therapeutic strategy is poised to change, yielding to highly quantitative measurements and associated calibrated drug administration. Therefore, the catalyst for change may come from personalized medication, which can classify individuals and measuring their disease burden so that a precise measured therapy may be offered.

With the development and widespread implementation of modern techniques in clinical diagnostics, it is imperative to consider the transition from "bench to bedside." Understanding the methodology and resulting data from an important diagnostic technique like qPCR can help determine which method of treatment to administer, at what dosages, and for what duration. Moreover, understanding the results can permit technique manipulations, facilitating better applications to translational medicine and diagnostics. A best-case scenario therefore could permit accurate disease assessment in terms of disease-associated molecules per disease-associated cell. Dimensionally, miniaturized formats are capable of handling cells as unique packages, and also possess the means (using continuous flow) to process large numbers of single cells to understand the distribution across a population of cells, and in terms of cell activity relative to specific location.

The use of a powerful technique like qPCR starts to facilitate a higher resolution understanding of the activity within many components in the body, related to both disease and healthy conditions, and their numerous interactions. This knowledge is invaluable for better modeling of function by systems biologists. From an enhanced elucidation of the body and its various interrelated components, systems biology and translational medicine, when combined, can significantly change the way modern medicine is viewed and diseases combated. As a result, through miniaturization, clinical biomarkers can be quantitatively assessed and patient diagnosis and treatment can excel to a more individualized level, permitting each patient to be treated based on his or her absolute personal needs.

The current outlook on personalized medicine still requires both technical and scientific advancements. The quality control and reliability of qPCR need to be overcome for trustworthy results in medical diagnostics. The speed, specificity, and reliability of this technique all can be improved from the miniaturization of qRT–PCR onto a microfluidic platform, advancing the use of translational medicine and personalized treatment through point-of-care techniques. Although the outlook for miniaturized qPCR includes a cost reduction, this is largely speculation. Initial costs may even increase as a result of the heightened cost of fabrication and assay development and optimization. Aside from this possibility, integration of μ–TAS with qPCR can provide infallible, precise multiplatform results along with integrated sample preparation, high sample throughput, less reagent and sample use, and faster analysis, as listed in Table 7–1. Furthermore, test frequency will increase exponentially because of the ability to repeatedly assess a patient's status at each doctor's visit.

Thoughts for the future of miniaturized quantitative clinical diagnostics include advancing to a technique capable of handling populations of single cells and single molecule analysis by eliminating the need for target amplification and potentially integrating techniques like various forms of spectroscopy, enabling direct molecular measurement.[103]

REFERENCES

1. Bustin SA (2002) Quantification of mRNA using real-time reverse transcription PCR (RT-PCR): trends and problems. *Journal of Molecular Endocrinology* **29**(1): 23–39.
2. Ramachandran C, Melnick SJ (1999) Multidrug resistance in human tumors – molecular diagnosis and clinical significance. *Journal of Molecular Diagnostics* **4**(2): 81–94.
3. Burrow GN (1988) From bench to bedside – the impact of the transfer of the new biology to clinical medicine. *Clinical and Investigative Medicine* **11**(4): 315–320.
4. Bustin SA (2000) Absolute quantification of mRNA using real-time reverse transcription polymerase chain reaction assays. *Journal of Molecular Endocrinology* **25**(2): 169–193.

5. Bustin SA, Nolan T (2004) Pitfalls of quantitative real-time reverse-transcription polymerase chain reaction. *Journal of Biomolecular Techniques* **15**(3): 155–166.
6. Ginzinger DG (2002) Gene quantification using real-time quantitative PCR: an emerging technology hits the mainstream. *Experimental Hematology* **30**(6): 503–512.
7. Halford WP, Falco VC, et al. (1999) The inherent quantitative capacity of the reverse transcription-polymerase chain reaction. *Analytical Biochemistry* **266**(2): 181–191.
8. Bustin SA, Benes V, et al. (2005) Quantitative real-time RT-PCR – a perspective. *Journal of Molecular Endocrinology* **34**(3): 597–601.
9. Mackay IM, Arden KE, et al. (2002) Real-time PCR in virology. *Nucleic Acids Research* **30**(6): 1292–1305.
10. Mackay J, Landt O (2007) Real-time PCR fluorescent chemistries. *Methods in Molecular Biology (Clifton, N.J.)* **353**: 237–261.
11. Dong QH, Zheng S, et al. (2005) Evaluation of ST13 gene expression in colorectal cancer patients. *Journal of Zhejiang University. Science B* **6**(12): 1170–1175.
12. Hill WE (1996) The polymerase chain reaction: applications for the detection of foodborne pathogens. *Critical Reviews in Food Science and Nutrition* **36**(1–2): 123–173.
13. Hoebeeck J, Speleman F, et al. (2007) Real-time quantitative PCR as an alternative to Southern blot or fluorescence in situ hybridization for detection of gene copy number changes. *Methods in Molecular Biology (Clifton, N.J.)* **353**: 205–226.
14. Jolkowska J, Derwich K, et al. (2007) Methods of minimal residual disease (MRD) detection in childhood haematological malignancies. *Journal of Applied Genetics* **48**(1): 77–83.
15. Ratcliff RM, Chang G, et al. (2007) Molecular diagnosis of medical viruses. *Current Issues in Molecular Biology* **9**(2): 87–102.
16. Stahlberg A, Hakansson J, et al. (2004) Properties of the reverse transcription reaction in mRNA quantification. *Clinical Chemistry* **50**(3): 509–515.
17. Sims CE, Allbritton NL (2007) Analysis of single mammalian cells on-chip. *Lab on a Chip* **7**(4): 423–440.
18. Schutten M, Niesters HG (2001) Clinical utility of viral quantification as a tool for disease monitoring. *Expert Review of Molecular Diagnostics* **1**(2): 153–162.
19. Kao CL, King CC, et al. (2005) Laboratory diagnosis of dengue virus infection: current and future perspectives in clinical diagnosis and public health. *Journal of Microbiology, Immunology, and Infection* **38**(1): 5–16.
20. Templeton KE, Scheltinga SA, et al. (2004) Rapid and sensitive method using multiplex real-time PCR for diagnosis of infections by influenza A and influenza B viruses, respiratory syncytial virus, and parainfluenza viruses 1, 2, 3, and 4. *Journal of Clinical Microbiology* **42**(4): 1564–1569.
21. Mulder J, McKinney N, et al. (1994) Rapid and simple PCR assay for quantitation of human immunodeficiency virus type 1 RNA in plasma: application to acute retroviral infection. *Journal of Clinical Microbiology* **32**(2): 292–300.
22. Sun R, Ku J, et al. (1998) Ultrasensitive reverse transcription-PCR assay for quantitation of human immunodeficiency virus type 1 RNA in plasma. *Journal of Clinical Microbiology* **36**(10): 2964–2969.
23. Palmer S, Wiegand AP, et al. (2003) New real-time reverse transcriptase-initiated PCR assay with single-copy sensitivity for human immunodeficiency virus type 1 RNA in plasma. *Journal of Clinical Microbiology* **41**(10): 4531–4536.
24. Candotti D, Temple J, et al. (2004) Multiplex real-time quantitative RT-PCR assay for hepatitis B virus, hepatitis C virus, and human immunodeficiency virus type 1. *Journal of Virological Methods* **118**(1): 39–47.
25. Eder M, Battmer K, et al. (1999) Monitoring of BCR-ABL expression using real-time RT-PCR in CML after bone marrow or peripheral blood stem cell transplantation. *Leukemia* **13**(9): 1383–1389.
26. Emig M, Saussele S, et al. (1999) Accurate and rapid analysis of residual disease in patients with CML using specific fluorescent hybridization probes for real time quantitative RT-PCR. *Leukemia* **13**(11): 1825–1832.

27. Gabert J (1999) Detection of recurrent translocations using real time PCR; assessment of the technique for diagnosis and detection of minimal residual disease. *Haematologica* **84** Suppl EHA-4: 107–109.
28. Hochhaus A, Reiter A, et al. (1998) Molecular monitoring of residual disease in chronic myelogenous leukemia patients after therapy. *Recent Results in Cancer Research* **144**: 36–45.
29. Olavarria E, Kanfer E, et al. (2001) Early detection of BCR-ABL transcripts by quantitative reverse transcriptase-polymerase chain reaction predicts outcome after allogeneic stem cell transplantation for chronic myeloid leukemia. *Blood* **97**(6): 1560–1565.
30. Szczepanski T. (2007) Why and how to quantify minimal residual disease in acute lymphoblastic leukemia? *Leukemia* **21**(4): 622–626.
31. Tysarowski A, Fabisiewicz A, et al. (2007) Usefulness of real-time PCR in long-term follow-up of follicular lymphoma patients. *Acta Biochimica Polonica* **54**(1): 135–142.
32. Park NJ, Zhou X, et al. (2007) Characterization of salivary RNA by cDNA library analysis. *Archives of Oral Biology* **52**(1): 30–35.
33. Muller MC, Hordt T, et al. (2004) Standardization of preanalytical factors for minimal residual disease analysis in chronic myelogenous leukemia. *Acta Haematologica* **112**(1–2): 30–33.
34. Raengsakulrach B, Nisalak A, et al. (2002) Comparison of four reverse transcription-polymerase chain reaction procedures for the detection of dengue virus in clinical specimens. *Journal of Virological Methods* **105**(2): 219–232.
35. Keilholz U, Willhauck M, et al. (1998) Reliability of reverse transcription-polymerase chain reaction (RT-PCR)-based assays for the detection of circulating tumour cells: a quality-assurance initiative of the EORTC Melanoma Cooperative Group. *European Journal of Cancer* **34**(5): 750–753.
36. Nolan T, Hands RE, et al. (2006) Quantification of mRNA using real-time RT-PCR. *Nature Protocols* **1**(3): 1559–1582.
37. Koh CG, Tan W, et al. (2003) Integrating polymerase chain reaction, valving, and electrophoresis in a plastic device for bacterial detection. *Analytical Chemistry* **75**(17): 4591–4598.
38. Afzal M (2003) Comparative evaluation of measles virus-specific RT-PCR methods through an international collaborative study. *Journal of Medical Virology* **70**: 171–176.
39. Niesters HG (2001) Quantitation of viral load using real-time amplification techniques. *Methods* **25**(4): 419–429.
40. Niesters HG (2004) Molecular and diagnostic clinical virology in real time. *Clinical Microbiology and Infection* **10**(1): 5–11.
41. Kwok S, Higuchi R (1989) Avoiding false positives with PCR. *Nature* **339**(6221): 237–238.
42. Bustin SA, Dorudi S (2004) Gene expression profiling for molecular staging and prognosis prediction in colorectal cancer. *Expert Review of Molecular Diagnostics* **4**(5): 599–607.
43. Bustin SA, Siddiqi S, et al. (2004) Quantification of cytokeratin 20, carcinoembryonic antigen and guanylyl cyclase C mRNA levels in lymph nodes may not predict treatment failure in colorectal cancer patients. *International Journal of Cancer* **108**(3): 412–417.
44. Schuster R, Max N, et al. (2004) Quantitative real-time RT-PCR for detection of disseminated tumor cells in peripheral blood of patients with colorectal cancer using different mRNA markers. *International Journal of Cancer* **108**(2): 219–227.
45. Akane A, Matsubara K, et al. (1994) Identification of the heme compound copurified with deoxyribonucleic acid (DNA) from bloodstains, a major inhibitor of polymerase chain reaction (PCR) amplification. *Journal of Forensic Science* **39**(2): 362–372.
46. Bustin SA (2005) Real-time, fluorescence-based quantitative PCR: a snapshot of current procedures and preferences. *Expert Review of Molecular Diagnostics* **5**(4): 493–498.

47. Wolffs P, Norling B, et al. (2005) Risk assessment of false-positive quantitative real-time PCR results in food, due to detection of DNA originating from dead cells. *Journal of Microbiological Methods* **60**(3): 315–323.
48. Bomjen G, Raina A, et al. (1996) Effect of storage of blood samples on DNA yield, quality and fingerprinting: a forensic approach. *Indian Journal of Experimental Biology* **34**(4): 384–386.
49. Bertolini E, Olmos A, et al. (2001) Single-step multiplex RT-PCR for simultaneous and colourimetric detection of six RNA viruses in olive trees. *Journal of Virological Methods* **96**(1): 33–41.
50. Curry J, McHale C, Smith MT (2002) Low efficiency of the Moloney murine leukemia virus reverse transcriptase during reverse transcription of rare t(8;21) fusion gene transcripts. *BioTechniques* **32**(4): 768, 770, 772, 754–755.
51. Day PJ (2006) Miniaturization applied to analysis of nucleic acids in heterogeneous tissues. *Expert Review of Molecular Diagnostics* **6**(1): 23–28.
52. Wittwer CT, Herrmann MG, et al. (1997) Continuous fluorescence monitoring of rapid cycle DNA amplification. *BioTechniques* **22**(1): 130–131, 134–138.
53. Yin JS, Shackel NA, Zekry A, McGuinness PH, Richards C, Putten KV, et al. (2001) Real-time reverse-transcriptase polymerase chain reaction (RT-PCR) for measurement of cytokine and growth factor mRNA expression with fluorogenic probes or SYBR Green I. *Immunology and Cell Biology* **79**: 213.
54. Ogura M, Mitsuhashi M (1994) Screening method for a large quantity of polymerase chain reaction products by measuring YOYO-1 fluorescence on 96-well polypropylene plates. *Analytical Biochemistry* **218**(2): 458–459.
55. Hoerndli FJ, Toigo M, et al. (2004) Reference genes identified in SH-SY5Y cells using custom-made gene arrays with validation by quantitative polymerase chain reaction. *Analytical Biochemistry* **335**(1): 30–41.
56. Greiner O, Day PJ, et al. (2001) Quantitative detection of *Streptococcus pneumoniae* in nasopharyngeal secretions by real-time PCR. *Journal of Clinical Microbiology* **39**(9): 3129–3134.
57. Greiner O, Day PJ, et al. (2003) Quantitative detection of *Moraxella catarrhalis* in nasopharyngeal secretions by real-time PCR. *Journal of Clinical Microbiology* **41**(4): 1386–1390.
58. Schulte TH, Bardell RL, et al. (2002) Microfluidic technologies in clinical diagnostics. *Clinica Chimica Acta* **321**(1–2): 1–10.
59. Sia SK, Linder V, et al. (2004) An integrated approach to a portable and low-cost immunoassay for resource-poor settings. *Angewandte Chemie (International ed. in English)* **43**(4): 498–502.
60. Thorsen T, Maerkl SJ, et al. (2002) Microfluidic large-scale integration. *Science* **298**(5593): 580–584.
61. Steigert J, Grumann M, et al. (2006) Fully integrated whole blood testing by real-time absorption measurement on a centrifugal platform. *Lab on a Chip* **6**(8): 1040–1044.
62. Steigert J, Grumann M, Dube M, Streule W, Riegger L, Brenner T, et al. (2006) Direct hemoglobin measurement on a centrifugal microfluidic platform for point-of-care diagnostics. *Sensors and Actuators A–Physical* **130–131**: 228–233.
63. Auroux PA, Iossifidis D, et al. (2002) Micro total analysis systems. 2. Analytical standard operations and applications. *Analytical Chemistry* **74**(12): 2637–2652.
64. Auroux PA, Koc Y, et al. (2004) Miniaturised nucleic acid analysis. *Lab on a Chip* **4**(6): 534–546.
65. Dittrich PS, Tachikawa K, et al. (2006) Micro total analysis systems. Latest advancements and trends. *Analytical Chemistry* **78**(12): 3887–3908.
66. Kricka LJ, Wilding P (2003) Microchip PCR. *Analytical and Bioanalytical Chemistry* **377**(5): 820–825.
67. Roper MG, Easley CJ, et al. (2005) Advances in polymerase chain reaction on microfluidic chips. *Analytical Chemistry* **77**(12): 3887–3893.

68. Vilkner T, Janasek D, et al. (2004) Micro total analysis systems. Recent developments. *Analytical Chemistry* **76**(12): 3373–3385.
69. Reyes DR, Iossifidis D, et al. (2002) Micro total analysis systems. 1. Introduction, theory, and technology. *Analytical Chemistry* **74**(12): 2623–2636.
70. Zhang CS, Xu JL, et al. (2006) PCR microfluidic devices for DNA amplification. *Biotechnology Advances* **24**(3): 243–284.
71. Wittwer CT, Garling DJ (1991) Rapid cycle DNA amplification: time and temperature optimization. *BioTechniques* **10**(1): 76–83.
72. Burns MA, Mastrangelo CH, et al. (1996) Microfabricated structures for integrated DNA analysis. *Proceedings of the National Academy of Sciences of the United States of America* **93**(11): 5556–5561.
73. Cheng J, Shoffner MA, et al. (1996) Chip PCR. 2. Investigation of different PCR amplification systems in microfabricated silicon-glass chips. *Nucleic Acids Research* **24**(2): 380–385.
74. Cheng J, Waters LC, et al. (1998) Degenerate oligonucleotide primed polymerase chain reaction and capillary electrophoretic analysis of human DNA on microchip-based devices. *Analytical Biochemistry* **257**(2): 101–106.
75. Kopp MU, de Mello AJ, et al. (1998) Chemical amplification: continuous-flow PCR on a chip. *Science* **280**(5366): 1046–1048.
76. Waters LC, Jacobson SC, et al. (1998) Microchip device for cell lysis, multiplex PCR amplification, and electrophoretic sizing. *Analytical Chemistry* **70**(1): 158–162.
77. Waters LC, Jacobson SC, et al. (1998) Multiple sample PCR amplification and electrophoretic analysis on a microchip. *Analytical Chemistry* **70**(24): 5172–5176.
78. Ueda M, Nakanishi H, et al. (2000) Imaging of a band for DNA fragment migrating in microchannel on integrated microchip. *Materials Science & Engineering C–Biomimetic and Supramolecular Systems* **12**(1–2): 33–36.
79. Chen L, Ren J (2004) High-throughput DNA analysis by microchip electrophoresis. *Combinatorial Chemistry & High Throughput Screening* **7**(1): 29–43.
80. Obeid PJ, Christopoulos TK (2003) Continuous-flow DNA and RNA amplification chip combined with laser-induced fluorescence detection. *Analytica Chimica Acta* **494**(1–2): 1–9.
81. Ibrahim MS, Lofts RS, et al. (1998) Real-time microchip PCR for detecting single-base differences in viral and human DNA. *Analytical Chemistry* **70**(9): 2013–2017.
82. Northrup MA, Benett B, et al. (1998) A miniature analytical instrument for nucleic acids based on micromachined silicon reaction chambers. *Analytical Chemistry* **70**(5): 918–922.
83. Cady NC, Stelick S, et al. (2005) Real-time PCR detection of Listeria monocytogenes using an integrated microfluidics platform. *Sensors and Actuators B–Chemical* **107**(1): 332–341.
84. Xiang Q, Xu B, et al. (2005) Real time PCR on disposable PDMS chip with a miniaturized thermal cycler. *Biomedical Microdevices* **7**(4): 273–279.
85. Hu GQ, Xiang Q, et al. (2006) Electrokinetically controlled real-time polymerase chain reaction in microchannel using Joule heating effect. *Analytica Chimica Acta* **557**(1–2): 146–151.
86. Nakayama T, Kurosawa Y, et al. (2006) Circumventing air bubbles in microfluidic systems and quantitative continuous-flow PCR applications. *Analytical and Bioanalytical Chemistry* **386**(5): 1327–1333.
87. Zhou ZM, Liu DY, et al. (2004) Determination of SARS-coronavirus by a microfluidic chip system. *Electrophoresis* **25**(17): 3032–3039.
88. Kaigala GV, Huskins RJ, et al. (2006) Automated screening using microfluidic chip-based PCR and product detection to assess risk of BK virus-associated nephropathy in renal transplant recipients. *Electrophoresis* **27**(19): 3753–3763.
89. Liu CN, Toriello NM, et al. (2006) Multichannel PCR-CE microdevice for genetic analysis. *Analytical Chemistry* **78**(15): 5474–5479.

90. Lee TM, Hsing IM, Lao AI, Carles MC (2000) A miniaturized DNA amplifier: Its application in traditional Chinese medicine. *Analytical Chemistry* **72**(17): 4242–4247.
91. Giordano BC, Copeland ER, et al. (2001) Towards dynamic coating of glass microchip chambers for amplifying DNA via the polymerase chain reaction. *Electrophoresis* **22**(2): 334–340.
92. Lee TM, Carles MC, et al. (2003) Microfabricated PCR-electrochemical device for simultaneous DNA amplification and detection. *Lab on a Chip* **3**(2): 100–105.
93. Liu RH, Yang JN, et al. (2004) Self-contained, fully integrated biochip for sample preparation, polymerase chain reaction amplification, and DNA microarray detection. *Analytical Chemistry* **76**(7): 1824–1831.
94. Lotspeich E, Schoene M, et al. (2007) Detection of disseminated tumor cells in the lymph nodes of colorectal cancer patients using a real-time polymerase chain reaction assay. *Langenbeck's Archives of Surgery* **392**(5): 559–566.
95. Snow M, McKay P, et al. (2006) Development, application and validation of a Taqman real-time RT-PCR assay for the detection of infectious salmon anaemia virus (ISAV) in Atlantic salmon (*Salmo salar*). *Developments in Biologicals* **126**: 133–145; discussion 325–326.
96. Oberg AN, Lindmark GE, et al. (2004) Detection of occult tumour cells in lymph nodes of colorectal cancer patients using real-time quantitative RT-PCR for CEA and CK20 mRNAS. *International Journal of Cancer* **111**(1): 101–110.
97. Tsimberidou AM, Jiang Y, et al. (2002) Quantitative real-time polymerase chain reaction for detection of circulating cells with t(14;18) in volunteer blood donors and patients with follicular lymphoma. *Leukemia & Lymphoma* **43**(8): 1589–1598.
98. Ghossein RA, Rosai J (1996) Polymerase chain reaction in the detection of micrometastases and circulating tumor cells. *Cancer* **78**(1): 10–16.
99. Dendukuri N, Khetani K, et al. (2007) Testing for HER2-positive breast cancer: a systematic review and cost-effectiveness analysis. *CMAJ: Canadian Medical Association Journal* **176**(10): 1429–1434.
100. Holodniy M (1994) Clinical application of reverse transcription-polymerase chain reaction for HIV infection. *Clinics in Laboratory Medicine* **14**(2): 335–349.
101. Kaplan JC, Kahn A, et al. (1992) Illegitimate transcription: its use in the study of inherited disease. *Human Mutation* **1**(5): 357–360.
102. Spitzack KD, Ugaz VM (2006) Polymerase chain reaction in miniaturized systems: big progress in little devices. *Methods in Molecular Biology (Clifton, N.J.)* **321**: 97–129.
103. Rasmussen A, Deckert V (2005) New dimension in nano-imaging: breaking through the diffraction limit with scanning near-field optical microscopy. *Analytical and Bioanalytical Chemistry* **381**(1): 165–172.

8 The road from qualitative to quantitative assay: What is next?

Michael W. Pfaffl

The PCR is widely used in many applications throughout the world. It has its secure place in the history of molecular biology as one of the most revolutionary methods ever. The principles of PCR are clear, but how can the reaction procedure be optimized to bring out the best in each assay? What is the status quo and what is next? Where are there areas for improvement?

INTRODUCTION

PCR is defined as a relatively simple heat-stable Taq polymerase–based technique, invented by Kary B. Mullis and coworkers,[1,2] who were awarded the Nobel Prize for chemistry in 1993 for this discovery. However, this terrain is contested, and many other scientists were instrumental in making PCR work in all kinds of deoxyribonucleic acid (DNA), ribonucleic acid (RNA), and protein (immuno quantitative PCR [qPCR])–based applications. Reverse transcription (RT) followed by PCR represents a powerful tool for messenger RNA (mRNA) quantification.[3–5] Nowadays, real-time RT–PCR is widely and increasingly used because of its high sensitivity, good reproducibility, and wide dynamic quantification range.[6,7] Today, quantitative real-time RT–PCR (qRT–PCR) represents the most sensitive method for the detection and quantification of gene expression levels. It has its tremendous advantages in elucidating small changes in mRNA expression levels in samples with low RNA concentrations, from limited tissue samples and in single cell analysis.[8,9] Sensitivity and reproducibility is a particular requirement of expression profiling, which focuses on the fully quantitative approach for mRNA quantification, rather than simply qualitative analysis.

The enormous potential for scientific and diagnostic assays makes a comprehensive understanding of the underlying principles of RT–qPCR mandatory. As a quantitative method, it suffers from accumulated problems arising during the amplification workflow in (1) the pre-PCR steps (tissue handling, RNA extraction, and storage), (2) the RT and PCR steps (RT and PCR enzyme, primer design, detection dye, plastic ware, sealing), and (3) the post-PCR steps (data acquisition, background correction, quantification method, efficiency correction, normalization, statistical testing, data visualization) (summarized by Pfaffl[10]). Importantly, the absolute fidelity of a qRT–PCR assay is associated with its "true" specificity, sensitivity, reproducibility, robustness, and correctness.[11]

This chapter explains the improvements in chemistry, hardware, and software over the last two decades; focuses on considerations of specificity, sensitivity, variability, reproducibility, and data analysis; and presents some new ideas for data analysis.

PRE-PCR STEPS

The so-called pre-PCR steps are important and influence the result of a quantitative assay in a substantial way.[12,13] The process of sampling, tissue handling, and storage, followed by RNA extraction, is important for a reliable and quantitative assay. The scientific community has recognized this in the recent past, and the preanalytical steps are now gaining more attention. The development of RNA integrity testing by innovative lab-on-a-chip capillary electrophoresis has made a particularly big step toward quality control. All pre-PCR steps up to the extracted total RNA can now be carefully controlled to preserve the quality and integrity of the RNA material. It is well known that mRNA is sensitive to degradation by postmortem processes and inadequate sample handling or storage.[14] For a reliable quantification we need high integrity RNA that should be preferentially free of any DNA or inhibitors.[15,16] To prevent any RNA degradation, we recommend the RNA*later*® (Ambion) and PAXgene™ systems (Qiagen), which were recently optimized for high-quality total mRNA and microRNA extraction.[17] The accuracy of gene expression evaluation is recognized to be influenced by the quantity and quality of starting RNA.[18] The RNA purity and integrity are the most determining factors for the overall success of RNA-based quantification. Starting with low-quality RNA may strongly compromise the results of downstream applications that are often labor intensive, time consuming, and highly expensive.[18,19] It is therefore important to use high-quality intact RNA, ideally with RNA integrity numbers higher than five[12] as a starting point in quantitative molecular biological as well as diagnostic applications. In clinical applications with unique and precious limited tissue material – such as samples obtained after surgery, by biopsy, or from single cell studies – a reliable RNA quality analysis is necessary.[20–22]

A second important parameter relating to the pre-PCR step is the RT. It is one of the most variable reaction steps in the entire quantification assay. Even today, after the development of recombinant enzyme types with various new properties, it is the major source of variability. Each reverse transcriptase enzyme has specific reaction conditions that have to be optimized for each application and primer pair. The reaction fidelity suffers from differences in RT efficiencies, resulting in highly variable amounts of synthesized complementary DNA (cDNA) copies.[13] For most quantitative applications, Moloney murine leukemia virus (MMLV) H^- RT is the enzyme of choice,[23,24] as its cDNA synthesis rate can be up to fifty-fold greater than that of avian myeloblastosis virus (AMV).[25,26] Newly available thermostable RT enzymes maintain their activity up to 70°C, thus relieving the amount of secondary RNA structure during RT and permitting increased specificity and efficiency of first primer annealing. Each of the enzymes used to generate cDNA differs significantly with respect to specificity as well as cDNA yield and variety. Consequently, it is important to realize that RT–PCR results are comparable only when the same priming strategy and reaction conditions are used.[13] In addition, by using mFold software,[27] the first primer binding site can be checked for better mRNA accessibility and the RT reaction step can be optimized to prevent any false priming.[22] To circumvent these high inter-assay variations, an internal quality control for cDNA synthesis can be used. These internally grown controls can be artificially, like alien RNA, or naturally occurring reference genes, like glyceraldehyde 3-phosphate dehydrogenase (GAPDH), albumin, actins, tubulins, cyclophilin, microglobulins, or ribosomal subunits (18S or 28S ribosomal RNA [rRNA]).[28,29]

In summary, the vast efforts made at improving the RT step in terms of enzyme development, protocols optimizing the preamplification step, and software improvements, as well as the RNA integrity testing, have resulted in substantial improvements to the standardization and reliability in the pre-PCR setup.

INVENTIONS MADE IN "ABSOLUTE" QUANTIFICATION ASSAYS

The fidelity of a quantification assay is measured by its specificity, low background fluorescence, steep fluorescence increase, high amplification efficiency, and high level plateau. The absolute dynamic range of the detectable fluorescence (maximal plateau minus background fluorescence) should be maximized in a quantitative assay. For single PCR product reactions with well-designed primers, intercalating dyes like SYBR® Green I work perfectly well, with spurious nonspecific background showing up only in very late cycles.[30,31] Among the real-time detection chemistry, SYBR® Green I and probe-based TaqMan® assays produce comparable quantification ranges and sensitivities, although SYBR® Green I detection is more precise and produces a more linear decay plot than do the TaqMan® probes.[32,33] Nowadays new intercalating and saturated dyes are available (SYBR® GreenER, SYTO® 9, EvaGreen®, LCGreen®, CHROMOFY) that give higher fluorescence readouts and reduce the risk of primer–dimer formation.[34] The new dyes have the added advantage that, at least in theory, the sensitivity of the assay should be increased, because C_t value acquisition can take place at earlier cycles.

Assay improvements are not solely due to improved dyes and chemistries, however. A whole range of new polymerase types and mixtures has been introduced to the market. In addition to single polymerase reaction mixes, multiple polymerase mixes are now available, such as combinations of the classical Taq polymerase and proofreading polymerases. "Hot start PCR" was already a topic in the early days of classical block PCR, when we worked with wax to prevent early reaction start-up at too-low temperatures. Combining PCR components at low temperatures often leads to nonspecific high backgrounds and low product yield. Certain PCR enzymes exhibit significant polymerase activity at the typical reaction setup temperatures lower than 25°C or during the ramping steps. Nonspecific primer annealing and extension at nonrestrictive temperatures produce undesirable products that are amplified throughout the remaining PCR cycles. Today the polymerase is usually activated via antibody blockage,[35,36] through chemical modifications of the enzyme, or by an inert ligand that detaches immediately from the active enzyme center of the polymerase when there is an increase in temperature. The inert ligand has the advantage that the activation step is unnecessary; furthermore it has a "Cold Stop" feature (Eppendorf®; 5-Prime, Germany): When the temperature drops beneath a critical threshold value during the primer-annealing step, the inert ligand "binds" onto the polymerase again and deactivates it. Again, improvements to the enzymes themselves are but one aspect of the improvements made to the qPCR assay. Significant efforts have been made to optimize buffer conditions to simplify the reaction setup. For example, "self-adjusting" magnesium (Mg^{2+}) buffers reduce the need for pipetting during PCR setup, with optimal Mg^{2+} concentration always present in the tube (Eppendorf®; 5-Prime, Germany).

Besides these chemical and enzymatic improvements, hardware, plastic ware, and cycling procedures have been improved significantly. Today the term "rapid cycling" is a synonym for quicker and better results. The heating and cooling performances of the blocks have been improved, allowing the shortening of the "ramping time" between single amplification cycles. Better block surface alloying and thinner tube materials have led to higher temperature uniformity and conduction while cycling. Therefore, the unspecific reaction times have been minimized, resulting in better PCR amplification performance.[37] Additional attention has been paid to seemingly minor items such as tube sealing: Instead of self-adhesive sealing foil that can result in poor seals of the reaction tube at plate borders, new automatic heat-sealing methods that use a glue-free and highly transparent foil guarantee tube-to-tube individual sealed reaction chambers,

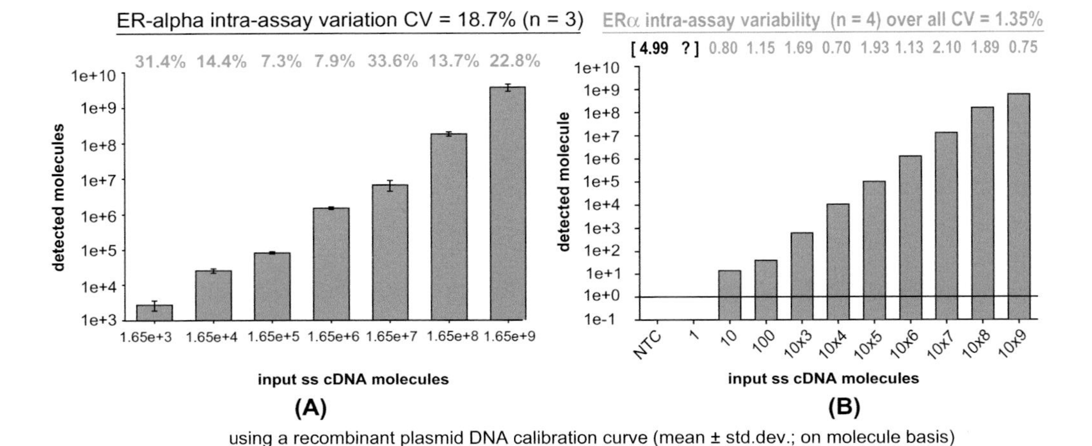

Figure 8–1. Standard curve variability performed with Estrogen Receptor alpha (ERα) single-stranded plasmid deoxyribonucleic acid (ssDNA). **(A)** Assay variability using 1,650 to 1.65 billion ERα ssDNA start molecules and a classical SYBR® Green I dye ($n = 3$). **(B)** One to one billion ERα start molecules using a new generation intercalating saturated dye ($n = 4$). Assay variability is indicated as a percentage. CV, coefficient of variation; cDNA, complementary DNA.

thus preventing any evaporation (Abgene, UK; 4-titude, UK; Eppendorf®, Germany).

The improvements made during the last ten years are nicely demonstrated on an estrogen receptor alpha (ERα) assay developed in 1997.[38,39] Both assays shown were run with the same plasmid DNA standard material using different kits and platforms, one in 1997 and the other in 2007 (Figure 8–1, a and b). On the left-hand side, assay variability is plotted using 1.65×10^3 to 1.65×10^9 ERα single-strand DNA (ssDNA) starting molecules and a classical SYBR® Green I dye ($n = 3$). An assay overall variability of 18.7% was derived in 1997. One decade later the standard material was run from 1 molecule to 1×10^9 starting molecules using a new generation intercalating saturated dye. The average variability in four replicates was 1.45%, which is remarkably low. Furthermore, the assay sensitivity was ten molecules per reaction tube.

Summarizing this, we can conclude that chemicals and hardware are made more sensitive and more reproducible while resulting in remarkable reductions in assay variability.

HOW THE RELATIVE QUANTIFICATION STRATEGY CHANGED

Alongside the "absolute" quantification according to a given standard curve, relative quantification has been of particular interest to all areas of physiological science. Relative quantification in qRT–PCR is easier to perform than the absolute assay setup, because a calibration curve is not necessary. It is based on the expression levels of a target gene versus one or more reference genes (sometimes called housekeeping or internal control genes). It is adequate for most purposes to investigate minor physiological changes in gene expression levels.[40,41] The units used to express relative quantities are irrelevant, and the relative quantities can be compared across multiple real-time RT–PCR experiments.[42] Relative quantification setup determines the changes in steady-state mRNA levels of a gene across multiple samples and expresses it relative to the levels of an internal control RNA. This reference gene can be coamplified in the same tube in a multiplex assay or can be amplified in a separate tube.[43] Therefore, relative quantification does not require standards with known concentrations, and the reference can be any transcript, as long as its sequence is known.[44]

The calculation of the expression changes will be measured by mathematical algorithms that are based on the "delta delta C_t method," established originally by Livak and Schmittgen.[45] Calculations rely on the comparison of the distinct cycle, such as threshold cycles (C_t) at a constant level of fluorescence or C_t acquisition according to established mathematic algorithms.[46,47] To date, several quantification models that calculate the relative expression ratio have been developed. Relative quantification models with and without efficiency correction use single or multiple reference genes for normalization and are available and published (summarized by Pfaffl[10]). According to such ratio calculation models, appropriate software applications were developed, such as LightCycler® Relative Quantification Software (Roche Applied Science),[48] QGene,[49] Relative Expression Software Tool (REST®),[50] SoFar,[51] Data Assimilation Research Testbed (DART),[52] qPCR–data analysis and management system (qPCR–DAMS),[53] and Qbase®.[54]

The application of such algorithms that calculate PCR efficiency on a single PCR run basis has been shown to be important for the generation of correct results.[18,47] Therefore, PCR efficiency corrections are being included in new relative quantification software (e.g., REST 2008; http://REST.gene-quantification.info/). It is desirable that the real-time qPCR software applications should calculate automatically the qPCR efficiency and implement it in proven relative quantification modules.[10]

WHAT ABOUT PCR EFFICIENCY?

All qPCR methods, absolute and relative, assume that the target and the sample amplify with similar efficiency,[45] but we know that is not the case! Unfortunately, unknown samples may contain substances that significantly reduce the efficiency of the RT[12] as well as in the PCR.[55] As discussed, sporadic RT and PCR inhibitors or different RNA/cDNA distributions can occur. A dilution series can be run on the unknown samples, and the inhibitory factors often can be diluted out, causing a nonlinear standard curve.[56,57] Individual samples can generate different fluorescence histories in real-time RT–PCR. The shapes of amplification curves differ in the steepness of any fluorescence increase and in the absolute fluorescence levels at plateau depending on background fluorescence levels. The PCR efficiency has a major impact on the overall fidelity as well as accuracy of the assay, and is critically influenced by PCR components. Efficiency evaluation is an essential marker in gene quantification procedure.[47]

A correction for efficiency, as performed in efficiency-corrected mathematical models, is strongly recommended and results in a more reliable estimation of the "real expression ratio" compared to no efficiency correction.[55] Small efficiency differences between target and reference gene generate false expression ratios, and the researcher over- or underestimates the "real and initial" mRNA amount present in the biological sample (LightCycler® Relative Quantification Software; Roche Applied Science).[48]

To conclude, quantitative efficiency corrections should be included in the automation and calculation procedure in relative quantification models, and are a major goal for the future in real-time PCR cycler and software development.

ASSAY VARIANCE AND HOW TO PERFORM A PROPER NORMALIZATION

It is important to realize that any measured variation in gene expression between subjects is caused by three sources: (1) processing variance that occurs while sampling and during the RT and PCR reactions, which must be minimized by using more replicates and by normalization with internal standards; (2) individual biological variance, which can be minimized by repeated measurements of RT and PCR reactions and by an additional normalization to an untreated control group; and (3) treatment variance.

The processing variance occurs while sampling, during RT, and during the PCR. This variance can be minimized by using multiple replicates and by normalization with internal standards, such as reference genes. The individual biological variance can be minimized by repeated measurements at RT and PCR levels and by an additional normalization to an untreated control group. In contrast, there is the treatment variation, explaining the phenotype or underlying phenomenon under investigation. This variance should be reduced by random sampling and by taking a large number of biological samples.

One major hurdle in real-time PCR gene expression studies is the removal of this experimentally induced nonbiological variation from the true biological variation. As shown before, we are on the right path, but there is still some undefined assay variability left. There are several strategies to remove experimentally induced variation, each with its own advantages and considerations.[58] We can reduce reaction noise through normalization by controlling as many of the confounding variables as possible.[29] Although most of these methods cannot completely reduce all variance sources, it has been shown to be very important to control all the sources of variation during the entire PCR process.[59] If one does not meticulously try to standardize each step, variation can and will be introduced in the results and cannot be fully eliminated by applying normalization by reference genes.[13]

Although the use of reference genes for normalization of gene expression levels is certainly the "gold standard," some new ideas for normalization have been recently developed.[58] The quality of normalized quantitative expression data cannot be better than the quality of the normalizer itself. Any variation in the normalizer will obscure real changes and produce artifactual changes.[44] Real-time RT–PCR–specific errors in the quantification of mRNA transcripts are easily compounded with any variation in the amount of starting material between the samples, for example, caused by sample-to-sample variation, variation in RNA integrity, RT efficiency differences, and cDNA sample loading variation.[18,24,25] Normalization of target gene expression levels must be performed to compensate for intra- and inter–RT–PCR variability (sample-to-sample and run-to-run variations). Therefore, data normalization by more than one reference gives much more reliable results.[60] Vandesompele and colleagues recommended using at least three nonregulated references to perform a proper normalization. A set of candidate references has to be quantified in all biological samples under investigation, and a reliable test to determine the most stable reference must be performed. This can be done by various software applets available: geNorm,[29] BestKeeper,[60] or Qbase software.[54] It still remains up to the individual investigator to choose appropriate reference gene(s) that are best for normalization in the particular experimental setting. Over the years a panel of optimal references have been reported, which are more or less stable under specific biological treatments. Also the idea of Global Pattern Recognition (GPR) was developed to evaluate expression changes in real-time PCR data.[61] By comparing the expression of each gene to every other gene in the array, a global pattern was established, and significant changes are identified and ranked. GPR makes use of biological replicates to extract significant changes in gene expression, providing an alternative to relative normalization in real-time PCR experiments.

To summarize, the normalization strategy using software applets is prerequisite for accurate quantification of RT–PCR expression profiling, which opens up the possibility of studying the biological relevance of even small mRNA expression differences. The proper normalization process revolutionized the relative quantification in real-time RT–PCR, and guided us to a more reliable result.

EXPRESSION PROFILING, qPCR BIOINFORMATICS, AND STATISTICAL ANALYSIS

In research and in clinical diagnostics, real-time qRT–PCR is the method of choice for expression profiling. Enormous amounts of expression C_t data are created. However, accurate and straightforward mathematical and statistical analysis of qPCR data and management of growing data sets have become the major hurdles to effective implementation.[62] Nowadays up to 384- and 1536-well applications are the standard in research, but in the near future high-throughput applications with multiple thousand PCR spots will generate huge amounts of data. Various qPCR data sets need to be grouped, standardized, normalized, and documented by intelligent software applications.[54] The main challenge remains the mathematical and statistical analysis of the enormous amount of data gained, as these functions are not included in the software provided.[49] The so-called bioinformatics and biostatistics on real-time RT–PCR experimental data are highly variable, because various procedures are possible, involving different ways of performing background correction, threshold settings, or expression normalization. The possibilities in performing data analysis are nearly infinite! Many questions arise: Which one is the right analysis method? Can I use my generated data? Which one gives the best results, in terms of significance? Which one gives realistic results, in terms of the biological question? Which statistical test is the right one?

Prior to normalization or statistical testing, real-time qPCR data should be analyzed by automated verification methods, such as Kinetic Outlier Detection (KOD), to detect outliers and samples with dissimilar efficiencies.[63,64]

Later statistical testing in mRNA gene quantification is nowadays mainly performed on the basis of classical standard parametric tests, such as analysis of variance or *t* tests (summarized by Pfaffl[10]). Parametric tests depend on assumptions, such as normality of distributions, the validity of which is unclear.[49,65] When performing relative quantification analysis, where the quantities of interest are derived from expression ratios, assay variances might be high, normal distributions might not be expected, and it is unclear how a parametric test could be applied.[50] Up to now two available software packages have supported statistical analysis of expression results: QGene[49] and REST.[50] Both work on the basis of Visual Basic applets on the basis of Excel (Microsoft). In QGene, rapid and menu-guided performance of frequently used parametric and nonparametric statistical tests is provided. In REST, permutation or randomization tests are applied as alternatives to more standard parametric tests for analyzing experimental data. Both tests have the advantage of making no distributional assumptions about the data, while remaining as powerful as more standard tests, and are instead based on our knowledge that treatments were randomly allocated.[66]

WHAT IS NEXT IN REAL-TIME PCR?

In the near future, new PCR applications and improvements will be developed, both on the chemical and the hardware sides. Very interesting is the invention of high-throughput applications – even more than 384-well applications[67] – and digital PCR. Digital PCR represents a powerful example of PCR and provides unprecedented opportunities for molecular diagnostics, either on DNA or RNA levels. The technique is to amplify single DNA or RNA templates from highly diluted samples, therefore generating PCR products that are derived from one template. Thus, digital PCR transforms the exponential and analog signals obtained from conventional PCR to linear digital signals, allowing statistical analysis of the PCR product. Digital PCR has been applied in various applications for mutant detection and will offer high convincing results in future molecular diagnostics.[68,69]

In this section I want to focus on the new data analysis methods and how these models will help us generate more useful information from multiple gene expression data.[70] First we need a powerful concept and, of course, a set of algorithms to

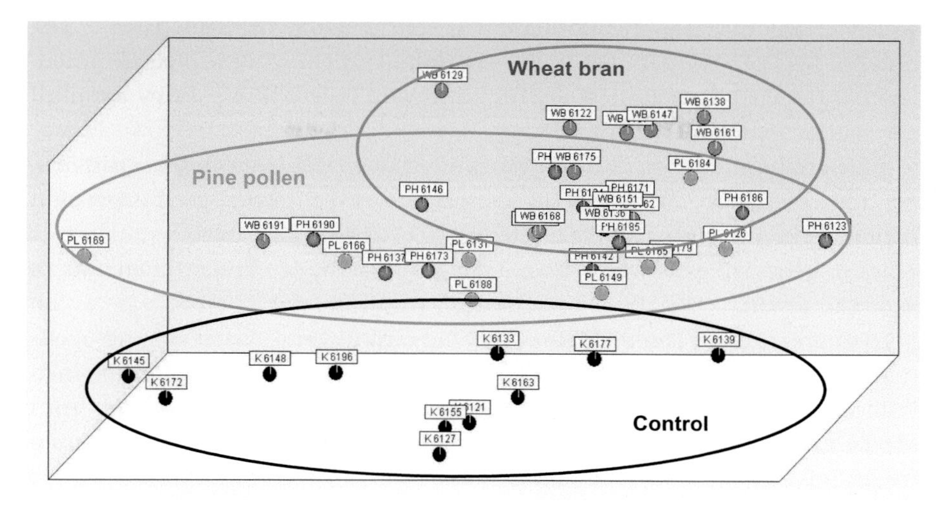

Figure 8–2. Multidimensional regression analysis via a three-dimensional scatter plot of daily intake, daily gain, and area of follicles in lymph node. Three different feeding regimens were investigated: wheat bran, pine pollen, and untreated control group. *See Color Plates.*

analyze extensive experimental expression data in parallel. Why? Suppose we pretend that our goal is to detect hidden interactions or correlations between genes. We may want to determine whether genes A and B are more influenced by our applied treatment than are genes C or D. A qPCR expression-profiling experiment generates a C_t value for each gene in each sample, thus recording the transcriptional activity of that gene in that particular sample. Although these data provide valuable and accurate information about the transcriptional response of the studied system, an even more powerful experimental design would incorporate an additional third parameter such as treatment time, applied treatment concentration, or type of treatment. Such studies generate so-called three-dimensional data sets (Figure 8–2) that are exceedingly informative and give more insight into the interaction of genes A and B over the parameter C.[71]

To analyze more data sets from an expression-profiling experiment, we need highly sophisticated algorithms, like cluster analysis,[72] which has been long established in the analysis of DNA array experiments, where thousands of data points have to be compared in parallel.[73,74] Gene expression clustering allows open-ended exploration of the data, without getting lost among the thousands of individual genes. Beyond simple visualization, there are also some important computational applications for gene clusters. The goal of clustering is to subdivide a set of items (in our case, genes) in such a way that similar items fall into the same cluster, whereas dissimilar items fall into different clusters.[70] The interpretation of clustering results will bring three general questions: (1) How do we decide what is similar – that is, which genes are similarly regulated? (2) How do we use this to cluster the items? (3) How do the different treatments cluster?

The fact that these questions often can be answered independently contributes to the bewildering variety of clustering algorithms. In hierarchical clustering, all information in the data is accounted for, but the data are analyzed sequentially, which means that not all information is considered at the same time.[70] The distance between two samples in the multidimensional space is typically calculated as the Euclidian shortest distance, by Ward's algorithm,[75] or by a ranked correlation approach.[76]

mRNA transcripts from different genes often share similar expression patterns. Ma and colleagues[77] developed an approach to reveal related gene expression patterns. The smoothing spline clustering (SSC) algorithm models natural properties of gene expression over time, taking into account natural differences in gene expression.

To summarize, the described three-way dimensional and cluster analysis opens the way to compare and interpret gene expression data in a multidimensional fashion. It creates gene groups, treatment groups, or groups of patients with similar mRNA regulation patterns and will give us much more information than will the classical gene-to-gene comparison.

GENE EXPRESSION AND MORE – THE SYSTEM BIOLOGY IDEA

Cluster analysis of gene expression data by three-dimensional data sets or by SSC is attractive, but we need even more sophisticated approaches. We do not simply wish to compare the gene expression data; what we are really interested in is the comparison between the applied treatment and the biology. This means incorporating a whole range of additional parameters, such as genetic, protein, and metabolic data sets from our samples (Figure 8–3). To visualize this, a nutrition study in 45 piglets will be presented.[78] Herein the gene expression data (C_t values) from various marker genes (apoptotic, cell-cycle, metabolic, pro- and anti-inflammatory markers), investigated in multiple organs (liver, stomach, jejunum, ileum, colon, lymph node, white blood cells), were implemented and compared with growth parameters (daily intake and daily gain, feed digestibility, feed conversion) as well as morphological data (length and width of villi, size of Peyer plates, various parameters from the lymph node morphology). Even more data sets, such as metabolic and bacterial counts in the gastrointestinal tract (GIT), will be implemented when available. All data were analyzed using GenEx software (http://www.multid.se).[79]

How is it possible to analyze hundreds of data sets that came from different measurement sources? How can we equilibrate all the data to make them comparable?

All data are measured by different analytical methods and therefore have different physical units. How we can bring these different data sets together and generate a complete readout to draw conclusions on treatment efficacy?

To do so, raw data should be autoscaled. Autoscaling is a well-established mathematical conversion that results in data sets of each parameter with the mean value of zero and the standard deviation of one (Figure 8–4). Autoscaling makes the expression data analysis robust.[70] Finally, all 107 data sets – that is, 107 different physiological parameters – from 45 animals underwent a parallel

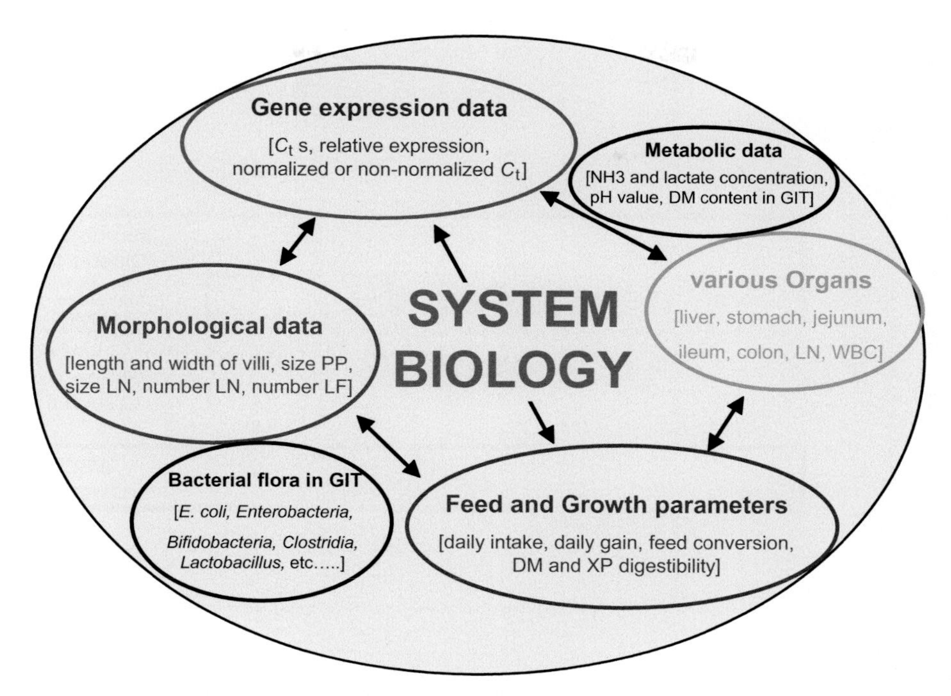

Figure 8–3. Multiple comparison of gene expression data in various organs, feed parameters, growth parameters, and morphological data (metabolic and bacterial) for a piglet feeding study for development of a system biology approach. *See Color Plates.*

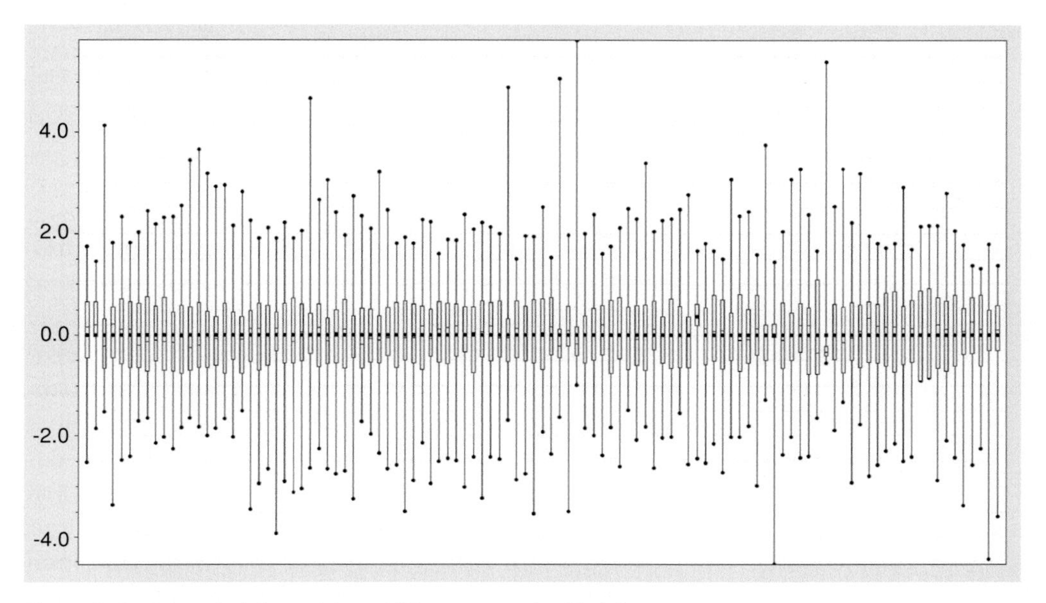

Figure 8–4. Autoscaled data set from 107 parameters in 45 piglets.

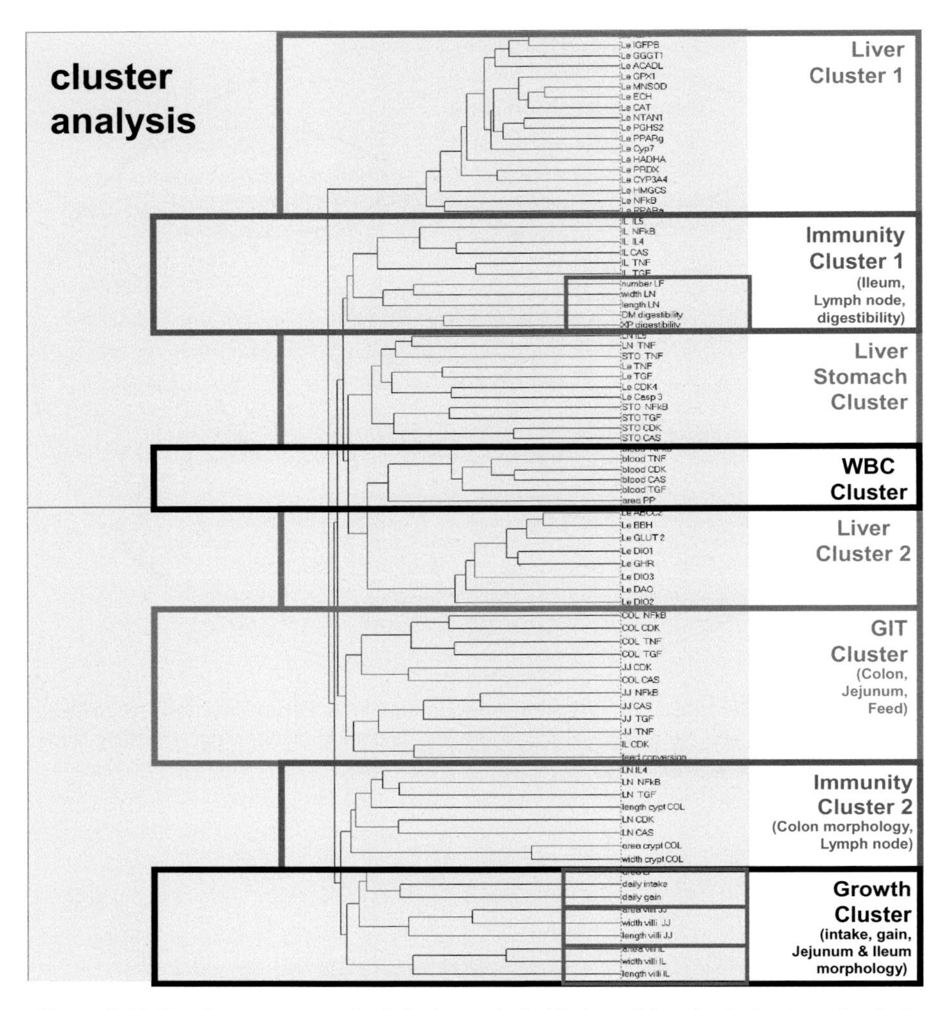

Figure 8–5. Dendrogram as result of cluster analysis. Various data sets cluster in main cluster and subcluster. WBC, white blood cell. *See Color Plates.*

cluster analysis. Figure 8–5 shows a dendrogram of the applied study. The dendrogram shows various main clusters, and of course subclusters, which correspond to genes expressed in distinct tissues, such as liver (Figure 8–6) or GIT, or belong to a functional group. As an example, many genes expressed in the liver group together, showing that gene expression is not solely regulated gene by gene. Furthermore, there is greater coherence between the individual tissues, and there is further regulation on the tissue level as well.

In immunity cluster 1, the cluster algorithm grouped the following parameters: immunological marker genes expressed in the ileum, lymph node relevant parameters, and feed parameters such as digestibility of dry matter and crude protein content (Figure 8–6). Here a direct conclusion about the overall correlation between gene expression data, morphological appearance in the GIT, and feed properties can be drawn. Furthermore, the growth cluster (Figure 8–7) functioned as a proof of concept. Within all 107 data sets the software conspicuously grouped

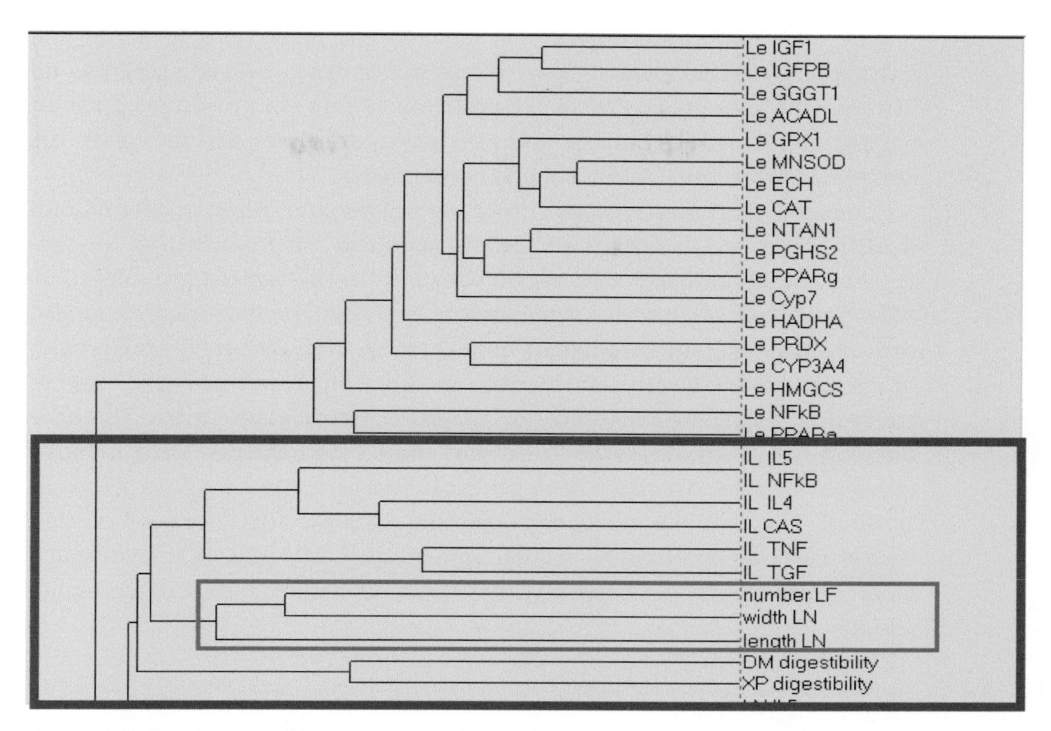

Figure 8–6. Dendrogram of liver and immunity subclusters. All liver genes cluster together in the upper part. Gene expression of immunological marker genes in the ileum, lymph node morphology, dry matter, and crude protein digestibility cluster clearly together (blue frame). *See Color Plates.*

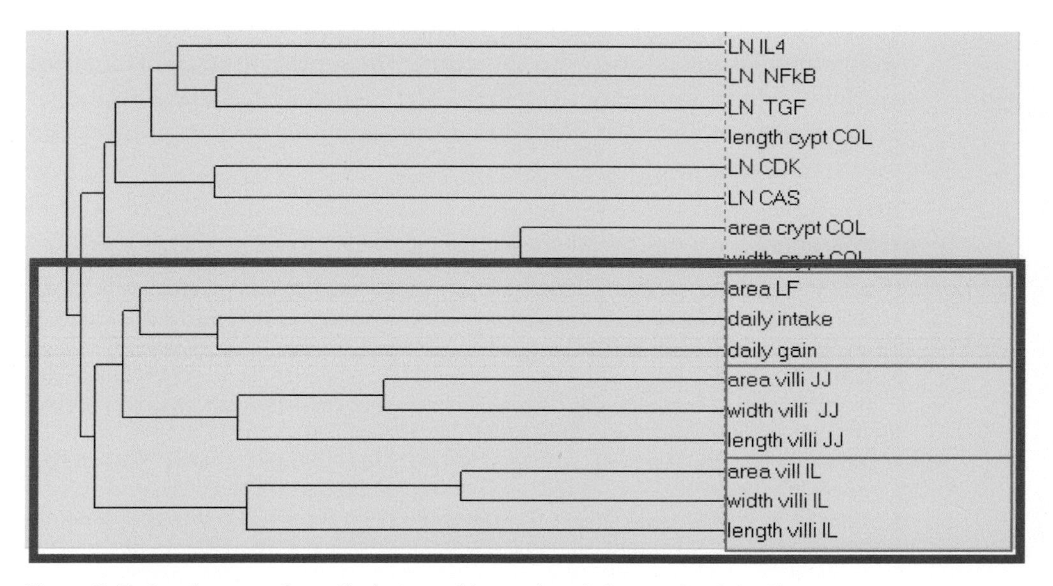

Figure 8–7. Dendrogram of growth cluster and immunity subclusters. *See Color Plates.*

side by side morphological data from jejunum and ileum (length, width, and area of the villi). The results show us that cluster analysis works and generates no fictitious and meaningless results. Importantly, within the growth cluster morphological data correlate highly with feed intake, daily gain, and the area of the lymph follicles in the ileal mesenterial lymph node.

Compared to cluster analysis, other algorithms are known to detect hidden structures between genes and other parameters. The idea behind the self-organizing map method[80] is to reflect variations in the expression profiles as a collection of cells, each with a representative expression profile, that are arranged to form a map with smooth changes in the profiles. When the expression profiles of the samples are located on the map, similar samples will be found close to each other.[70] In some situations the detailed expression pattern also can have prognostic value. Traditionally expression profiles are measured using microarrays, by which the expression of all genes can be assessed in a single experiment. However, the quality of microarray expression data usually is not good enough for detailed classification and accurate prognosis.[70] Real-time PCR gives much more information, is more sensitive, has a wider dynamic range, and has higher reproducibility.[10,42]

CONCLUSION

During the past two decades, important advances have been introduced, making quantification much more reliable. Improvements have been made in preanalytical steps, detection chemistry, applied dyes, quantification strategy, software application, and instrumentation. These improvements have led to the development of sensitive and stable assays whereby mRNA transcripts can be quantified in high throughput and precisely in a short time. The benefits in terms of increased sensitivity, reduced variability, reduced risk of contamination, increased throughput by automation, and meaningful data interpretation are obvious, even beyond gene expression data.

REFERENCES

1. Mullis K, Faloona F, Scharf S, Saiki R, Horn G, Erlich H (1986) Specific enzymatic amplification of DNA in vitro: the polymerase chain reaction. *Cold Spring Harbor Symposia on Quantitative Biology* **51**: 263–273.
2. Saiki RK, Scharf S, Faloona F, Mullis KB, Horn GT, Erlich HA, et al. (1985) Enzymatic amplification of beta-globin genomic sequences and restriction site analysis for diagnosis of sickle cell anemia. *Science* **230**(4732): 1350–1354.
3. Heid CA, Stevens J, Livak KJ, Williams PM (1996) Real time quantitative PCR. *Genome Research* **6**: 986–993.
4. Higuchi R, Fockler C, Dollinger G, Watson R (1993) Kinetic PCR analysis: real-time monitoring of DNA amplification reactions. *Biotechnology* **11**(9): 1026–1030.
5. Holland PM, Abramson RD, Watson R, Gelfand DH (1991) Detection of specific polymerase chain reaction product by utilizing the 5′-3′ exonuclease activity of *Thermus aquaticus* DNA polymerase. *Proceedings of the National Academy of Sciences of the United States of America* **88**(16): 7276–7280.

6. Ding C, Cantor CR (2004) Quantitative analysis of nucleic acids – the last few years of progress. *Journal of Biochemistry and Molecular Biology* **37**(1): 1–10.
7. Pfaffl MW, Hageleit M (2001) Validities of mRNA quantification using recombinant RNA and recombinant DNA external calibration curves in real-time RT-PCR. *Biotechnology Letters* **23**: 275–282.
8. Lockey C, Otto E, Long Z (1998) Real-time fluorescence detection of a single DNA molecule. *BioTechniques* **24**: 744–746.
9. Steuerwald N, Cohen J, Herrera RJ, Brenner CA (1999) Analysis of gene expression in single oocytes and embryos by real-time rapid cycle fluorescence monitored RT-PCR. *Molecular Human Reproduction* **5**: 1034–1039.
10. Pfaffl MW (2004) Quantification strategies in real-time RT-PCR. In: SA Bustin (ed), The Real-Time PCR Encyclopaedia A-Z of Quantitative PCR, pages 87–120. La Jolla, CA: International University Line.
11. Wittwer CT, Garling DJ (1991) Rapid cycle DNA amplification: time and temperature optimization. *BioTechniques* **10**: 76–83.
12. Fleige S, Walf V, Huch S, Prgomet C, Sehm J, Pfaffl MW (2006) Comparison of relative mRNA quantification models and the impact of RNA integrity in quantitative real-time RT-PCR. *Biotechnology Letters* **28**(19): 1601–1613.
13. Stahlberg A, Hakansson J, Xian X, Semb H, Kubista M (2004) Properties of the reverse transcription reaction in mRNA quantification. *Clinical Chemistry* **50**(3): 509–515.
14. Perez-Novo CA, Claeys C, Speleman F, Cauwenberge PV, Bachert C, Vandesompele J (2005) Impact of RNA quality on reference gene expression stability. *BioTechniques* **39**: 52–56.
15. Mannhalter C, Koizar D, Mitterbauer G (2000) Evaluation of RNA isolation methods and reference genes for RT-PCR analyses of rare target RNA. *Clinical Chemistry and Laboratory Medicine* **38**: 171–177.
16. Swift GH, Peyton MJ, MacDonald RJ (2000) Assessment of RNA quality by semi-quantitative RT-PCR of multiple regions of a long ubiquitous mRNA. *BioTechniques* **28**(3): 524–531.
17. Kruhoffer M, Dyrskjot L, Voss T, Lindberg RL, Wyrich R, Thykjaer T, et al. (2007) Isolation of microarray-grade total RNA, microRNA, and DNA from a single PAXgene blood RNA tube. *The Journal of Molecular Diagnostics* **9**(4): 452–458.
18. Fleige S, Pfaffl MW (2007) RNA integrity and the effect on the real-time qRT-PCR performance. *Molecular Aspects of Medicine* **27**(2–3): 126–139.
19. Raeymarkers L (1993) Quantitative PCR: theoretical consideration with practical implication. *Analytical Biochemistry* **214**: 582–585.
20. Bengtsson M, Ståhlberg A, Rorsman P, Kubista M (2005) Gene expression profiling in single cells from the pancreatic islets of Langerhans reveals lognormal distribution of mRNA levels. *Genome Research* **15**(10): 1388–1392.
21. Freeman TC, Lee K, Richardson PJ (1999) Analysis of gene expression in single cells. *Current Opinions in Biotechnology* **10**(6): 579–582.
22. Nolan T, Hands RE, Bustin SA (2006) Quantification of mRNA using real-time RT-PCR. *Nature Protocols* **1**(3): 1559–1582.
23. Freeman WM, Vrana SL, Vrana KE (1996) Use of elevated reverse transcription reaction temperatures in RT-PCR. *BioTechniques* **20**(5): 782–783.
24. Wong L, Pearson H, Fletcher A, Marquis CP, Mahler S (1998) Comparison of the efficiency of Moloney murine leukaemia virus (M-MuLV) reverse transcriptase, RNase H–M-MuLV reverse transcriptase and avian myeloblastoma leukaemia virus (AMV) reverse transcriptase for the amplification of human immunoglobulin genes. *Biotechnology Techniques* **12**(6): 485–489.
25. Stahlberg A, Kubista M, Pfaffl M (2004) Comparison of reverse transcriptases in gene expression analysis. *Clinical Chemistry* **50**(9): 1678–1680.
26. Stangegaard M, Dufva IH, Dufva M (2006) Reverse transcription using random pentadecamer primers increases yield and quality of resulting cDNA. *BioTechniques* **40**(5): 649–657.

27. Zuker M (2003) *m*-Fold web server for nucleic acid folding and hybridization prediction. *Nucleic Acids Research* **31**(13): 3406–3415.
28. Thellin O, Zorzi W, Lakaye B, De Borman B, Coumans B, Hennen G, et al. (1999) Housekeeping genes as internal standards: use and limits. *Journal of Biotechnology* **75**: 291–295.
29. Vandesompele J, De Preter K, Pattyn F, Poppe B, Van Roy N, De Paepe A, et al. (2002) Accurate normalization of real-time quantitative RT-PCR data by geometric averaging of multiple internal control genes. *Genome Biology* **3**(7): RESEARCH0034.
30. Pfaffl MW (2001) Development and validation of an externally standardised quantitative insulin like growth factor-1 (IGF-1) RT-PCR using LightCycler SYBR® Green I technology. In: S Meuer, C Wittwer, and K Nakagawara (eds), Rapid Cycle Real-Time PCR, Methods and Applications, pages 281–291. Heidelberg: Springer Press.
31. Reist M, Pfaffl MW, Morel C, Meylan M, Hirsbrunner G, Blum JW, et al. (2003) Quantitative mRNA analysis of bovine 5-HT receptor subtypes in brain, abomasum, and intestine by real-time PCR. *Journal of Receptors and Signal Transduction* **23**(4): 271–287.
32. Reynisson E, Josefsen MH, Krause M, Hoorfar J (2005) Evaluation of probe chemistries and platforms to improve the detection limit of real-time PCR. *Journal of Microbiological Methods* **66**(2): 206–216.
33. Schmittgen TD, Zakrajsek BA, Mills AG, Gorn V, Singer MJ, Reed MW (2000) Quantitative reverse transcription-polymerase chain reaction to study mRNA decay: comparison of endpoint and real-time methods. *Analytical Biochemistry* **285**(2): 194–204.
34. Blow N (2007) PCR's next frontier. *Nature Methods* **4**: 869–875.
35. Kellogg DE, Rybalkin I, Chen S, Mukhamedova N, Vlasik T, Siebert PD, et al. (1994) TaqStart antibody: "hot start" PCR facilitated by a neutralizing monoclonal antibody directed against Taq DNA polymerase. *BioTechniques* **16**(6): 1134–1137.
36. Sharkey DJ, Scalice ER, Christy KG Jr, Atwood SM, Daiss JL (1994) Antibodies as thermolabile switches: high temperature triggering for the polymerase chain reaction. *Biotechnology (NY)* **12**(5): 506–509.
37. Reiter M, Pfaffl MW (2008) Effects of plate position, plate type and sealing systems on real-time PCR results. *Biotechnology & Biotechnological Equipment* **22**: 824–828.
38. Schams D, Kohlenberg S, Amselgruber W, Berisha B, Pfaffl MW, Sinowatz F (2003) Expression and localisation of oestrogen and progesterone receptors in the bovine mammary gland during development, function and involution. *Journal of Endocrinology* **177**: 305–317.
39. Berisha B, Pfaffl MW, Schams D (2002) Expression of steroid receptors in the bovine ovary during estrous cycle and pregnancy. *Endocrine* **17**(3): 207–214.
40. Inderwies T, Pfaffl MW, Meyer HHD, Blum JW, Bruckmaier RM (2003) Detection and quantification of mRNA expression of α- and β-adrenergic receptor subtypes in the bovine mammary gland of dairy cows. *Domestic Animal Endocrinology* **24**: 123–135.
41. Livak KJ (2001) ABI Prism 7700 Sequence detection System User Bulletin #2. Relative quantification of gene expression. Accessed July 22, 2009 and available at: http://docs.appliedbiosystems.com/pebiodocs/04303859.pdf.
42. Bustin SA (2000) Absolute quantification of mRNA using real-time reverse transcription polymerase chain reaction assays. *Journal of Molecular Endocrinology* **25**: 169–193.
43. Wittwer CT, Herrmann MG, Gundry CN, Elenitoba-Johnson KS (2001) Real-time multiplex PCR assays. *Methods* **25**(4): 430–442.
44. Bustin SA (2002) Quantification of mRNA using real-time RT-PCR. Trends and problems. *Journal of Molecular Endocrinology* **29**(1): 23–39.
45. Livak KJ, Schmittgen TD (2001) Analysis of relative gene expression data using real-time quantitative PCR and the 2′ [-delta delta C(T)] method. *Methods* **25**(4): 402–408.
46. Tichopad A, Dzidic A, Pfaffl MW (2002) Improving quantitative real-time RT-PCR reproducibility by boosting primer-linked amplification efficiency. *Biotechnology Letters* **24**: 2053–2056.

47. Tichopad A, Dilger M, Schwarz G, Pfaffl MW (2003) Standardized determination of real-time PCR efficiency from a single reaction setup. *Nucleic Acids Research* **31**(20): e122.
48. LightCycler Relative Quantification Software. (2001) Version 1.0, Roche Applied Science, Roche Molecular Biochemicals.
49. Muller PY, Janovjak H, Miserez AR, Dobbie Z (2002) Processing of gene expression data generated by quantitative real-time RT-PCR. *BioTechniques* **32**(6): 1372–1378.
50. Pfaffl MW, Horgan GW, Dempfle L (2002) Relative expression software tool (REST) for group-wise comparison and statistical analysis of relative expression results in real-time PCR. *Nucleic Acids Research* **30**(9): e36.
51. Wilhelm J, Pingoud A, Hahn M (2003) SoFAR – Validation of an algorithm for automatic quantification of nucleic acid copy numbers by real-time polymerase chain reaction. *Analytical Biochemistry* **317**(2): 218–225.
52. Peirson SN, Butler JN, Foster RG (2003) Experimental validation of novel and conventional approaches to quantitative real-time PCR data analysis. *Nucleic Acids Research* **31**(14): e73.
53. Jin N, He K, Liu L (2006) qPCR-DAMS: a database tool to analyze, manage, and store both relative and absolute quantitative real-time PCR data. *Physiological Genomics* **25**(3): 525–527.
54. Hellemans J, Mortier G, De Paepe A, Speleman F, Vandesompele J (2007) qBase relative quantification framework and software for management and automated analysis of real-time quantitative PCR data. *Genome Biology* **8**(2): R19.
55. Pfaffl MW (2001) A new mathematical model for relative quantification in real-time RT-PCR. *Nucleic Acids Research* **29**: e35.
56. Rasmussen R (2001) Quantification on the LightCycler. In: S Meuer, C Wittwer, and K Nakagawara (eds), Rapid Cycle Real-Time PCR, Methods and Applications, pages 21–34. Heidelberg: Springer Press.
57. Morrison TB, Weis JJ, Wittwer CT (1998) Quantification of low-copy transcripts by continuous SYBR Green I monitoring during amplification. *BioTechniques* **24**(6): 954–962.
58. Huggett J, Dheda K, Bustin S, Zumla A (2005). Real-time RT-PCR normalisation; strategies and considerations. *Genes and Immunity* **6**: 279–284.
59. Bengtsson M, Hemberg M, Rorsman P, Stahlberg A (2008) Quantification of mRNA in single cells and modelling of RT-qPCR induced noise. *BMC Molecular Biology* **9**: 63.
60. Pfaffl MW, Tichopad A, Prgomet C, Neuvians TP (2004) Determination of stable housekeeping genes, differentially regulated target genes and sample integrity: BestKeeper – Excel-based tool using pair-wise correlations. *Biotechnology Letters* 26: 509–515.
61. Akilesh S, Shaffer DJ, Roopenian D (2003) Customized molecular phenotyping by quantitative gene expression and pattern recognition analysis. *Genome Research* **13**(7): 1719–1727.
62. Yuan JS, Reed A, Chen F, Stewart CN Jr (2006) Statistical analysis of real-time PCR data. *BMC Bioinformatics* **7**: 85.
63. Bar T, Stahlberg A, Muszta A, Kubista M (2003) Kinetic Outlier Detection (KOD) in real-time PCR. *Nucleic Acids Research* **31**(17): e105.
64. Burns MJ, Nixon GJ, Foy CA, Harris N (2005) Standardisation of data from real-time quantitative PCR methods – evaluation of outliers and comparison of calibration curves. *BMC Biotechnology* **7**: 31–37.
65. Sheskin D (2000) Handbook of Parametric & Nonparametric Statistical Procedures. Boca Raton, FL: CRC Press.
66. Manly B (1997) Randomization, Bootstrap and Monte Carlo Methods in Biology. London: Chapman & Hall.
67. Morrison T, Hurley J, Garcia J, Yoder K, Katz A, Roberts D, et al. (2006) Nanoliter high throughput quantitative PCR. *Nucleic Acids Research* **34**(18): e123.

68. Pohl G, Shih IeM. (2004) Principle and applications of digital PCR. *Expert Review of Molecular Diagnostics* **4**(1): 41–47.
69. Vogelstein B, Kinzler KW (1999) Digital PCR. *Proceedings of the National Academy of Sciences of the United States of America* **96**(16): 9236–9241.
70. Kubista M, Andrade JM, Bengtsson M, Forootan A, Jonak J, Lind K, et al. (2006) The real-time polymerase chain reaction. *Molecular Aspects of Medicine* **2–3**: 95–125.
71. Smilde A, Bro R, Geladi P (2004) MultiWay Analysis. Hoboken, NJ: John Wiley & Sons Ltd.
72. Bansal M, Belcastro V, Ambesi-Impiombato A, di Bernardo D (2007) How to infer gene networks from expression profiles. *Molecular Systems Biology* **3**: 78.
73. D'Haeseleer P (2005) How does gene expression clustering work? *Nature Biotechnology* **23**: 1499–1501.
74. Tavazoie S, Hughes JD, Campbell MJ, Cho RJ, Church GM (1999) Systematic determination of genetic network architecture. *Nature Genetics* **22**: 281–285.
75. Ward JH (1963) Hierarchical grouping to optimize an objective function. *Journal of the American Statistical Association* **58**(301): 236–244.
76. Tichopad A, Pecen L, Pfaffl MW (2006) Distribution-insensitive cluster analysis in SAS on real-time PCR gene expression data of steadily expressed genes. *Computer Methods and Programs in Biomedicine* **82**: 44–50.
77. Ma P, Castillo-Davis CI, Zhong W, Liu JS (2006) A data-driven clustering method for time course gene expression data. *Nucleic Acids Research* **34**: 1261–1269.
78. Schedle K, Pfaffl MW, Plitzner C, Meyer HHD, Windisch W (2008) Effect of insoluble fibre on intestinal morphology and mRNA expression pattern of inflammatory, cell cycle and growth marker genes in a piglet model using cluster analysis. *Archives of Animal Nutrition* **62**(6): 427–438.
79. GenEx software, MultiD, Göteburg, Sweden.
80. Kohonen T (1995) Self-Organizing Maps, Springer Series in Information Sciences. Vol. 30, 3rd ed. Berlin, Heidelberg, New York: Springer.

9 Taking control of the polymerase chain reaction

Tania Nolan, Tanya Novak, and Jim Huggett

INTRODUCTION

All living organisms use nucleic acid to store the genetic code. In most cases, this is in the form of deoxyribonucleic acid (DNA), although some viruses use a ribonucleic acid (RNA) molecule. DNA is used as the template for production of various RNA molecules. These have several functions, including regulation of the transcription of messenger RNA (mRNA), which is an intermediary molecule used in turn as the template for the production of proteins. It is proteins that are generally considered to be the active molecules of the cell. Of course there are many exceptions to this general pathway or "Central Dogma," and complex regulatory mechanisms are constantly being elucidated. Nonetheless it serves as a starting point for our discussion on the use of the polymerase chain reaction (PCR), because the study of these genetic materials is critical for our understanding of most aspects of life science. PCR is currently the cornerstone tool for the study of both DNA and (indirectly) RNA.

DNA ANALYSIS IN THE PRE-PCR ERA

Within human and other eukaryotic cells, DNA is compacted and organized into a number of chromosomes. Cytogenetic studies of entire chromosomes use banding patterns resulting from Giemsa staining (G banding) as structural markers. By the 1950s the techniques relating to G banding were sophisticated enough for the human karyotype (chromosome complement) to be defined as forty-six chromosomes that are arranged as twenty-two matching pairs and two sex-related chromosomes. In 1959, Jerome Lejeune et al.[1] discovered that an additional chromosome, later accepted as number 21 (trisomy 21), was consistent with Down's syndrome. Over the next two years, chromosomal abnormalities associated with human disorders were reported at almost one per month. Early in this phase of chromosomal investigation, Nowell and Hungerford[2,3] observed the presence of a minute chromosome fraction in patients suffering from chronic myeloid leukemia (CML). In a remarkably short report they concluded, "The findings suggest a causal relationship between the chromosome abnormality observed and chronic granulocytic leukaemia." Later this was referred to as the Philadelphia 1 chromosome (Ph; because the observation was made at University of Pennsylvania School of Medicine in Philadelphia), and thirteen years later Jane Rowley[4] demonstrated that the minute Ph chromosome was actually the remaining fraction of chromosome 22 with a portion of the longer arm missing. This portion of chromosome 22 was found translocated to the lower arm of chromosome 9, and a small fragment from chromosome 9 is translocated to the remainder of chromosome 22. This translocation event encodes a fused protein complex referred to as BCR-ABL that prevents normal regulation of a tyrosine kinase activity resulting in the malignant changes observed in CML. An understanding of these genetic changes is now used in the diagnosis, treatment, and management of CML patients. This observation provides a classic example of a rearrangement of the DNA sequences of the genome resulting in a clinical disorder.

There is tremendous variability in the absolute order of the individual bases that comprise the DNA molecules within a normal, complete genome. The variability is to such a degree that every individual will have a different DNA sequence. It has long been taught that because identical twins originate from a single fertilization event, they begin life with an identical genomic DNA sequence. However, it was also clear that identical twins do differ phenotypically to a greater or lesser degree, and it was believed that these differences were due to differences in epigenetic control of gene expression and not to DNA sequence variations. During a lifetime, twins do accumulate mutations and so become genetically distinct, although these variations are not usually at a rate that makes the sequences readily distinguishable. To further investigate the causes of the observed phenotypic differences, Bruder et al.[5] investigated the genetic sequence of nineteen pairs of twins. They selected pairs of twins in which only one sibling showed signs of dementia or Parkinson's disease. These were considered to be phenotypic variations, but the disease states have been associated with the presence of multiple copies of specific genomic DNA sequences. Interestingly, sequence analysis revealed clear genetic differences in the number of copies of these specific sequences between the twins. At this stage it is unclear whether these changes occur at the embryonic level, as the twins age, or both. It is clear, however, that identical twins do not always have identical genomic DNA sequences. Although some DNA regions, such as those that code for structural elements, remain highly conserved, often even between diverse organisms, there are also approximately 100 hypervariable repeats (mini satellite DNA) that contain repeated DNA sequences. The number of repeats of a given element varies between unrelated individuals. Repeated sequence patterns may contain 20 to 100 base pairs and are called variable number tandem repeats (VNTRs). Analysis and comparison of the length of these VNTRs form the basis of "DNA fingerprinting" that was developed by Professor Alec Jeffreys of Leicester University in 1984.[6] The probability that two unrelated individuals have exactly the same pattern of VNTRs depends upon the particular hypervariable region. In one system of analysis the region targeted yields up to thirty-six different-sized DNA repeat sequences for each individual. In theory, if random variability is assumed, the probability of two unrelated individuals sharing the same pattern for all thirty-six elements is approximately 0.25^{36} (5×10^{21}) or 1 in 5,000 billion billion, which is more than the number of grains of sand on all of the beaches on our planet. It is clear then that these patterns of repeated elements could be used to characterize the DNA from a given individual; this perceived genetic variability is the basis of VNTR-based forensic examination.

The first criminal case to make use of VNTR-based DNA evidence was in the conviction of Colin Pitchfork for the murder of two schoolgirls in the small town of Narborough in Leicestershire, UK. Lynda Mann was murdered in 1983 when she was fifteen years old. Forensic examination of a semen sample taken from her body revealed that her attacker was a person with type A blood. Two years later Dawn Ashworth, also fifteen years old, was strangled and sexually assaulted. Forensic analysis revealed that her attacker had the same blood type and enzyme profile as Lynda's murderer. The prime suspect was a local man who revealed unreleased details about Dawn Ashworth and eventually confessed to her murder but emphatically denied killing Lynda Mann. The police were convinced that the same person had murdered both girls, so they requested the, then-novel, DNA fingerprint analysis from Alec Jeffreys. DNA extracted from the semen samples isolated from both murder victims was compared to DNA extracted from a blood sample from the suspect. This analysis proved that both girls were indeed killed by the same person; however, it was also sufficient to demonstrate that the suspect was not the murderer. This suspect was the first person to be cleared of a criminal charge through the use of DNA profiling. The clearing of the suspect was followed by the world's first forensic DNA screen. Approximately 5,000 men from the surrounding villages were asked to provide blood or saliva samples. DNA fingerprint analysis was carried out on men who had the same blood type and enzyme profile as the killer. Even in these early days of DNA forensic analysis, the murderer clearly recognized the power of the technology and tried to escape conviction by persuading another man to give blood in his place. The switch was identified after the impostor was overheard in a bar telling others about the false sample. Colin Pitchfork was arrested and found guilty after his DNA profile was matched to the DNA from the semen samples from the murder victims.

Although clearly a powerful technique, the original DNA fingerprinting protocol was a labor-intensive and lengthy process. The sample DNA was subjected to restriction enzymatic digestion, and these fragments were then resolved through an agarose gel prior to transfer onto nitrocellulose membrane using Southern blotting. When the fragments were immobilized onto the membrane they were detected using radiolabeled DNA probes and the bound radioactive signal identified after exposure to x-ray film. The life science research community was using these familiar techniques, irradiating laboratories as a matter of course, and investing weeks of work to generate clones of interesting sequences. In contrast to a routine research lab, in the case of a forensic examination, crime scene material can be limited and there may be insufficient material to complete a thorough analysis. This limitation could lead to far fewer than thirty-six elements being examined, which would make the estimated probability of 1 in 5×10^{21} unreliable. In addition, the assumption of a random sequence distribution is flawed because the VNTR distribution is dependent upon genetic inheritance and therefore is not randomly distributed across all of the human population. As well as familial associations there is also a clear bias associated with race and population.

The power of the VNTR technique and its use to assist in both crime scene and immigration disputes demonstrated that there was a requirement for a powerful DNA-based forensic test, preferably applicable to as little material as possible.

PCR

As described for VNTR analysis and routine molecular biological studies, conventional DNA analysis techniques – including Southern blotting, northern blotting, and even cloning – were lengthy and cumbersome and usually involved at least one olfactory challenging step. In those good old days before simple DNA analysis, fathers were left to decide whether they believed mothers-to-be about the paternity of the infant with only blood typing available to address fatherhood conflicts. In the UK, the Child Support Agency was still a twinkle in the eye of an aspiring politician. Families were torn apart in immigration battles, unable to categorically prove that legal immigrants were related to persons desiring entry to UK and we can only postulate about the potential miscarriages of justice around the world. Many areas of life science research, genetic analysis, clinical diagnosis, and forensic studies were hindered by the lack of sufficient target material for study.

It is widely reported that, in 1969, Kjell Kleppe gave a presentation at a Gordon Conference in New Hampshire in which he demonstrated a reaction and amplified a nucleic acid target in front of a live audience. In 1971, Kleppe and the Nobel laureate Gobind Khorana[7] published studies including a description of techniques that could be considered to be the basic principles of a method of nucleic acid replication. In a well-quoted passage he writes,

> The principles for extensive synthesis of the duplexed tRNA genes which emerge from the present work are the following. The DNA duplex would be denatured to

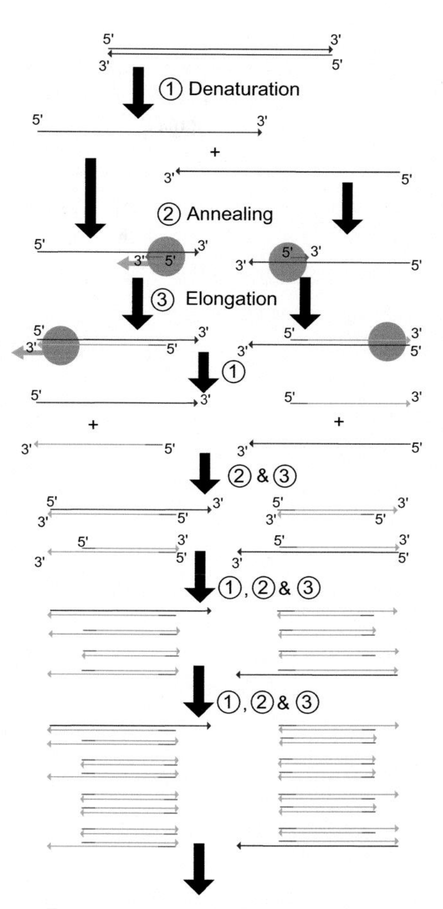

Figure 9–1. The polymerase chain reaction (PCR). The first step of the PCR is to separate the strands of target deoxyribonucleic acid (DNA). Short PCR primers are then annealed to specific sequences and elongated using a DNA polymerase to incorporate deoxynucleotide triphosphates (dNTPs). In theory each target strand is duplicated with each cycle, resulting in a logarithmic increase in the region encompassed by the primers. *See Color Plates.*

form single strands. This denaturation step would be carried out in the presence of a sufficiently large excess of the two appropriate primers. Upon cooling, one would hope to obtain two structures, each containing the full length of the template strand appropriately complexed with the primer. DNA polymerase will be added to complete the process of repair replication. Two molecules of the original duplex should result. The whole cycle could be repeated, there being added every time a fresh dose of the enzyme. It is however, possible that upon cooling after denaturation of the DNA duplex, renaturation to form the original duplex would predominate over the template-primer complex formation. If this tendency could not be circumvented by adjusting the concentrations of the primers, clearly one would have to resort to the separation of the strands and then carry out repair replication. After every cycle of repair replication, the process of strand separation would have to be repeated. Experiments based on these lines of thought are in progress.

This is a clear description of the forerunner to the process now recognized as PCR (Figure 9–1).

Kary Mullis, a Nobel laureate for the invention of PCR, recognized the pioneering work of Kleppe and Khorana; "He [Kleppe] almost had it. He saw the problems but didn't realise how fast things happen [in the PCR]."[8] Kleppe and coworkers were ahead of their time: Oligonucleotide synthesis was a painful and expensive process performed by organic chemists, and in 1971 synthetic oligos were not available as simply, rapidly, and inexpensively as in 1984 or certainly as now.

Mullis described the road to his discovery and development of PCR in his Nobel Prize acceptance speech. "With two oligonucleotides, DNA polymerase, and the four nucleosidetriphosphates I could make as much of a DNA sequence as I wanted and I could make it on a fragment of a specific size that I could distinguish easily."[9]

Originally, Mullis assumed that he could add the primers to denatured DNA and that these would be extended. Then the extension products would come unwound from their templates, be primed again, and the process of extension repeated. Unfortunately, this does not occur by simple diffusion. After the first round of strand copying, the DNA must be heated to almost boiling to denature the newly formed, double-stranded DNA, allowing the strands to separate to provide new single-stranded templates for the next round of amplification (Figure 9–1). This limitation caused the original PCR technique to be slow and labor intensive. The enzyme used was the Klenow fragment of DNA polymerase I extracted from the bacterium *Escherichia coli*. Unfortunately, heating the reaction to denature the DNA also irreversibly inactivated the polymerase, so more enzyme was required at the start of each cycle (as predicted by Kleppe in 1971). The critical development leading to the success of the PCR technique was the concept of using a thermally stable DNA polymerase that could tolerate the high temperature of the repeated denaturation steps. The most commonly used DNA polymerase (Taq polymerase) is extracted from the bacterium *Thermus aquaticus*, which lives in thermal hot springs and is resistant to permanent inactivation by exposure to high temperature.[10,11] When used in PCR, several rounds of complete amplification can be carried out without opening the reaction tube. An additional benefit to using this enzyme was that the DNA synthesis step could be performed at a higher temperature than with the original reactions using the Klenow fragment, which synthesizes at 37°C. This simple improvement was sufficient to bring about the desperately needed increase in replication fidelity. Early attempts at PCR resulted in approximately 200,000-fold amplification of the target sequence, but only approximately 1% of the PCR product was the desired fragment so products were detected using a DNA probe in a system analogous to Southern blotting. The use of thermostable DNA polymerase reduced nonspecific product formation, allowing the products to be detected directly on ethidium bromide–stained agarose gels. In addition, whereas Klenow-mediated synthesis was restricted to amplification of fragments up to 400 bp, Taq could support amplification of much larger products. Initially a brief description of PCR was included in an article describing detection of the mutation causing sickle cell anemia[12] and then in greater detail in subsequent publications.[13,14] By 1989, the

PCR technique was being used in all areas of modern biological sciences research, including clinical and diagnostic studies and, ironically, detection of human immunodeficiency virus (HIV) in acquired immune deficiency syndrome (AIDS) patients.

The awarding of the Nobel Prize to Kary Mullis for the discovery of PCR is a clear indication that adoption of the PCR technique was sufficient to revolutionize life science and associated fields. The impact was certainly way beyond the stated aim of Mullis to increase the market for DNA oligonucleotides (although that was certainly achieved).

THE MODERN PCR

The standard PCR is a deceptively simple process: Template DNA from a source of interest is combined at low concentration with a forward and a reverse oligonucleotide primer in a reaction buffer containing variations on a basic composition consisting of ammonium sulfate, Tris, ethylenediaminetetraacetic acid (EDTA), bovine serum albumin (BSA), β-mercaptoethanol, deoxynucleotide triphosphates (dNTPs), $MgCl_2$, KCl, NaCl, and DNA polymerase. The absolute optimum buffer composition is dependent upon the DNA polymerase used; different enzymes can affect PCR efficiency and therefore product yield.[15] It is generally accepted that Taq DNA polymerase performs optimally in a basic buffer of 50 mM KCl and 10 mM Tris-HCl, pH 8.3 (measured at room temperature). Some enzymes have a requirement for added protein (BSA is usually added, when required). Although dNTPs are the standard substrate for DNA polymerases, modified substrates may be incorporated such as digoxigenin–2′-deoxyuridine 5′-triphosphate (dUTP), biotin-11-dUTP, dUTP, c7-deaza-dGTP, and fluorescently labeled dNTPs. In a standard PCR, the concentration of dNTPs is included in equimolar ratios, usually 200 μM (or up to 500 μM) of each dNTP, with a higher concentration if a longer product is to be amplified. However, incorporation of unbalanced dNTP concentrations can be used as a technique for promotion of base misincorporation resulting in random mutagenesis. Many commercially available buffers may also contain PCR enhancers such as single-stranded binding protein, betaine, formamide, and dimethyl sulfoxide. The presence of detergent improves the activity of some enzymes, presumably by reducing aggregation.[16]

The salt concentration within the buffer affects the melting temperature (T_m) of the primer–template duplex and is required for primer annealing. Concentrations of KCl or NaCl greater than 50 mM can be inhibitory, whereas $MgCl_2$ is required as a cofactor for DNA polymerase. The $MgCl_2$ concentration should be optimized for each primer–template. The most influential factor affecting free magnesium ions is the concentration of dNTPs in the reaction, so the magnesium ion concentration must exceed the dNTP concentration. Typical reactions contain 1.5 mM $MgCl_2$ in the presence of 0.8 mM dNTPs, resulting in approximately 0.7 mM free magnesium. Optimization results in significant differences in the efficiency of the reaction and in the yield of PCR product (Figure 9–2).

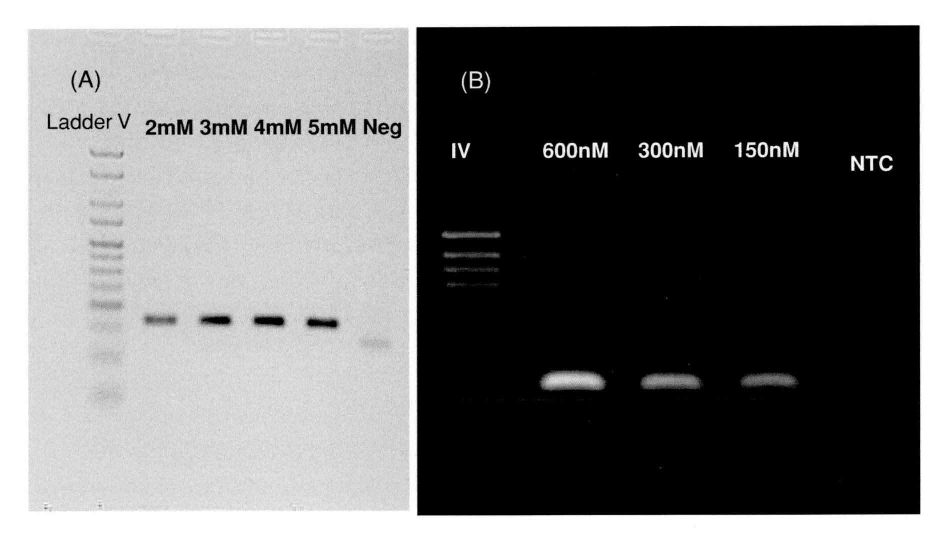

Figure 9–2. Polymerase chain reaction (PCR) optimization. Differences in the biochemical conditions can result in extreme differences in PCR yield. Each component contributes to the efficiency of the reaction. **(A)** Identical reactions were run under different $MgCl_2$ concentrations, ranging from 2 mM to 5 mM. An increase in product yield is evident in reactions containing 3 mM and 4 mM $MgCl_2$ and less in those reactions run in 2 mM and 5 mM $MgCl_2$. **(B)** Identical reactions were run using different concentrations of primers. In each case the forward and reverse concentrations are equally matched. In this example, it is apparent that higher primer concentrations result in an increase in PCR product yield. NTC, no template control.

Initially the target DNA molecules are denatured by incubation at 95°C for 10 minutes (complementary DNA [cDNA]) to 15 minutes (genomic DNA [gDNA]). The oligonucleotide primers are then hybridized to the single-stranded targets by cooling the reaction to approximately 55°C (determined from primer design and empirically). Elongation by the DNA polymerase then proceeds, usually at 72°C. The period of elongation is determined by the length of the desired product and the processivity rate of the DNA polymerase used.[17] The process of denaturation, annealing, and elongation is repeated for the desired number of cycles. In subsequent rounds of amplification, the initial denaturation stage is reduced to approximately 30 seconds because the relatively short PCR product (amplicon) requires less denaturation than total gDNA or cDNA. The absolute period of denaturation will largely depend upon the length of the amplicon and the base composition – longer guanine–cytosine (GC)-rich sequences will require a longer denaturation than shorter adenine-thymine (AT)-rich sequences.

Ideally the template may be present at any concentration from a single copy to approximately 10^{11} copies. High concentration of template will inhibit the reaction, resulting in reduced yield. Low initial concentration can result in lack of detection of amplified product if the final yield is extremely low. In addition, in the absence of specific target, nonspecific primer hybridization, including primer dimerization, may occur. Determination of the most appropriate target concentration from a cDNA sample may require the testing of several dilutions. As a guide, for a medium to highly expressed gene, including the equivalent of 0.5 μL

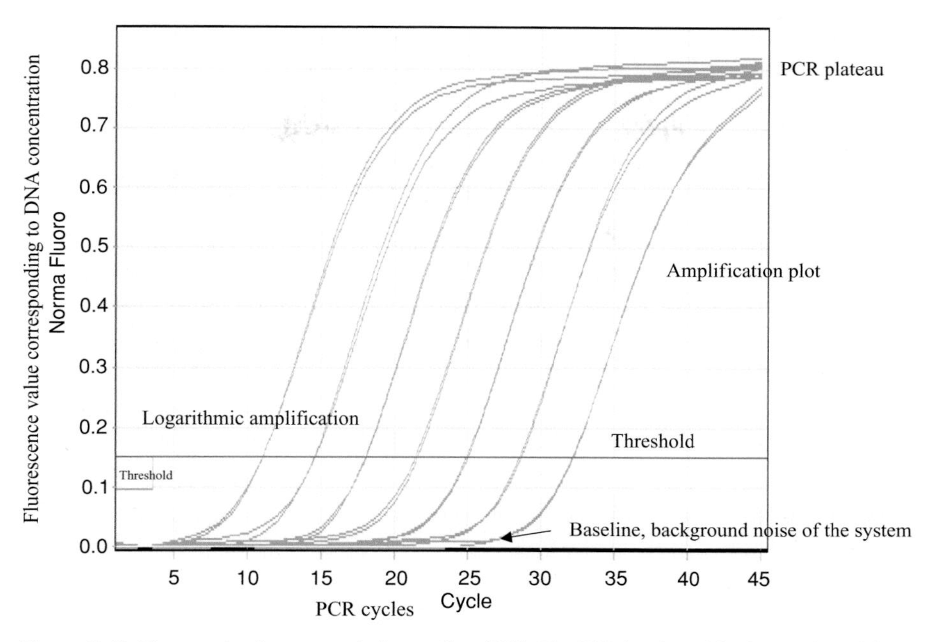

Figure 9–3. Phases of polymerase chain reaction (PCR). The PCR is a logarithmic reaction. When using real-time quantitative PCR (qPCR), a fluorescent label is added to the PCR that associates with the amplicon at each cycle. This addition allows the increase in deoxyribonucleic acid (DNA) concentration to be monitored. Also, the phases of PCR can be visualized. During early cycles the changes in DNA concentration cannot be determined because the signals are below the detection limit (baseline) of the qPCR system. When sufficient amplicon molecules have been produced, the PCR logarithmic amplification is observed until the PCR is limited and reaches the plateau phase. In the experiment shown, a fold serial dilution of template DNA was amplified. With each dilution there is a corresponding delay in the amplification plot indicating that more cycles are required to produce enough DNA (fluorescent signal) to be detected. Measurements are conventionally taken relative to the threshold. *See Color Plates*.

of a cDNA synthesis (from a total cDNA synthesis preparation of approximately 500 ng of total RNA in 20 μL of total reaction volume) in a 25-μL PCR should be sufficient to see a PCR product on an ethidium bromide–stained gel after 30 cycles of amplification. An approximate copy number of a given target in gDNA can be determined using the approximation of the genome size of the organism and the average base pair mass of 650 daltons.[18]

The number of PCR amplification cycles should be optimized with respect to the starting concentration of the target DNA. Approximately 40 to 45 cycles are generally required to amplify 50 target molecules and 25 to 30 cycles to amplify 10^5 molecules to the same concentration. The final concentration is restricted by the plateau effect (Figure 9–3), which occurs in the later cycles of the PCR when the exponential accumulation of product ceases. Various reasons for the plateau effect have been postulated: degradation of reactants (dNTPs, enzyme); reactant depletion (primers, dNTPs); end-product inhibition (pyrophosphate formation); competition for reactants by nonspecific products; or competition for primer binding by reannealing of concentrated product. However,it is clear that the most likely explanation for the prevention of exponential growth is nonspecific binding of DNA polymerase to amplification products.[19] These amplification products may be the specific target product or the result of amplification of nonspecific products formed as a result of primer dimerization.[20]

Because it is clear that accumulation of nonspecific amplification products is detrimental to PCR, various methods have been suggested for prevention of nonspecific priming. It is believed that nonspecific priming may occur as a result of nonspecific hybridization at low temperatures during the setup of the reaction. As soon as the reaction temperature is raised to the elongation temperature, these 3′ matches serve as suitable templates for elongation. One method to avoid this, and to increase specificity, is to use a hot start protocol. Originally, hot start protocols involved setting up the reaction but physically separating the enzyme from the PCR components at low temperature. One method was to raise the temperature of the reaction and then remove the tubes from the PCR block and add the enzyme. Another innovation was to layer wax on the surface and add the enzyme on top of it. As the wax melted at higher temperatures, the enzyme joined the rest of the components and the reaction would proceed. As can be imagined, neither of these solutions is ideal. Removing tubes and adding an almost random volume of enzyme tube by tube created endless opportunities for error. The wax method required extra time for the wax to melt, and then extracting the final reaction through the hardened wax was a job best performed by persons with the most delicate fingers. The most practical system is to use an enzyme that is inactive at low temperatures and becomes active only after incubation at raised temperatures. Enzymes are inactivated either by chemical modification or by binding an antibody to the active site of the enzyme. Chemical modifications require lengthy activation steps that could result in depurination and breakage of the target DNA,[21] whereas effective antibody inactivation may require multiple antibodies. A more recent suggestion is to use cold-sensitive mutant such as Klentaq.[22] In addition to the adoption of a hot start protocol amplification, specificity should be addressed at the assay design stage and also during assay optimization.

When considering optimization of any technique, it is difficult to avoid the sense that we have entered the "Groundhog Day" film set (recall that the plot is roughly that a man finds himself living through the same day over and over and over . . .). Similarly, when optimizing a PCR it is clear that in a system with multiple variables, each time an optimized condition is determined, it is dependent upon the other factors for the reaction remaining stable. There is the risk of reoptimizing each factor over and over. To avoid circular arguments, it is important to determine an optimization protocol and remain focused on the aim of the optimization procedure. In many cases that aim is to produce a linear reaction with high reaction efficiency such that estimates of template quantity can be made.

Primer optimization serves to drive the kinetics of binding the primers to the specific template sequence. Annealing is a kinetic result of the annealing temperature of the reaction and also the concentration of the primers. As a starting condition, an approximation of annealing temperature can be made of approximately T_a (annealing temperature) $\sim T_m - 5°C$,[23] taking the T_m for the primer with the lowest T_m. The T_m for a short oligo can be calculated easily by the approximation:

$$T_m = 4(\text{number G} + \text{number C residues}) + 2(\text{number A} + \text{number T residues})$$

However, applying the optimal T_a will result in higher specificity and yield. Using too low a T_a results in nonspecific priming, and too high a T_a results in inefficient priming and elongation. When this T_a fails, different temperatures must be tested in steps of 0.5°C. For longer products (>1 kb), Rychlik et al.[24] take the concept of T_a optimization a step further and recommend that the T_a be increased by 1°C at alternate cycles. Alternatively, a simpler option is to use a temperature gradient PCR block. As an alternative, or in addition, primer concentration may be optimized using a primer concentration matrix in which all primer concentrations from 50 nM to 600 nM are tested against each other and the conditions producing the highest concentration of specific template are selected (Figure 9–2).[25]

QUANTITATIVE PCR

Even with the best optimization, the plateau phase reached in later cycles of the PCR prevents quantification using estimations of end point product yield, without further indirect measurements.

The growing requirement to assess gene expression patterns and gene copy number required a modification to the basic PCR technique such that measurements could be taken earlier in the process, prior to the plateau. The challenge was that insufficient DNA is produced during earlier cycles to detect when using conventional techniques. Higuchi et al.[26] recognized that the process of PCR could be tracked by including a fluorescent label in the reaction that would associate with the accumulating PCR product. As the PCR product increases, the intensity of the signal also increases (Figure 9–3). In current quantitative PCR (qPCR) technology, these signals are generated by inclusion of either fluorescent DNA binding dyes or additional oligonucleotide probes. Fluorescent DNA binding dyes (such as SYBR® Green I and related derivatives, BEBO, and BOXTO[27,28]) are included in the PCR buffer along with DNA primers. As the target is amplified, the dye binds to the DNA product and adopts an alternative conformation. This conformational change results in a change from low to high fluorescent emission. The increased fluorescence intensity is monitored during the reaction and can be used to quantify the DNA target. As described previously, primers are apt to dimerize or misprime, particularly in the absence of specific target. Primer dimerization results in formation of nonspecific PCR product that is also detected by DNA binding dyes, alongside the specific product. Optimization can reduce this effect, but when quantifying low copy numbers it is preferable to use a specific form of detection. Greater specificity is achieved by introduction of an additional oligo probe situated between the two primers. This oligo probe is labeled and, in most cases, also quenched. Various probe options are available, but the most popular are the linear hydrolysis probe (referred to colloquially as TaqMan® probes),[29] Molecular Beacons,[30] Scorpions™ probes,[31] and LightCycler® probes.[32] Linear hydrolysis probes are oligos with a fluorescent label on the 5′ end and a quencher molecule on the 3′ end. These probes are designed to have an annealing temperature 7 to 10°C greater than that of the primers; therefore, in the reaction the probe hybridizes to the target sequence before the primers. On further cooling, the primers hybridize and the new strand is elongated until the Taq polymerase reaches the 5′ of the probe. The probe is then cleaved by 5′ to 3′ exonuclease activity of the enzyme, releasing the fluorescent label. In this way a fluorescent label should be released with each amplicon synthesized. In reality, between 4% and 47% of amplicons are detected using the linear probe system. For this reason, alternative probe systems may be used to increase detection sensitivity. Different probe methods provide different sensitivities of detection because of greater efficiency in separating the fluorescent label from the quencher.[33] It has been observed that incorporation of the locked nucleic acid modified nucleotides residues into a linear, hydrolysis probe can increase detection sensitivity by up to tenfold (T Nolan, unpublished observation, 2006). Locked Nucleic Acids are the result of a modification to the backbone of the nucleotide such that a methylene bridge is formed between positions 2 and 4 of the sugar molecule. This bridge causes the resulting DNA to be more rigid. The increase in sensitivity is most likely because of increased stability of probe binding, thus promoting cleavage and reducing displacement.

Molecular Beacons is a structured probe system with a loop region complementary to the target sequence and an additional double-stranded stem sequence that holds the label and quencher in close proximity. In the presence of the specific target, the kinetics of binding of the probe to target is stronger than the stem, so the Molecular Beacon opens and the label is separated from the quencher. The close positioning of the fluorophore to the quencher results in a low background signal and therefore higher relative signal to noise.

Scorpion probes are also a structured detection system. These combine the forward primer and detection probe into a single molecule, also holding the fluorophore and quencher in close proximity with a stem structure analogous to the Molecular Beacon system. Initially the primer region hybridizes and elongates from the single-stranded target. After melting away the template, the Scorpion opens and the probe region hybridizes to the target region, separating the label from the quencher.

Tracking the increase in fluorescence as the PCR progresses reveals that the reaction can be defined by a series of phases. The background noise of the detection system prior to sufficient amplification signal is recorded as the **baseline**, which is followed by a period of **logarithmic amplification** during which the detected signal is correlated to an increase in DNA concentration until the **plateau phase** is reached (Figure 9–3). It is inaccurate to take data measurements close to the background noise of the system, so one method of analysis is to define a constant florescence intensity, correlating to a constant DNA concentration, and determine the number of cycles required to reach this threshold setting.

Alternative systems are used to analyze the curve shape using a variety of algorithms that aim to predict the cycle at which a positive signal occurs. Over time, each of these systems has been used to produce a cycle number used for quantification. It has been proposed that when communicating these values a common term is adopted, the C_q (quantification cycle).[34]

The C_q difference between equivalent dilutions of template material remains constant, so the system can be used to make quantitative assessments of input DNA concentration. This being the case, there are two main options for quantification. A calibration curve can be constructed from DNA of known concentration or relative dilution. The C_q for each dilution is plotted relative to the log of the input DNA concentration (PCR is a logarithmic reaction), and measurements of the concentration of material in the samples are taken relative to this calibration series. Alternatively, it is possible to make a relative measurement of the apparent quantity of the target sequence in sample A relative to sample B by making assumptions about the efficiency of the reaction and therefore the C_q differences.[35,36] Using this technology it is possible to estimate quantity to as low as a single copy or to quantify targets from a single cell.

The sensitivity of the assay is in part determined by the assay design. Much has been written about the design and optimization of qPCR assays, and general guidelines can be found in Nolan et al.[25]

THE PCR REVOLUTION

As we have seen, the replicative power and, therefore, detection sensitivity of PCR are phenomenal. In theory, a single template molecule should be replicated during each PCR cycle. Assuming absolute, perfect replication at each cycle, a single template molecule would yield 2^{30} or 1,073,741,824 template molecules in a tube after 30 rounds of amplification. The number of applications for the technique is probably somewhere close to this too (but we will not go into detail here, for all of this information is in this chapter)!

Variations to the basic technique include the possibility of modifying the primer sequences so that they contain additional sequences at the 5′ end that match restriction enzyme sites. This technique allows PCR products to be cloned, regardless of the sequence of the fragment between the primers and therefore without prior knowledge of the entire sequence. Primers that were designed to be specific matches to a given gene in a target organism are used to isolate the homologous gene sequences from related organisms by altering the PCR protocol to tolerate initial priming with primers that are not exact matches (i.e., degenerate). These approaches enable fragments of unknown sequence to be cloned and subsequently sequenced, or the PCR product could be sequenced directly (depending upon the sequencing protocol and fragment length). Such approaches led to the explosion in sequencing projects and ultimately to entire genome sequencing.

The availability of vast amounts of sequence enabled the identification of submicroscopic genetic markers that are characteristic of malignancies in a manner comparable to the Ph chromosome as described previously. Using PCR, these genetic markers can be tracked and used to monitor patient relapse, along with effectiveness of drug and transplant treatments. Infectious diseases such as HIV, tuberculosis, and malaria can be detected with tremendous sensitivity and treatment efficacy monitored. In other industries, such as the production of fine foods, PCR is used to quantify proportions of, for example, genuine caviar or virgin olive oil in a product and to identify products bulked out with lower quality material. The proportion of material from genetically modified organisms in foods and even the quality of beer are also measured using qPCR. The applications of PCR are apparently endless.

The development of PCR has led to improvements in forensic investigation techniques such that it is now possible to use minute samples. One modern form of DNA profiling examines "short tandem repeats" or STRs. These are repetitive DNA sequences of 2 to 6 bases long that are repeated in tandem and occur at different chromosomal loci. The number of repeated sequence blocks varies within the population, from four to forty depending on the STR. Each variation is referred to as the allele for that specific marker. Each person has the potential for two alleles of each marker, one inherited from each parent. These two alleles for a particular marker may be identical, if both parents had the same form, or differ such that each different marker was inherited from each parent. The U.S. Federal Bureau of Investigation (FBI) identification system, called CODIS (Combined DNA Indexing System), has become the standard DNA profiling system in use in the United States and uses a set of thirteen 4-base repeated sequences. For example, the STR defined as D7S820 is located on chromosome 7 and consists of six to fifteen repeated *GATA* elements. Each of these STR loci has multiple alleles in the population. After isolation of DNA from samples, the regions specifying the thirteen target STR sequences are amplified using specific PCR primers. For STR analysis, the primers are labeled with a fluorescent dye so that, during amplification, the fluorescent label becomes incorporated into the product and is detected by instrumentation that can then define the fragment length associated with the labeled molecule.

Improvements in the sensitivity of forensic detection have led to an increasing number of convictions after reopening "cold cases" (and therefore TV series such as "New Tricks" and "Cold Case") as well as exonerations of persons found guilty of crimes that they were innocent of committing. In the United States, the focus of the Innocence Project is to review convictions, the majority being rape, and to challenge these convictions based on DNA analysis. These re-examinations of criminal cases are dependent on improved techniques in analysis of crime scene evidence and DNA matches to sequences of convicted criminals that are stored in police databases. The exponential increase in solving old cases is the main justification for the requirement for national DNA databases. In the United States, many states allow DNA sample data to be retained even in the absence of a conviction. From September 2007, all states in the United States have laws that require convicted sex offenders to submit to DNA testing; forty-four states require convicted criminals to submit samples for DNA analysis, and the remainder have restricted regulations with respect to DNA sampling of persons arrested and convicted. In the United Kingdom, all suspects and arrestees (including persons accused of begging and being drunk and disorderly) can be forced to provide a DNA sample. Traffic wardens in Edinburgh and public transport staff in Scotland generally are to be issued with DNA kits in an effort to track down people who attack or even spit at them. However, the European Court of Human Rights has recently ruled that the indiscriminate storage of DNA or fingerprint data from persons not convicted of a criminal offense is unlawful and contravenes the right for privacy under the European Convention on Human Rights. Of course, there will always be persons who will try to beat the system. Back in 2000, Dr. John Schneeberger was found accused of drugging and sexually assaulting female patients. To evade arrest, he sliced open his arm and inserted a plastic tube containing the blood of a male patient. On each of three occasions (in 1992, 1993, and 1996) when blood samples were taken, the blood from the tube was actually drawn. This led to the conclusion that he could not be the rapist because the blood DNA did not match recovered semen DNA. Presumably he had also seen the film "Gattaca" in which the genetically inferior hero inserts the blood of a genetically superior male into his fingertip to produce as a sample when he is required to give blood to prove his identity. (I am sorry if we spoiled the plot!) The truth won out, and Dr. Schneeberger was charged and convicted after the police analyzed samples from his hair and found a perfect match. He was also convicted of obstructing justice!

Whereas the previous examples of PCR applications rely upon the tremendous sensitivity of the technique, the specificity afforded by the technique has also opened up new possibilities. Paternity and identification testing have been revolutionized using the fragment comparison tests described. DNA matching has been used for identification of victims of crimes and also disasters such as the terrorist attacks of 9/11 in New York. In the clinic, there are examples of differences in the severity of diseases being due to differences in the causative agent. A classic example is provided by the defined link between infection by the human papilloma virus (HPV) and cervical cancer, where infection with subtypes HPV16 or HPV18 results in a greater severity of disease than other subtypes. The power of the technique is further exemplified by its use to detect infections such as *Mycobacterium tuberculosis* and *Plasmodium falciparum* malaria in trace amounts of DNA extracted from ancient Egyptian mummies.[37,38]

As well as being used for the diagnosis of disease, there is a growing trend toward personalized medicine. Single base pair variations, referred to as single nucleotide polymorphisms (SNPs), are changes to the original (or wild type) genomic DNA sequence that occur through mistakes in the natural process of DNA replication and are considered to be the driving force behind evolution. Unfortunately, many of these mistakes result in genetic disease states such as cystic fibrosis or sickle cell anemia. It has been recognized that these also are associated with disposition toward diseases such as cardiac disorders, diabetes, bipolar disorder, schizophrenia, and even dyslexia. There is the suggestion that these polymorphisms and repeated sequence variations are involved in personality traits such as the need for thrills to promote a sense of excitement. A personalized map of these polymorphisms and associated personality traits could be used to encourage targeted healthy living and specific forms of education and life management as well as early diagnosis that could reduce the childhood and adolescent psychological pain of an undefined disorder.

In addition to genomic sequence analysis, it has long been believed that investigating specific mRNA sequences can be informative about the biology of the cell. Investigating gene quantity changes between normal tissues and diseased cells, or looking for changes in gene expression at different times of day or night or in response to drug treatments, is being used to understand how regulation gene expression is a part of the complex system of control of cellular processes.

Measuring mRNA requires an additional step to convert RNA to a DNA template suitable for PCR amplification. This process is carried out using a reverse transcriptase enzyme and extension from one or more oligo primers. Priming of the reverse transcription (RT) may be from a sequence specific primer, from a series of random primers that hybridize along the length of the mRNA, or from a primer directed toward a tract of adenosines that are added to the 3′ end of most messenger RNA sequences, the polyA tail. After elongation from the primer, a double-stranded hybrid of RNA and DNA, called cDNA, is produced. cDNA is then a perfect template for PCR, and relative quantities of specific templates could be determined by carefully controlling the amplification in a process of semiquantitative PCR or qPCR (as described earlier in this chapter).

CONTROLLING THE REVOLUTIONARIES

The very power of the PCR technique could also be considered a drawback; under favorable conditions, a single target DNA molecule will amplify even if it was not actually the intended target sequence. In the absence of a specific target, primers will often settle for a best match or self-dimerize, still resulting in an amplified product. The film "Gatacca" provides a futuristic vision of a genetically defined population and also a science fiction approach to sample collection by illustrating that a single hair root revealed the desired information when collected along with the rubbish in an entire building of several hundred workers. This challenge does provide a glimpse into the real challenges faced by routine molecular biologists in either clinical or routine settings, as well as by forensic scientists and officers at a scene of a crime.

One of the most extreme crime scenes was created on August 16, 1998, when 220 people were injured and 29 killed after a 255-kg car bomb exploded in the small town of Omagh in Northern Ireland. After a grueling police investigation lasting almost ten years, the chief suspect was brought to trial for the murders in one of Britain's largest murder trials. The critical factor on which the trial hung was the DNA evidence. The prosecution claimed similarities between the bombs used in this and other attacks and that DNA from the accused had been found on the timing mechanisms of these previously used bombs. Justice Reg Weir was unimpressed and delivered a not guilty verdict on December 20, 2007.

He considered that the teams concerned with analyzing the samples did not take "appropriate DNA protective precautions."[39] The case rested on the analysis of traces of DNA extracted from minute samples typically as a result of surface contact rather than, for example, from blood stains. The techniques used pushed PCR to the limit for analysis of low-copy-number samples,[40,41] and the judgment was a recognition of the extreme care required when handling samples and performing PCR. Following the conviction of Mark Dixie for the murder of eighteen-year-old Sally Anne Bowman, Detective Superintendent Stuart Cundy said, "It is my opinion that a national DNA register – *with all its appropriate safeguards* – could have identified Sally Anne's murderer within 24 hours. Instead it took nearly nine months before Mark Dixie was identified, and almost two-and-a-half years for justice to be done."

"WITH ALL ITS APPROPRIATE SAFEGUARDS"

These "appropriate safeguards" are a critical subject to be considered by all scientists using PCR and associated techniques. As demonstrated by the previous examples, PCR can produce data that can be revolutionary or can be highly misleading, maybe even depending on the desired interpretation. Like the music of Mozart, the PCR could be considered to be too simple for beginners and too complex for experts. Remaining with the analogy, every component must be carefully controlled to optimize the final performance. When starting a project requiring PCR, it is important to ensure that instrumentation is functioning correctly. PCR machines are notoriously variable both between instruments and within a single block. It is not unheard of to have a reaction that functions only in the central wells and not around the edges, due to lack of thermal uniformity. Also, micropipettes are subject to constant abuse. This ill treatment results in their dispensing inaccurate volumes, causing variable component concentrations in the reactions, and leads to suboptimal, if not failed, experiments. Pipette calibration is a simple, inexpensive way to pass a Friday afternoon, and regular servicing could coincide neatly with literature review periods.

To protect against uncertainties due to reaction variability, it is important to include a series of controls alongside all samples for an experiment: test, study, report, just as would be recommended as best practice for any scientific procedure. The concept of including controls appears to be out of fashion, and persons who insist on them are regarded as being a small step from a cozy chair in a corner of the room, gazing into the fire and reminiscing about the good old days. As the anonymous seventeenth-century nun prays, "Lord, Thou knowest better than I know myself, that I am growing older and will someday be old.... Release me from craving to straighten out everybody's affairs (*experiments*).... With my vast store of wisdom, it seems a pity not to use it all, but Thou knowest Lord that I want a few friends at the end." With this risk in mind, it is appropriate to discuss the value of scientific controls in the context of PCR.

The choice of controls and whether to include them largely depend upon the nature of the study and should be independent of how many samples need running by tomorrow, how much money the reagents cost, or how much time there is before the next grant body review meeting. Although these may be factors to consider, they should not interfere with a desire for science that is beyond reproach. Some controls certainly should be considered obligatory and honest, and original data should be presented for inspection, especially when the results of the study or analysis lead to life-or-death decisions.[34,42] It is worth recognizing that the simple disregard for the data from controls in a study claiming to link autism to the measles, mumps, and rubella (MMR) vaccination leads to the distress of parents of children affected by the disorder. In 2002, it was reported that qPCR had been used to identify measles virus in the intestinal tissue of children reported with gut disorders and developmental delay.[43] This observation was used as the basis for claims of an association between MMR vaccination and a new form of autism. However, detailed examination of the data reveals that inconsistencies in the results for the control samples show that the findings are unreliable. For example, the measles virus exists only in the RNA form; therefore, an RT step is required prior to PCR to generate a cDNA template. In the original experiments there was approximately the same quantity of measles virus detected in samples following an experimental procedure either including or excluding the RT step. This aberrant result provided a clear indication that the experimental setup was vulnerable to detection of DNA contamination and, therefore, that the apparent detection of measles virus was due to DNA contamination. Since doubts about vaccination arose, there has been a steady decrease in MMR uptake and recently the death of a child not protected against measles.[42] This is a striking example of the absolute necessity for extreme caution tending to obsession over the rigor of control data.

The PCR can be considered to have two major contributing sets of factors: variability pertaining to the reaction itself and that due to the samples and gene target of interest. Each source of variability must be validated and controlled for. The process of assay optimization, discussed earlier, serves to minimize variability due to the assay. However, it is critical that the validity of the assay is checked each time it is run by simply including a positive control alongside the samples to ensure greater confidence in reporting negative results. In addition, because PCR is remarkably sensitive, it is also imperative to include a negative control: a reaction containing all components with the exception of DNA/cDNA template. A positive signal from this sample is indicative of contamination or nonspecific amplification, and therefore the data from associated samples should be further investigated at least and potentially be disregarded (if of dubious reliability). When the target is cDNA, a further control containing all components of the RT step (with the exception of the reverse transcriptase enzyme) is absolutely critical. This control is especially important when the target cDNA is present at low copy and/or the gene sequence is present in multiple copies. Should gDNA be present in the RNA sample, these sequences also will amplify and be interpreted as a false-positive signal. In many cases, this can be avoided at the design stage by ensuring that the primers span an intron/exon boundary; but the design specificity should also be tested experimentally.

The significance of variability due to the sample quality cannot be underestimated. The choice of sample source is usually determined by the experiment, but it is clear that not all samples are created equal. The biological variability between individuals may mask variability due to the test condition. Similarly, the variability between cells within a tissue sample may blur localized differences in mRNA concentration. Hence the tried and tested recommendation that all experiments are designed with reference to a statistician; seek advice on the minimum number of samples required and the associated control or non–test condition samples. When the response to this inquiry is several hundred clinical samples with an even larger problem being the requirement for several normal or reference samples, it becomes apparent that accurate statistics are set to detract from a good biological story. So it is that probability assessments are provided based on analysis of just a handful of samples. Fewer samples are required when the test and reference are from the same biological source. A classic example is provided by experimental designs requiring the testing of tumor material and comparison of expression profiles to those in adjacent normal tissue. This design requires the assumption that the adjacent normal tissue in a cancer patient is in no way different from the normal tissue of a nonsuffering individual. It is apparent that this assumption is unsafe. In a study to compare the expression profiles in breast tumor material, RNA was extracted from both biopsies of adjacent normal tissue and also from tissue from patients undergoing reduction mammoplasty. The copy number of a range of genes was compared in these samples and tumor samples derived from cancers of varying degrees of severity. There were clear differences in the expression profiles of genes associated with the tumor when the adjacent normal tissues were compared to the tumor biopsy material. In addition, it is clear that the genetic profile of the normal tissue was clearly distinct from that of the pathologically normal material from the cancer patients. This difference may be due to sampling in that the adjacent material may have been genetically but not pathologically affected, or it could be that in these patients there was a disposition to tumor formation that was evident in the normal tissue (T Nolan, unpublished data, 2001. Figure 9–4). Hence the definition of "normal" must be clarified and even tested prior to experiment design.

Having established the choice and number of samples to work with, the next step is to extract the target nucleic acid from the biological sample. This extraction is considerably more challenging if the sample is a fragment of clothing at a murder scene than if it is a fresh collection of cultured cells. Specific procedures are required for DNA or RNA extraction depending on the sample; these procedures should be optimized accordingly. When the focus of the procedure is to amplify and then analyze a target of interest, it is clearly advantageous to actually have that target in the sample to begin with to avoid reporting a false negative.

One method to monitor the effectiveness of the purification procedure is to introduce an additional template into the samples prior to extraction. This control will proceed through the purification steps and can be used to track the efficiency of extraction procedures. There are some technical difficulties in detecting a purification control alongside the sample; when a distinguishable DNA sample

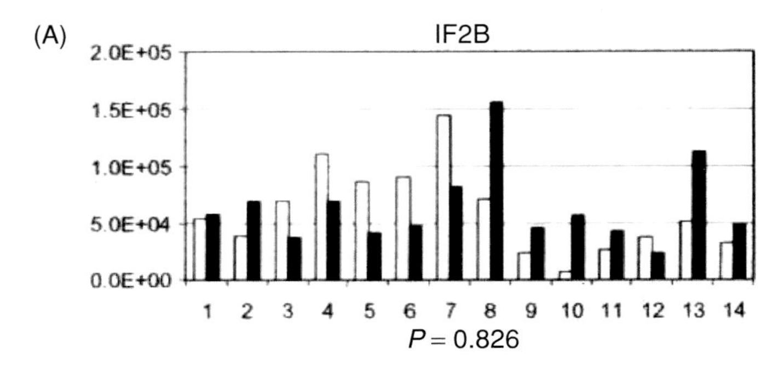
(A)
IF2B
2.0E+05
1.5E+05
1.0E+05
5.0E+04
0.0E+00
1 2 3 4 5 6 7 8 9 10 11 12 13 14
P = 0.826

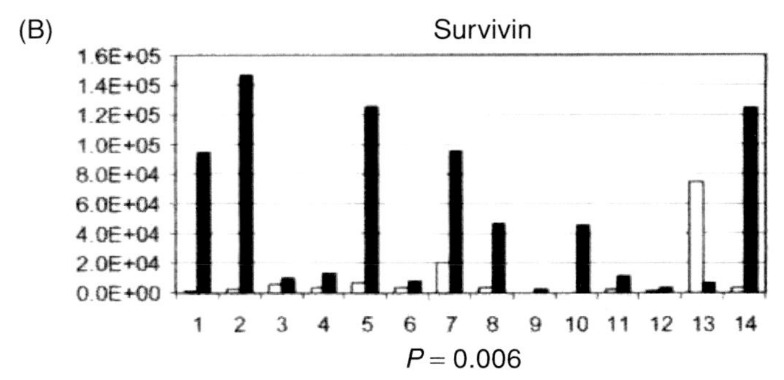
(B)
Survivin
1.6E+05
1.4E+05
1.2E+05
1.0E+05
8.0E+04
6.0E+04
4.0E+04
2.0E+04
0.0E+00
1 2 3 4 5 6 7 8 9 10 11 12 13 14
P = 0.006

Normal
Tumor

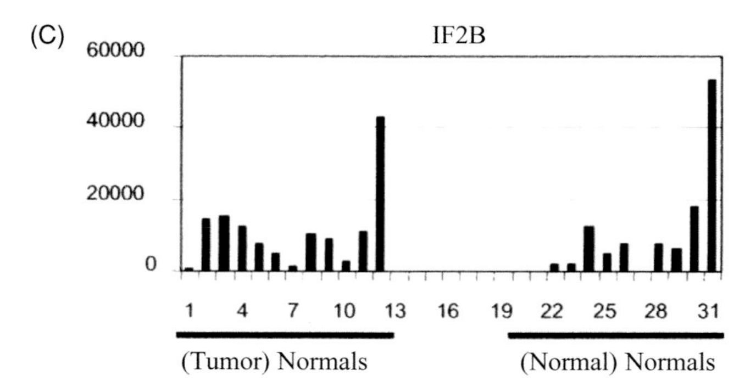
(C)
IF2B
60000
40000
20000
0
1 4 7 10 13 16 19 22 25 28 31
(Tumor) Normals
(Normal) Normals

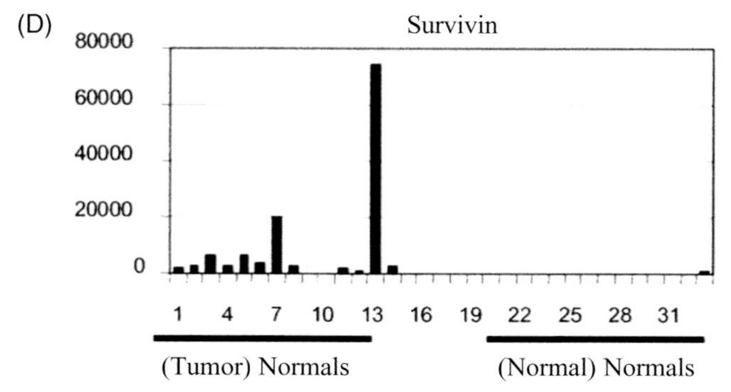
(D)
Survivin
80000
60000
40000
20000
0
1 4 7 10 13 16 19 22 25 28 31
(Tumor) Normals
(Normal) Normals

is added, additional primers are required to detect a diagnostic sequence and, ideally, the control PCR is performed alongside the PCR for the target sequence. Running multiple PCR amplifications adds to the complexity of the reaction and can result in less efficient amplification of either or both assays. An undected inefficient assay would lead the scientist to conclude that the extraction had been less efficient than was actually true. To use this approach, it is critical that the PCR for all templates is well optimized. An alternative is to add a synthetic DNA molecule that would function as a PCR target (much to the delight of Kary Mullis, no doubt) or a plasmid containing a modified version of the target sequence. Ideally it would be created such that the ends of the target were recognized by the same primers as the target of interest, but the total product would differ in size or sequence such that it could be distinguished either by gel electrophoresis or by a specific qPCR detection probe.

In addition, the purity of the sample must be considered. Factors remaining in the DNA/RNA sample that may effect downstream enzymatic reactions will also effect the apparent quantification. If a sample contains a PCR inhibitor or enhancer, the assay will be effected such that an apparently higher or lower copy number than reality would be estimated.[44] Even commercial extraction procedures can leave behind such material, so it is advisable to perform a routine quality check of all samples, particularly those samples extracted from clinical or environmental sources.[45]

Another factor contributing to sample quality is the state of the nucleic acid and how degraded it may be. When quantifying mRNA targets in particular, it is clear that the data are inaccurate when the sample is degraded.[46,47] At the very least, it is advisable to determine the quality of the samples and to refrain from using degraded material, or adopt extreme caution when comparing samples of different quality.

An additional consideration is to control exactly how much of the specific sample has been included in the amplification reaction. In the simplest cases, a template loading control is required to ensure that the sample is present and to confirm a negative test for the target gene of interest with greater confidence. For any quantitative measures to be made using qPCR one or more normalization controls are used. In this situation it is important to validate that the same

Figure 9–4. *(facing page).* **Comparison of gene expression between tumor and matched normal samples and clinically normal material from cancer patients and noncancer patients**

The expression of the genes IF2B and Survivin was compared in clinical samples derived from breast tumors and adjacent normal tissue from the same patient. The copy number measured on the X-axis in the tumor sample is represented on the bar graph as solid black and of the normal tissue is in open clear. The Y-axis defines the patient such that a single patient has data for both the tumor and normal tissue (**A, B**). There is a significant increase in expression of Survivin (**A**) in tumor samples but no difference in the expression of IF2B (**B**).

The expression of these genes was also compared in pathologically normal tissue excised from tissue surrounding the breast tumor (tumor normals) to breast tissue derived from reduction mammaplasty surgery where there was no evidence of tumor (normal normals). There is no difference in the expression of IF2B (**C**), however there is a significantly higher expression of Survivin in the normal tissue derived from breast cancer patients (**D**).

amount of sample has been included in each reaction prior to target amplification. Consider a situation in which the quantity of gene *x* is measured in samples 1 to 10. If an equal amount of sample DNA is included in each reaction, any change in the detection of gene *x* quantity must be due to genuine differences in the target concentration. In contrast, if different amounts of sample are loaded into each reaction, differences in the quantity of gene *x* detected would be a combination of the difference in starting copy number and in the biological differences between samples. Appropriate normalization of qPCR data, in particular, remains the subject of heated discussions requiring several visits to coffee bars or similar refreshment houses.

The controls described are often used to give a statement of success or failure, to prevent inappropriate data interpretation. When an experiment does not have appropriate controls, how reliable can the data be considered to be? A challenging extreme of this situation would be the analysis of the blood stain on the Turin Shroud. In the book "The DNA of God?"[48] the results of a DNA analysis of shroud blood samples are revealed. The PCR laboratory at the University of Texas tested for and identified three key genes. The β-globin gene on chromosome 11 yielded a sequence of 268 base pairs. The first 80 of these are

AGCCAAGGAC AGGT <u>CCAAT</u> GTCATCACTT TCCTAAGCCA GTGCCA
AGACCTCACC CTGTGGAGCC ACACCCTAGG GTTGGCCAAT CTACTCCCAG.

Apart from the short underlined string, all 268 base pairs matched the control (HUMHBB221) used for testing for this gene. Of course, there are always small differences when comparing DNA of different individuals. That is the whole basis of DNA identification. So what does this sequence mean? What do we now know about the blood stain from the Turin Shroud? In the absence of any realistic controls, pretty well nothing more than the stain probably contained human β-globin gDNA.

CONCLUSION

PCR is an apparently simple investigative tool with far-reaching applications across all areas of life science. The deceptive simplicity with which data can be generated has led to abuse of the technique and a generation of rainforests' worth of questionable literature. As PCR-based methods are increasingly being used in applications of growing importance to human life, such as forensics and diagnosis, it is critical that standard operating procedures are reviewed and that scrupulous scientific practice is adopted. The power of the use of PCR controls is in the ability to differentiate the difference between rare, meaningful and abundant, meaningless data.

REFERENCES

1. Lejeune J, Gautier M, Turpin R (1959) Étude des chromosomes somatiques de neuf enfants mongoliens. *Comptes Rendus Hebdomadaires des Séances de l'Académie des Sciences* **248**(11): 1721–1722.

2. Nowell PC, Hungerford DA (1960) A minute chromosome in human chronic granulocytic leukemia. *Science* **142**: 1497.
3. Nowell PC, Hungerford DA (1960) Chromosome studies on normal and leukemic human leukocytes. *Journal of the National Cancer Institute* **25**: 85–110.
4. Rowley JD (1973) A new consistent chromosomal abnormality in chronic myelogenous leukaemia identified by quinacrine fluorescence and Giemsa staining. *Nature* **243**: 290–293.
5. Bruder CE, Piotrowski A, Gijsbers AA, Andersson R, Erickson S, de Ståhl TD, et al. (2008) Phenotypically concordant and discordant monozygotic twins display different DNA copy-number-variation profiles. *American Journal of Human Genetics* **82**(3): 763–771.
6. Jeffreys A, Wilson V, Thein SL (1985) Hypervariable minisatellite regions in human DNA. *Nature* **314**: 67–73.
7. Kleppe K, Ohtsuka E, Kleppe R, Molineuz I, Khorana HG (1971) Studies on polynucleotides XCVL. Repair replication of short synthetic DNAs as catalysed by DNA polymerases. *Journal of Molecular Biology* **56**: 341–361.
8. Mullis KB, Ferré F, Gibbs R (1994) The Polymerase Chain Reaction: A Textbook. Boston: Birkhäuser.
9. Mullis KB (1997) The Polymerase Chain Reaction (Nobel Prize acceptance speech, 1993). In: BG Malmström (ed), Nobel Lectures, Chemistry 1991–1995. Singapore: World Scientific Publishing Co.
10. Saiki R, Gelfand D, Stoffel S, Scharf S, Higuchi R, Horn G, et al. (1988) Primer-directed enzymatic amplification of DNA with a thermostable DNA polymerase. *Science* 239: 487–491.
11. Lawyer F, Stoffer S, Saiki R, Chang S, Landre P, Abramson R, et al. (1993) High-level expression, purification, and enzymatic characterization of full-length *Thermus aquaticus* DNA polymerase and a truncated form deficient in 5′ to 3′ exonuclease activity. *PCR Methods and Applications* 2: 275–287.
12. Saiki R, Scharf S, Faloona F, Mullis K, Horn G, Erlich H (1985) Enzymatic amplification of beta-globin genomic sequences and restriction site analysis for diagnosis of sickle cell anemia. *Science* **230**: 1350–1354.
13. Mullis K, Faloona F, Scharf S, Saiki R, Horn G, Erlich H (1986) Specific enzymatic amplification of DNA in vitro: the polymerase chain reaction. *Cold Spring Harbor Symposium in Quantitative Biology* **51**: 263–273.
14. Mullis K, Faloona F (1987) Specific synthesis of DNA in vitro via a polymerase-catalyzed chain reaction. *Methods in Enzymology* **155**: 335–350.
15. Wolffs P, Grage H, Hagberg O, Rådström P (2004) Impact of DNA polymerases and their buffer systems on quantitative real time PCR. *Journal of Clinical Microbiology* **42**(Jan): 408–411.
16. Chakrabarti R (2004) Novel PCR enhancing compounds and their modes of action. In: T Weissensteiner, HG Griffin (eds), PCR Technology, Current Innovations. Second edition. Boca Raton: CRC Press.
17. Takagi M, Nishioka M, Kakihara H, Kitabayashi M, Inoue H, Kawakami B, et al. (1997) Characterization of DNA polymerase from Pyrococcus sp. strain KOD1 and its application to PCR. *Applied and Environmental Microbiology* **63**(11): 4504–4510.
18. New England BioLabs, Inc. Nucleic acid data. Available at: http://www.neb.com/nebecomm/tech_reference/general_data/nucleic_acid_data.asp.
19. Morrison C, Gannon F (1994) The impact of the PCR plateau phase on quantitative PCR. *Biochimica et Biophysica Acta* **1219**: 493–498.
20. Kainz P (2000) The PCR plateau phase – towards an understanding of its limitations. *Biochimica et Biophysica Acta* **1494**: 23–27.
21. Lindahl T (1993) Instability and decay of the primary structure of DNA. *Nature* **362**(6422): 709–715.
22. Kermekchiev MB, Tzekov A, Barnes W (2003) Cold-sensitive mutants of Taq DNA polymerase provide a hot start for PCR. *Nucleic Acids Research* **31**(21): 6139–6147.
23. Innis MA, Gelfand DH (1990) Optimization of PCRs. In: MA Innis, DH Gelfand, JJ Sninsky and TJ White (eds), PCR Protocols. New York: Academic Press.

24. Rychlik W, Spencer WJ, Rhoads RE (1990) Optimization of the annealing temperature for DNA amplification *in vitro*. *Nucleic Acids Research* **18**(21): 6409–6412.
25. Nolan T, Hands RE, Bustin SA (2006) Quantification of mRNA using real-time RT-PCR. *Nature Protocols* **1**(3): 1559–1582.
26. Higuchi R, Fockler C, Dollinger G, Watson R (1993) Kinetic PCR analysis: real-time monitoring of DNA amplification reactions. *BioTechnology* **11**: 1026–1030.
27. Bengtsson M, Karlsson K, Westman G, Kubista M (2003) A minor groove binding asymmetric cyanine reporter dye for real time PCR. *Nucleic Acids Research* **31**(8): 1–5.
28. Ahmad A (2007) BOXTO as a real-time thermal cycling reporter dye. *Journal of Biosciences* **32**: 229–239.
29. Holland PM, Abramson RD, Watson R, Gelfand DH (1991) Detection of specific polymerase chain reaction product by utilizing the 5′–3′ exonuclease activity of *Thermus aquaticus* DNA polymerase. *Proceedings of the National Academy of Sciences of the United States of America* **88**(16): 7276–7280.
30. Tyagi S, Kramer FR (1996) Molecular Beacons: probes that fluoresce upon hybridization. *Nature Biotechnology* **14**: 303–308.
31. Whitcombe D, Theaker J, Guy SP, Brown T, Little S (1999) Detection of PCR products using self-probing amplicons and fluorescence. *Nature Biotechnology* **17**: 804–807.
32. Wittwer CT, Herrmann MG, Moss AA, Rasmussen RP (1997) Continuous fluorescence monitoring of rapid cycle DNA amplification. *BioTechniques* **22**(1): 130–131, 134–138.
33. Wang L, Blasic JR Jr, Holden MJ, Pires R (2005) Sensitivity comparison of real-time PCR probe designs on a model DNA plasmid. *Analytical Biochemistry* **344**: 257–265.
34. Bustin S, Garson J, Hellemans J, Huggett J, Kubista M, Mueller R, et al. (2009) The MIQE Guidelines: Minimum Information for Publication of Quantitative Real-Time PCR Experiments. *Clinical Chemistry*, **55**(4): 611–622.
35. Livak KJ, Schmittgen TD (2001) Analysis of relative gene expression data using real-time quantitative PCR and the 2(-Delta Delta C(T)) method. *Methods* **25**(4): 402–408.
36. Pfaffl MW (2001) A new mathematical model for relative quantification in real-time RT–PCR. *Nucleic Acids Research* **29**(9): e45.
37. Nerlich AG, Haas CJ, Zink A, Szeimies U, Hagedorn HG (1997) Molecular evidence for tuberculosis in an ancient Egyptian mummy. *Lancet* **350**(9088): 1404.
38. Nerlich AG, Schraut B, Dittrich S, Jelinek T, Zink A (2008) *Plasmodium falciparum* in Ancient Egypt. *Emerging Infectious Diseases* **14**(8): 1317–1319.
39. Court Proceedings R vs Sean Hoey (2008). Available at: http://www.xproexperts.co.uk/newsletters/feb08/R%20v%20Hoey.pdf.
40. Lowe A, Murray C, Richardson P, Wivell R, Gill P, Tully G, et al. (2003) Use of low copy number DNA in forensic inference. *International Congress Series* **1239**: 799–801.
41. van Oorschot, RAH, Treadwell S, Beaurepaire J, Holding NL, Mitchell RJ (2005) Beware of the possibility of fingerprinting techniques transferring DNA. *Journal of Forensic Sciences* **50**(6): 1417–1422.
42. Bustin SA (2008) Real-time quantitative PCR – opportunities and pitfalls. *European Pharmaceutical Review* (**4**): 18–23.
43. Uhlmann V, Martin CM, Sheils O, Pilkington L, Silva I, Killalea A, et al. (2002) Potential viral pathogenic mechanism for new variant inflammatory bowel disease. *Molecular Pathology* **55**(2): 84–90.
44. Huggett JF, Novak T, Garson JA, Green C, Morris-Jones SD, Miller RF, et al. (2008) Differential susceptibility of PCR reactions to inhibitors: an important and unrecognised phenomenon. *BMC Research Notes* **28**(1): 70.
45. Nolan T, Hands RE, Ogunkolade BW, Bustin SA (2006) SPUD; a qPCR assay for the detection of inhibitors in nucleic acid preparations. *Analytical Biochemistry* **351**: 308–310.
46. Fleige S, Pfaffl M (2006) RNA integrity and the effect on real-time qRT-PCR performance. *Molecular Aspects of Medicine* **27**: 126–139.
47. Pérez-Novo CA, Claeys C, Speleman F, van Cauwenberge P, Bachert C, Vandesompele J (2005) Impact of RNA quality on reference gene expression stability. *BioTechniques* **39**(1): 52, 54, 56.
48. Garza-Valdés LA (1998) The DNA of God? London: Hodder and Stoughton.

II Applications

10 Polymerase chain reaction–based methods for the detection of solid tumor cancer cells for clinical diagnostic and prognostic assays

Susan A. Burchill

The biological process of metastases requires multiple individual steps to successfully establish a solid tumor at a secondary site. Tumor cell(s) need to migrate through and from the primary tumor mass, intravasate into and survive within the hemopoietic or lymphatic vascular systems, extravasate from these systems into secondary tissues and initiate proliferation and angiogenesis. Multiple molecular and microenvironment factors influence these processes, defining which tumor cells survive, spread to distant organs and give rise to metastases.[1] Despite considerable advances in the treatment of solid cancers, metastases remain the major clinical challenge for successful treatment. The classic view is that metastatic spread is a late process in disease progression. However, the prognosis for patients with small or even undetectable primary tumors is still limited by metastatic relapse, sometimes long after removal of the primary tumor. This has led to the hypothesis that primary tumors may shed tumor cells at an early stage, resulting in the dissemination of tumor cells to distant sites and development of metastases. Interestingly the bone marrow (BM) and lymph nodes (LN) seem to be common homing tissues for many different disseminating tumor cells, and so the early detection and characterization of tumor cells in these compartments could help guide treatment decisions before the onset of overt metastases, as well as in the setting of advanced disease. Unfortunately the number of tumor cells in these sites is usually small and they are not detected using current technologies.

THE CLINICAL NEED TO ACCURATELY DETECT METASTATIC DISEASE

The accurate and sensitive detection of metastatic disease is fundamental to the initial staging of patients at diagnosis, and usually defines the treatment that patients will receive. For most cancers, the presence of metastases at diagnosis is an indicator of poor outcome, these patients requiring more aggressive therapy. Although metastatic disease detected by conventional methods is one of the most powerful markers of poor prognosis for many cancers, some patients with apparently localized disease have rapidly progressing cancer of which they subsequently die. This suggests these patients have minimal-metastatic disease (MD) that is not detected by current routine methods used in staging of patients. MD includes minimal residual disease (MRD); disease that cannot be detected by routine procedures after treatment and/or surgical removal of the primary tumor. Tumor cells in the bone marrow are frequently referred to as disseminated tumor cells, and those in peripheral blood as circulating tumor cells. MD, if left unchecked, may give rise to metastases and disease progression, and in some cases could persist in a dormant state for many years within the BM from which it may subsequently recirculate to initiate metastases decades after the initial diagnosis. More sensitive and specific methods for the detection of this disease will improve understanding of its role in the disease course, and may provide better strategies for improved stratification of some patients and their successful treatment.

In patients with metastatic disease identified by conventional methods, detection of MD may better define patient response by analysis of BM and/or peripheral blood (PB) during successive cycles of chemotherapy. If MD is present at the end of therapy, patients may be offered more intensive or novel therapeutics; alternatively, those patients with no evidence of metastases may be spared further cytotoxic therapy and consequently escape complications associated with that treatment. Such methods may detect relapse before overt metastases occur, resulting in a redefinition of disease-free. Accurate measurement of MD in peripheral blood stem cell (PBSC) harvests or autologous BM for transplantation may also be important if the reinfusion of tumor cells leads to relapse,[2–4] although in the absence of randomized trials showing improved disease-free survival after successful tumor cell purging, this remains controversial. Accurate assessment of disease status is also essential to monitor patients on and off therapy, and might also be exploited to objectively and rapidly evaluate the efficacy of novel therapeutics targeting MD and MRD.[5–8]

The development of distant metastases frequently occurs through the hemopoietic system. Since the collection of BM aspirates is part of the diagnostic procedure for the management of many cancers, BM is an attractive compartment to study MD. However, collecting BM is still rather invasive and time consuming; it is also difficult to standardize the quality of aspirates. Therefore a number of studies have focused on analyzing MD in PB, which is easy to collect, minimally invasive, and readily accessible for repeated sampling. However, monitoring MD in PB may not necessarily reflect disease status in the BM, and in some cancers may not be clinically relevant. There are a few studies where MD has been compared in BM and PB collected at the same time-points, these studies usually demonstrating that the frequency of tumor cell detection in BM is consistently higher than in PB from the same patient at the same time-point.[9] This is consistent with the hypothesis that the BM provides an appropriate microenvironment for accumulation and survival of MD, whereas PB may simply be a vehicle for dissemination of disease. Although the development of distant metastases frequently occurs through BM and PB, tumor cells can also disseminate in the lymphatic system and lodge in the LN where their presence is often an indicator of poor outcome. This emphasizes the importance of evaluating the prognostic power of MD detected in different compartments (PB, BM, or LN) for individual cancers, across prospective clinical studies to define what is clinically informative.

CONVENTIONAL METHODS AND LIMITATIONS

Initial diagnostic procedure typically depends on imaging and morphological examination of BM and/or LN for infiltrating tumor cells. In addition to defining the extent of a primary tumor mass, imaging by computed tomography (CT) and magnetic resonance imaging (MRI) also identifies the extent of other tissue involvement. For example, in Ewing's sarcoma these methods can identify the boundaries of primary tumor within the long bone of a leg, the extent of additional bony involvement, and frequency of soft tissue metastases.[10] Using cytology (after Romanovsky staining of BM smears) it is possible to identify a BM infiltration of more than 10% tumor cells, which for many patients at diagnosis is informative and predictive of outcome. Therefore, these methods are useful for the assessment of disease in the majority of patients at diagnosis, as most present with overt metastatic disease. These methods may be equally informative at the time of clinical relapse when extensive disease is present. However, for those patients with low level metastases (<10% infiltrating tumor cells) at diagnosis and for most patients on therapy, these methods are suboptimal.

For more than two decades, immunocytology and immunohistochemistry have been investigated as tools to stage and monitor disease in BM and LN; the specificity and sensitivity of these methods being dependent on the availability of antibodies to tumor-associated antigens that are expressed on all target tumor cells but not in the normal cells of the BM or LN. Where suitable antibodies are available, these methods can be robust and potentially clinically informative. For example, immunocytology using antibodies to GD2 (a cell-surface antigen expressed by neuroblastoma [NBL] cells) has successfully identified 1 tumor cell in 1×10^6 normal BM cells from children with NBL.[11–16] However, interpretation of such assays is subjective and can lead to inter-observer variation. This is a particular challenge when identifying a single or clumps of 2–10 NBL cells, when cells do not produce GD2[17] or if antigen that is shed by NBL cells is taken up by macrophages (which will stain positive with alkaline phosphatase and potentially be identified as false positives). These challenges can be overcome by combining standardized immunocytology with morphological criteria,[18] although this remains rather subjective. Therefore, the development of more sensitive methods for the unequivocal, objective detection of MD is essential to allow a robust evaluation of its clinical significance and utility.

APPLICATION OF PCR

PCR[19] has made an enormous impact on the sensitivity and specificity of MD assessment, allowing the accurate detection of a single tumor cell in up to 1×10^7 white blood cells. This compares with a sensitivity of 1 tumor cell in 1×10^2 detected by cytology, and 1×10^6 detected by immunocytology. However, although the amplification of tumor-specific or tumor-associated messenger ribonucleic acid (mRNA) by reverse transcriptase–polymerase chain reaction (RT–PCR) has been used to detect MD burden in a number of different cancers, the clinical utility of this method is still largely unresolved. Importantly the presence of tumor-specific or tumor-associated mRNA is thought to reflect the presence of the disseminating tumor cell or cells that have the capacity to metastasize. This is in contrast to the detection of circulating free nucleic acid in plasma, serum, or urine that may be useful in cancer detection, prognostication, and monitoring,[20–22] but reflects tumor turnover and mass. It is important to remember, however, that although the tumor cell detected by RT-PCR has the capacity to metastasize it will not necessarily go on to form secondary disease, as this is dependent on other biological processes.[1]

Amplification of tumor-specific abnormalities

PCR-based amplification of tumor-specific mRNA to detect MD has been most powerful in hematological malignancies where consistent, well-characterized molecular abnormalities have been described. These studies have resulted in the introduction of new interventions to target this disease as an essential part of current trials, the results of which will determine whether MRD-based treatment is associated with improved outcome in hematological malignancies.[23–27]

For most solid cancers, tumor-specific gene rearrangements have not been described, but where they have they provide a robust target for RT-PCR to assess the clinical significance of MD. For example, the mRNA products of the nonrandom chromosome translocations between the *EWS* gene on chromosome 22q12 and members of the *ETS* gene family in the Ewing's sarcoma family of tumors (ESFT[28]) have successfully been used to detect MD by RT–PCR. The presence of EWS–ETS mRNA in BM is reported to be an indicator of poor prognosis[29,30] and may detect MD in BM from patients with apparently localized disease identified by more conventional methods.[31] Some studies have suggested that the presence of EWS–ETS fusion transcripts in PB is not clinically informative,[32] although this has been challenged.[30,33] The clinical significance of MD detected by RT–PCR in PBSC is also controversial, in some studies being associated with relapse,[34] whereas in others it does not predict event-free or over-all survival.[35]

Interestingly, the collection of PBSC harvest after two courses of chemotherapy is reported to have reduced MD compared to collection in earlier courses, suggesting the use of later harvests may minimize the potential risk of secondary disease from reinfused tumor cells.[36] Although some of these studies imply an association between the presence of tumor cells in PBSC and poor outcome, they do not demonstrate that the reinfused tumor cell is responsible for the relapse. The presence of clinically significant MD detected by RT–PCR for gene rearrangements in solid cancers has also been exploited in alveolar rhabdomysarcoma[37–39] and desmoplastic small round cell tumors[38]; other tumor specific fusion transcripts have been identified in soft tissue cancers that could be exploited for detection of MD.[28]

More recently, recurrent gene fusions involving the 5′ untranslated region of the androgen-regulated gene *TMPRSS2* and the *ETS* family members *ERG, ETV1,* and *ETV4* have been described in a high proportion of prostate cancers, suggesting they too may be exploited as a tool to improve staging of this difficult and complex disease.[40,41] These fusion transcripts have already been detected by RT–PCR in the urine of patients with clinically localized prostate cancer,[42] although whether this is clinically relevant remains to be seen. Furthermore, the diversity in structure of the TMPRSS2–ERG transcripts may limit their rapid exploitation for therapeutic advantage as routine diagnostic and monitoring tools; so far 14 distinct hybrid transcripts have been described.[43]

Amplification of tumor-associated wild-type mRNA

Unfortunately, for most solid tumors, specific gene abnormalities that can be exploited to detect MD have not been identified. In these cases amplification of tumor-associated wild-type mRNA by RT–PCR has been used to detect MD burden.[44,45] Optimal sensitivity and specificity require the identification of a target mRNA expressed in all tumor cells but not in the cells of the normal compartment to be studied; the choice of tumor-associated mRNA target may be compartment dependent as some targets are differentially expressed in the normal hemopoietic or lymph node cells. Expression of a target mRNA for the detection of MD by RT–PCR should be stable, and its expression be unaffected by chemotherapeutic agents so that RT–PCR can be used to provide an accurate assessment of MD throughout treatment and disease course. Ideally it should be encoded by a gene with introns, so that primers for amplification can be designed across an exon–exon junction to selectively amplify complementary DNA (cDNA) generated from mRNA, and not contaminating genomic DNA; including a reverse transcriptase negative control is useful to confirm the specificity of amplification from mRNA.

Since many solid tumors arise from epithelial tissues several groups have attempted to identify a generic marker that might be used to detect all epithelially derived cancers,[44] although this has thus far remained elusive. Many groups have explored using cytokeratins as targets, cytokeratin 19 (CK19) being the most frequently investigated, despite low specificity[46,47] and a high number of identified pseudogenes.[48] Although cytokeratin 20 may be more specific,[49] the restricted expression profile of this cytokeratin may limit its general applicability. The heterogeneity of mRNA expression across different cancer cells, even among cells of the same tumor, and low level expression of some targets in normal cells suggests that increased sensitivity and specificity of RT–PCR detection of MD might best be achieved by exploiting multiple mRNA targets.[50–54]

The application of quantitative (Q)RT–PCR has increased the scope and potential for MD monitoring by providing assays with a wider linear dynamic range, superior sensitivity and objective interpretation of results. These properties are enhanced by good intra- and inter-assay reproducibility, and the generation of a permanent quantitative record of the data that can be reviewed independently. Additional attractions include high throughput capacity, speed, and elimination of lengthy post-PCR handling steps, reducing the risk of potential carryover contamination.[16,55–56] A further obvious advantage of QRT–PCR over more traditional qualitative RT–PCR is that it allows a precise quantification of a single or multiple mRNA(s) in small clinical samples. This is not a direct measure of absolute cell number, since the level of a target mRNA per cell may vary; however, it does allow an objective accurate measure of mRNA content within and across clinical samples. This is particularly important for defining the clinically relevant level of MD and MRD at diagnosis and during disease course respectively. For accurate reporting the selection of an optimal reference mRNA against which the test mRNA(s) can be normalized is essential[57,58]; beta-2 microglobulin (B2M) is frequently selected as the standard house-keeping gene to report expression of MD in hematopoietic cells.[58,59] The ability to accurately detect the level of mRNA transcripts is certainly valuable to assess sample quality, informing the development of optimal methods for collection, storage, transport, and preparation of clinical samples.

Since initial proof-of-principal experiments using tyrosinase mRNA as a target to detect melanoma cells in PB were published,[60] RT–PCR for wild-type mRNA has been used to detect MD in a number of different cancer cell types by many investigators worldwide. These studies have usually focused on the detection of systemic disease in the hemopoietic compartments (BM or PB). The application of RT–PCR for tyrosine hydroxylase (TH), prostate specific antigen (PSA), and tyrosinase mRNA to detect MD and MRD has been extensively studied in NBL, prostate cancer, and melanoma respectively. In each case RT–PCR for these targets has been shown in model systems to detect tumor cells with increased sensitivity than more conventional methods, although the clinical value and application of these targets to detect clinically significant disease has been variable. In the remainder of this chapter, these three cancers and targets have been used to demonstrate the advantages and challenges when utilizing RT–PCR to detect MD and MRD, and to emphasize the importance of evaluating specific targets and the clinical utility of this technology in different clinical settings.

TH mRNA in NBL

NBL is one of the most common solid tumors of childhood, accounting for 5%–10% of all cancers in patients up to the age of 15 years; ~15% of all cancer deaths in children are due to NBL. The presence of metastatic disease (identified by imaging and cytology) at diagnosis is an indicator of poor prognosis in children with this disease; most present over the age of 1 year with disseminated disease (International Neuroblastoma Staging System [INSS] stage 4; 40–50% of NBL).

Because catecholamines are produced by NBL, the first enzyme in the catecholamine synthesis pathway, TH, has been used as an mRNA target for the detection of NBL by RT–PCR. Although other targets for the detection of NB cells by RT–PCR have been evaluated, TH mRNA is currently the single most widely used target.[61–65] Using this methodology, researchers can specifically detect a single NBL cell in 1×10^6 normal cells. The success of TH as a target for detection of MD is attributed to its ubiquitous expression in NBL cells and lack of expression in hemopoietic cells.[53,62,66–68] Using RT–PCR for TH mRNA, clinically significant disease has been detected in PB[61,63,68–70] and BM[63,68,69,71–74] from children with NBL at diagnosis, on therapy, on follow-up, and at relapse. Furthermore, the technique has been used to detect NBL cells in PBSC harvests from children with high-risk disease,[36,71,75–77] although the clinical significance of this disease is currently controversial.

Even though there is a relatively large literature demonstrating that RT–PCR for TH mRNA detects NBL cells in BM, PB, and PBSC harvests from children with NBL, the clinical utility of this method remains unclear, reflecting the small number of patients studied, absence of quality control, lack of uniform methodology, and inconsistency of reporting between studies.[78] Therefore, the clinical significance of RT–PCR for TH mRNA in BM, PB, and PBSC is currently being evaluated in a large prospective clinical trial (HR-NBL1/ESIOP; www.siopen-r-net.org), with appropriate standardization and quality control between participating countries and laboratories.[79,80]

It has been suggested that increased sensitivity and specificity of NBL detection might be achieved using a panel of targets to overcome the heterogeneity of NBL.[53,54,81] However, there is currently no consensus on which markers are clinically reliable[78]; therefore PB and BM from children entered into HR-NBL1/ESIOP are being evaluated by QRT–PCR for multiple validated markers to develop the best model for detection of clinically relevant MD in children with NBL.[53] Within this study the clinical utility of detecting NBL cells using QRT–PCR will be compared to imaging, BM cytology, and BM immunocytology for the assessment of disease status. In addition, the efficacy of MRD treatment strategies exploiting 13-*cis* retinoic acid and anti-GD2 monoclonal antibody therapy, to which children in this trial are randomized, is being evaluated.

PSA mRNA in prostate cancer

Prostate cancer is now the most common malignancy affecting men in the UK; the incidence is 72.7 cases per 100,000 (age-standardized rate; Cancer Research UK). It is also the second most common cause of cancer-associated death in men; death rate = 27.3 deaths per 100,000 (age-standardized rate; Cancer Research UK). PSA mRNA has been detected by RT–PCR in preoperative blood samples in many studies where its expression has been linked to capsular penetration, seminal vesicle involvement, positive surgical margins, and biochemical relapse[82–85]; however, these observations remain controversial.[86–91] These conflicting data most likely reflect the expression of PSA mRNA in blood samples from healthy volunteers.[86,87,92–95] However, the optimal management of patients with localized prostate cancer currently presents a significant clinical dilemma, reflecting the need for improved staging and prognostic stratification of patients. Using a real-time QRT–PCR to detect PSA mRNA in PB it is possible to distinguish patients with metastatic prostate cancer from healthy volunteers with a specificity of 95% and a sensitivity of 68%,[96] suggesting that this assay may have a role in the assessment and monitoring of patients with metastatic prostate cancer. However, this assay did not identify patients with localized pathologically higher grade disease (T3 and T4 lesions or nodal involvement) from healthy volunteers; these patients are not usually amenable to complete surgical excision and have a higher rate of subsequent relapse. Failure to detect tumor cells in PB using QRT–PCR is disappointing and contrasts with some reports using nonquantitative RT–PCR.[82,83,85]

Currently the clinical utility of RT–PCR in prostate cancer remains undefined, reflecting low specificity of PSA as a target for detection of MD. However, using QRT–PCR to accurately measure the level of PSA mRNA in clinical samples may be more informative, if a clinically relevant cut-off is defined that will allow discrimination of those patients with low level PSA mRNA in PB or BM reflecting expression in normal leukocytes and those that have higher expression due to the presence of contaminating prostate cancer cells. The independent prognostic value of detecting prostate cancer cells using QRT–PCR for PSA, alone or in combination with QRT–PCR for TMPRSS2-ETS fusion transcripts (see Amplification of Tumor-Specific Abnormalities), should be compared to that of serum PSA protein, which is commonly elevated in patients with large tumor mass(es) and is frequently used to confirm prostate cancer in asymptomatic patients, monitor the effects of treatment on tumor mass, and predict relapse.[97]

Tyrosinase mRNA in malignant melanoma

Cutaneous malignant melanoma is the rarest form of skin cancer, although it accounts for nearly 80% of skin cancer–related deaths. Patients who present with systemic or lymphatic metastases have a reduced survival; 5% of patients with stage IV disease survive for 5 years compared to 90% of those with disease localized to the primary tumor site at the time of diagnosis. Patients with stage III disease represent an enigmatic group in which approximately half of the patients will succumb to their disease in 5 years, suggesting that these patients may have metastatic disease at diagnosis that is currently undetected. Consequently, recent studies have focused on the application of RT–PCR to distinguish patients with primary melanoma of low metastatic potential from those with high metastatic potential disease.

RT–PCR for tyrosinase (the first enzyme regulating the synthesis of melanin) was the first mRNA target used to detect melanoma cells in PB.[60] This assay has been the most frequently used to detect MD in melanoma, probably reflecting the primer sequences that were optimized to limit amplification of genomic DNA by designing primers to span intron boundaries.

Many similar studies have followed,[98–102] the presence of the circulating tumor cells detected by RT–PCR in PB having been shown to be of prognostic value.[100,103] However, much of the early literature is confusing as suboptimal handling of samples and methodological differences[104] led to false-positive and false-negative results. Variability in the frequency of tumor cell detection in PB is most striking in patients at diagnosis with stage IV melanoma (distant metastases), where the reported frequency of melanoma cell detection ranges from 0% to 100%.[105] However, the inconsistency in detection rate is not only restricted to patients with stage IV disease; in stage I/II patients with localized disease the range is 0% to 53% and in stage III patients with regional metastases is 0% to 82%.[102,106,107] This variability is not due to the detection of tyrosinase in PB from healthy volunteers,[98,100,101] although it could reflect heterogeneity of tyrosinase expression in primary and metastatic lesions.[108–111] This might be overcome by using additional targets for the detection of MD in melanoma.[112–115] Quality assurance in Europe and North America will resolve methodological differences and facilitate a comparison of results from different laboratories and countries. While patients with persistent MD may have a higher risk of relapse, not all patients will develop recurrence during follow-up. The presence of tyrosinase mRNA in PB from patients with stage IV disease in long-term remission[98] is consistent with the hypothesis that tumor cells may persist in a dormant state for many years. The possibility that this cell might subsequently be reactivated and lead to late relapses emphasizes the need for a more accurate evaluation and characterization of MD and its role in the metastatic process.

Metastatic melanoma is most often first detected by the presence of LN metastases rather than systemic disease in BM or PB, making the LN potentially the preferred optimal compartment for initial staging studies. The value of RT–PCR for tyrosinase to detect melanoma cells in LN has been evaluated in a small number of patients, although there is currently no consensus on its clinical utility.[116–118] In one study of patients with stage I and II disease, the frequency of melanoma cell detection was 68% by RT–PCR compared to 38% detected by histology,[119] and in a more recent study the presence of melanoma cells detected by RT–PCR in LN has been shown to correlate with a worse overall and event-free survival,[120] suggesting that the RT–PCR status of LN may have clinically relevant prognostic power. However, detection of tyrosinase mRNA by RT–PCR in LN is reported in other studies not to increase the likelihood of disease recurrence above conventional pathology.[121–123] The choice of target to detect melanoma cells in the LN must be carefully considered, as when sampling through the skin it is possible to contaminate the LN with melanocytes, benign naevi, or Schwann cells, which also produce melanin and express many of the targets being considered to detect melanoma cells by RT–PCR (including tyrosinase). This could contribute to a high false-positive detection rate and limit an evaluation of clinical significance; again QRT–PCR for multiple melanoma mRNA targets may overcome such sampling problems.

SUMMARY AND THE FUTURE

Tumor-specific or tumor-associated mRNAs can be identified to allow detection of the vast majority of solid tumor cells. The ideal mRNA for detecting MD using RT–PCR is highly expressed in all target tumor cells, not expressed in BM, PB, PBSC, or LN, and stably expressed throughout the life span of the tumor cell.

Where true cancer-specific mRNA targets have not been identified, amplification of rare transcripts in non-cancer cells has led to some skepticism about the clinical value of this methodology. However the introduction of QRT–PCR, allowing an accurate quantitative assessment of transcript number, has informed the identification of cut-off values in non-cancer controls to permit the definition of transcripts that are of tumor origin. QRT–PCR for mRNA transcripts can detect clinically relevant tumor cells with increased sensitivity and specificity compared to most other methods. However, the presence of tumor cells detected by RT–PCR is not always predictive of outcome. For example in BM from some children less than 1 year old with stage 4s NBL, tumor cells are frequently detected at diagnosis but later resolve as the stage 4s disease regresses, reflecting the unusual biology of this disease.[124] Examples such as this demonstrate the importance of establishing when and in which patients RT–PCR for MD is clinically informative. It also emphasizes that the presence of a tumor cell in BM, PB, or PBSC does not mean it will metastasize, just that it has the capacity to disseminate; genetic and molecular properties of the tumor cell(s) and their interaction with the tumor microenvironment that are not usually accounted for when detecting MD by QRT–PCR play critical roles in the metastatic disease process.[125,126] Although RT–PCR for tumor-specific or tumor-associated mRNA may be clinically informative, as with any assay that may have clinical impact it is import that robust standard operating procedures (SOPs) for sample handling, processing, and analysis are established to ensure accurate and reliable results. However, large prospective clinical outcome studies must first demonstrate the need for this technology for patient benefit, and justify the necessary investment for its introduction into clinical practice. For the future, greater understanding of the genes and molecules involved in the process of tumor cell dissemination and survival in different physiological compartments is critical to improve prognostication and identify targets for the development of novel therapeutics. The role of metastasis-associated microRNAs and metastasis-suppressor genes in the development and homing of MD might be particularly fruitful.

REFERENCES

1. Eccles SA, Welch DR (2007) Metastasis: recent discoveries and novel treatment strategies. *Lancet* **369**: 1742–1757.
2. Brenner MK, Rill DR, Moen RC, Krance RA, Heslop HE, Mirro J Jr, et al. (1994) Gene marking and autologous bone marrow transplantation. *Annals of the New York Academy of Sciences* **716**: 204–214.
3. Brenner MK (1995) The contribution of marker gene studies to hemopoietic stem cell therapies. *Stem Cells* **13**: 453–461.

4. Brenner M (1998) Use of gene marking technologies in oncology. *Forum (Genova)* **8**: 342–353.
5. Cheung IY, Lo Piccolo MS, Kushner BH, Cheung NK (2003a) Quantitation of GD2 synthase mRNA by real-time reverse transcriptase polymerase chain reaction: clinical utility in evaluating adjuvant therapy in neuroblastoma. *Journal of Clinical Oncology* **21**: 1087–1093.
6. Cheung IY, Lo Piccolo MS, Kushner BH, Cheung NK (2003b) Early molecular response of marrow disease to biologic therapy is highly prognostic in neuroblastoma. *Journal of Clinical Oncology* **21**: 3853–3858.
7. Hess G, Bunjes D, Siegert W, Schwerdtfeger R, Ledderose G, Wassmann B, et al. (2005) Sustained complete molecular remissions after treatment with imatinib-mesylate in patients with failure after allogeneic stem cell transplantation for chronic myelogenous leukemia: results of a prospective phase II open-label multicenter study. *Journal of Clinical Oncology* **23**: 7583–7593.
8. Lo-Coco F, Ammatuna E (2007) Front line clinical trials and minimal residual disease monitoring in acute promyelocytic leukemia. *Current Topics in Microbiology and Immunology* **313**: 145–156.
9. Benoy IH, Elst H, Philips M, Wuyts H, Van Dam P, Scharpé S, et al. (2006) Real-time RT-PCR detection of disseminated tumour cells in bone marrow has superior prognostic significance in comparison with circulating tumour cells in patients with breast cancer. *British Journal of Cancer* **94**: 672–680.
10. Lewis I, Burchill SA, Souhami R (2002) Ewing's sarcoma and the Ewing family of tumors. In: RL Souhami, I Tannock, P Hohenberger, and J-C Horiot, (eds), Oxford Textbook of Oncology. Second edition. Chapter 16, pages 2539–2551. Oxford, UK: Oxford University Press.
11. Favrot MC, Frappaz D, Maritaz O, Phiip I, Fontaniere B, Gentilhomme O, et al. (1986) Histological, cytological and immunological analyses are complementary for the detection of neuroblastoma cells in bone marrow. *British Journal of Cancer* **54**: 637–641.
12. Cheung NK, Van Hoff DD, Strandjord SE, Coccia PF (1986) Detection of neuroblastoma cells in bone marrow using GD2 specific monoclonal antibodies. *Journal of Clinical Oncology* **4**: 363–369.
13. Rogers DW, Treleaven JG, Kemshead JT, Pritchard J (1989) Monoclonal antibodies for detecting bone marrow invasion by neuroblastoma. *Journal of Clinical Pathology* **42**: 422–426.
14. Carey PJ, Thomas L, Buckle G, Reid MM (1990) Immunocytochemical examination of bone marrow in disseminated neuroblastoma. *Journal of Clinical Pathology* **43**: 9–12.
15. Corrias MV, Faulkner LB, Pistorio A, Rosanda C, Callea F, Piccolo MS, et al. (2004) Detection of neuroblastoma cells in bone marrow and peripheral blood by different techniques: accuracy and relationship with clinical features of patients. *Clinical Cancer Research* **10**: 7978–7985.
16. Beiske K, Ambros PK, Burchill SA, Cheung IY, Swerts K (2005) Detecting minimal residual disease in neuroblastoma patients – the present state of the art. *Cancer Letters* **228**: 229–240.
17. Schumacher-Kuckelkorn R, Hero B, Ernestus K, Berthold F (2005) Lacking immunocytological GD2 expression in neuroblastoma: report of 3 cases. *Pediatric Blood Cancer* **45**: 195–201.
18. Swerts K, Ambros PF, Brouzes C, Navarro JM, Gross N, Rampling D, et al. (2005) Standardization of the immunocytochemical detection of neuroblastoma cells in bone marrow. *Journal of Histochemistry and Cytochemistry* **53**: 1433–1440.
19. Saiki RK, Bugawan TL, Horn GT, Mullis KB, Erlich HA (1986) Analysis of enzymatically amplified beta-globin and HLA-DQ alpha DNA with allele-specific oligonucleotide probes. *Nature* **324**: 163–166.
20. Taback B, Hoon DS (2004) Circulating nucleic acids in plasma and serum: past, present and future. *Current Opinion in Molecular Therapeutics* **6**: 273–278.

21. Zeerleder S (2006) The struggle to detect circulating DNA. *Critical Care* **10**: 142.
22. O'Driscoll L (2007) Extracellular nucleic acids and their potential as diagnostic, prognostic and predictive biomarkers. *Anticancer Research* **27**: 1257–1265.
23. Foroni L, Harrison CJ, Hoffbrand AV, Potter MN (1999) Investigation of minimal residual disease in childhood and adult acute lymphoblastic leukaemia by molecular analysis. *British Journal of Haematology* **105**: 7–24.
24. Faderl S, Talpaz M, Kantarjian HM, Estrov Z (1999) Should polymerase chain reaction analysis to detect minimal residual disease in patients with chronic myelogenous leukemia be used in clinical decision making? *Blood* **93**: 2755–2759.
25. Roman J, Alvarez MA, Torres A (2000) Molecular basis for therapeutic decisions in chronic myeloid leukemia patients after allogeneic bone marrow transplantation. *Haematologica* **85**: 1072–1082.
26. Hochhaus A, Weisser A, La Rosee P, Emig M, Muller MC, Saussele S, et al. (2000) Detection and quantification of residual disease in chronic myelogenous leukemia. *Leukemia* **14**: 998–1005.
27. Campana D, Neale GA, Coustan-Smith E, Pui CH (2001) Detection of minimal residual disease in acute lymphoblastic leukemia: the St. Jude experience. *Leukemia* **15**: 278–279.
28. Burchill SA (2008) Molecular abnormalities in Ewing's sarcoma. *Expert Review of Anticancer Therapy* **8**: 1675–1687.
29. Zoubek A, Ladenstein R, Windhager R, Amann G, Fischmeister G, Kager L, et al. (1998) Predictive potential of testing for bone marrow involvement in Ewing tumor patients by RT-PCR: a preliminary evaluation. *International Journal of Cancer* **79**: 56–60.
30. Schleiermacher G, Peter M, Oberlin O, Philip T, Rubie H, Mechinaud F (2003) Increased risk of systemic relapses associated with bone marrow micrometastasis and circulating tumor cells in localized ewing tumor. *Journal of Clinical Oncology* **21**: 85–91.
31. Avigad S, Cohen IJ, Zilberstein J, Liberzon E, Goshen Y, Ash S, et al. (2004) The predictive potential of molecular detection in the nonmetastatic Ewing family of tumors. *Cancer* **100**: 1053–1058.
32. Fagnou C, Michon J, Peter M, Bernoux A, Oberlin O, Zucker JM (1998) Presence of tumor cells in bone marrow but not in blood is associated with adverse prognosis in patients with Ewing's tumor. Société Française d'Oncologie Pédiatrique. *Journal of Clinical Oncology* **16**: 1707–1711.
33. de Alava E, Lozano MD, Patino A, Sierrasesumaga L, Pardo-Mindan FJ (1998) Ewing family tumors: potential prognostic value of reverse-transcriptase polymerase chain reaction detection of minimal residual disease in peripheral blood samples. *Diagnostic Molecular Pathology* **7**: 152–157.
34. Yaniv I, Cohen IJ, Stein J, Zilberstein J, Liberzon E, Atlas O, et al. (2004) Tumor cells are present in stem cell harvests of Ewings sarcoma patients and their persistence following transplantation is associated with relapse. *Pediatric Blood Cancer* **42**: 404–409.
35. Vermeulen J, Ballet S, Oberlin O, Peter M, Pierron G, Longavenne E, et al. (2006) Incidence and prognostic value of tumour cells detected by RT-PCR in peripheral blood stem cell collections from patients with Ewing tumour. *British Journal of Cancer* **95**: 1326–1333.
36. Burchill S, Picton S, Wheeldon J, Kinsey S, Lashford L, Lewis I (2003) Reduced tumor load in peripheral blood after treatment with G-CSF and chemotherapy in children with tumors of the Ewing sarcoma family but not neuroblastoma. *Blood* **102**: 3459–3460.
37. Thomson B, Hawkins D, Felgenhauer J, Radich J (1999) RT-PCR evaluation of peripheral blood, bone marrow and peripheral blood stem cells in children and adolescents undergoing VACIME chemotherapy for Ewing's sarcoma and alveolar rhabdomyosarcoma. *Bone Marrow Transplantation* **24**: 527–533.

38. Athale UH, Shurtleff SA, Jenkins JJ, Poquette CA, Tan M, Downing JR, et al. (2001) Use of reverse transcriptase polymerase chain reaction for diagnosis and staging of alveolar rhabdomyosarcoma, Ewing sarcoma family of tumors, and desmoplastic small round cell tumor. *Journal of Pediatric Hematology and Oncology* **23**: 99–104.
39. Gallego S, Llort A, Roma J, Sabado C, Gros L, de Toledo JS (2006) Detection of bone marrow micrometastasis and microcirculating disease in rhabdomyosarcoma by a real-time RT-PCR assay. *Journal of Cancer Research and Clinical Oncology* **132**: 356–362.
40. Schlomm T, Erbersdobler A, Mirlacher M, Sauter G (2007) Molecular staging of prostate cancer in the year 2007. *World Journal of Urology* **25**: 19–30.
41. Mao X, Shaw G, James SY, Purkis P, Kudahetti SC, Tsigani T, et al. (2008) Detection of TMPRSS2:ERG fusion gene in circulating prostate cancer cells. *Asian Journal of Andrology* **10**:467–473.
42. Laxman B, Tomlins SA, Mehra R, Morris DS, Wang L, Helgeson BE, et al. (2006) Noninvasive detection of TMPRSS2:ERG fusion transcripts in the urine of men with prostate cancer. *Neoplasia* **8**: 885–888.
43. Clark J, Merson S, Jhavar S, Flohr P, Edwards S, Foster CS, et al. (2007) Diversity of TMPRSS2-ERG fusion transcripts in the human prostate. *Oncogene* **26**: 2667–2673.
44. Burchill SA, Selby P (1999) Early detection of small volume disease using molecular technology. *Cancer Topics* **10**: 1–4.
45. Burchill SA, Selby PJ (2000) Molecular detection of low-level disease in patients with cancer. *Journal of Pathology* **190**: 6–14.
46. Burchill SA, Bradbury MF, Pittman K, Southgate J, Smith B, Selby P (1995) Detection of epithelial cancer cells in peripheral blood by reverse transcriptase-polymerase chain reaction. *British Journal of Cancer* **71**: 278–281.
47. Ko Y, Grünewald E, Totzke G, Klinz M, Fronhoffs S, Gouni-Berthold I, et al. (2000) High percentage of false-positive results of cytokeratin 19 RT-PCR in blood: a model for the analysis of illegitimate gene expression. *Oncology* **59**: 81–88.
48. Savtchenko ES, Schiff TA, Jiang CK, Freedberg IM, Blumenberg M (1988) Embryonic expression of the human 40-kD keratin: evidence from a processed pseudogene sequence. *American Journal of Human Genetics* **43**: 630–637.
49. Wyld DK, Selby P, Perren TJ, Jonas SK, Allen-Mersh TG, Wheeldon J, et al. (1998) Detection of colorectal cancer cells in peripheral blood by reverse-transcriptase polymerase chain reaction for cytokeratin 20. *International Journal of Cancer* **79**: 288–293.
50. Cheung IY, Barber D, Cheung N (1998) Detection of microscopic neuroblastoma in marrow by histology, immunocytology and reverse transcription-PCR of multiple molecular markers. *Clinical Cancer Research* **4**: 2801–2805.
51. Medic S, Pearce RL, Heenan PJ, Ziman M (2007) Molecular markers of circulating melanoma cells. *Pigment Cell Research* **20**: 80–91.
52. Xi L, Nicastri DG, El-Hefnawy T, Hughes SJ, Luketich JD, Godfrey TE (2007) Optimal markers for real-time quantitative reverse transcription PCR detection of circulating tumor cells from melanoma, breast, colon, esophageal, head and neck, and lung cancers. *Clinical Chemistry* **53**: 1206–1215.
53. Viprey VF, Lastowska MA, Corrias MV, Swerts K, Jackson MS, Burchill SA (2008) Minimal disease monitoring by QRT-PCR: guidelines for identification and systematic validation of molecular markers prior to evaluation in prospective clinical trials. *The Journal of Pathology* **126**: 245–252.
54. Stutterheim J, Gerritsen A, Zappeij-Kannegieter L, Yalcin B, Dee R, van Noesel MM, et al. (2009) Detecting minimal residual disease in neuroblastoma: the superiority of a panel of real-time quantitative PCR markers. *Clinical Chemistry* **55**: 1316–1326.
55. van Houten VM, Tabor MP, van den Brekel MW, Denkers F, Wishaupt RG, Kummer JA, et al. (2000) Molecular assays for the diagnosis of minimal residual head-and-neck cancer: methods, reliability, pitfalls, and solutions. *Clinical Cancer Research* **6**: 3803–3816.

56. Benoy IH, Elst H, Van Dam P, Scharpé S, Van Marck E, Vermeulen PB, et al. (2006) Detection of circulating tumour cells in blood by quantitative real-time RT-PCR: effect of pre-analytical time. *Clinical Chemistry and Laboratory Medicine* **44**: 1082–1087.
57. Beillard E, Pallisgaard N, van Der Velden VH, Bi W, Dee R, van der Schoot E, et al. (2003) Evaluation of candidate control genes for diagnosis and residual disease detection in leukemic patients using 'real-time' quantitative reverse-transcriptase polymerase chain reaction (RQ-PCR) – a Europe against cancer program. *Leukemia* **17**: 2474–2486.
58. Viprey VF, Corrias MV, Kagedal B, Oltra S, Swerts K, Vicha A, et al. (2007) Standardisation of operating procedures for the detection of minimal disease by QRT-PCR in children with neuroblastoma: quality assurance on behalf of SIOPEN-R-NET. *European Journal of Cancer* **43**: 341–350.
59. Livak KJ, Schmittgen TD (2001) Analysis of relative gene expression data using real-time quantitative PCR and the 2(-Delta Delta C(T)) Method. *Methods* **25**: 402–408.
60. Smith B, Selby P, Southgate J, Pittman K, Bradley C, Blair GE (1991) Detection of melanoma cells in peripheral blood by means of reverse transcriptase and polymerase chain reaction. *Lancet* **338**: 1227–1229.
61. Burchill SA, Bradbury FM, Lewis IJ (1995) Early clinical evaluation of reverse transcriptase-polymerase chain reaction (RT-PCR) for tyrosine hydroxylase. *European Journal of Cancer* **31**: 553–556.
62. Miyajima Y, Kato K, Numata S, Kudo K, Horibe K (1995) Detection of neuroblastoma cells in bone marrow and peripheral blood at diagnosis by the reverse transcriptase-polymerase chain reaction for tyrosine hydroxylase mRNA. *Cancer* **75**: 2757–2761.
63. Kuroda T, Saeki M, Nakano M, Mizutani S (1997) Clinical application of minimal residual neuroblastoma cell detection by reverse transcriptase-polymerase chain reaction. *Journal of Pediatric Surgery* **32**: 69–72.
64. Gilbert J, Norris MD, Marshall GM, Haber M (1997) Low specificity of PGP9.5 expression for detection of micrometastatic neuroblastoma. *British Journal of Cancer* **75**: 1779–1781.
65. Cheung IY, Cheung NK (2001a) Detection of microscopic disease: comparing histology, immunocytology, and RT-PCR of tyrosine hydroxylase, GAGE, and MAGE. *Medical and Pediatric Oncology* **36**: 210–212.
66. Burchill SA, Bradbury FM, Smith B, Lewis IJ, Selby P (1994) Neuroblastoma cell detection by reverse transcriptase-polymerase chain reaction (RT-PCR) for tyrosine hydroxylase mRNA. *International Journal of Cancer* **57**: 671–675.
67. Cheung IY, Cheung NK (2001) Quantitation of marrow disease in neuroblastoma by real-time reverse transcription-PCR. *Clinical Cancer Research* **7**: 1698–1705.
68. Träger C, Vernby A, Kullman A, Ora I, Kogner P, Kågedal B (2008) mRNAs of tyrosine hydroxylase and dopa decarboxylase but not of GD2 synthase are specific for neuroblastoma minimal disease and predicts outcome for children with high-risk disease when measured at diagnosis. *International Journal of Cancer* **123**: 2849–2855.
69. Shono K, Tajiri T, Fujii Y, Suita S (2000) Clinical implications of minimal disease in the bone marrow and peripheral blood in neuroblastoma. *Journal of Pediatric Surgery* **35**: 1415–1420.
70. Burchill SA, Lewis IJ, Abrams KR, Riley R, Imeson J, Pearson AD, et al. (2001) Circulating neuroblastoma cells detected by reverse transcriptase polymerase chain reaction for tyrosine hydroxylase mRNA are an independent poor prognostic indicator. *Journal of Clinical Oncology* **19**: 1795–1801.
71. Miyajima Y, Horibe K, Fukuda M, Matsumoto K, Numata S, Mori H, et al. (1996) Sequential detection of tumor cells in the peripheral blood and bone marrow of patients with stage IV neuroblastoma by the reverse transcription-polymerase chain reaction for tyrosine hydroxylase mRNA. *Cancer* **77**: 1214–1219.
72. Seeger RC, Reynolds CP, Gallego R, Stram DO, Gerbing RB, Matthay KK (2000) Quantitative tumor cell content of bone marrow and blood as a predictor of outcome in

stage IV neuroblastoma: a Children's Cancer Group Study. *Journal of Clinical Oncology* **18**: 4067–4076.

73. Horibe K, Fukuda M, Miyajima Y, Matsumoto K, Kondo M, Inaba J, et al. (2001) Outcome prediction by molecular detection of minimal residual disease in bone marrow for advanced neuroblastoma. *Medical and Pediatric Oncology* **36**: 203–204.
74. Fukuda M, Miyajima Y, Miyashita Y, Horibe K (2001) Disease outcome may be predicted by molecular detection of minimal residual disease in bone marrow in advanced neuroblastoma: a pilot study. *Journal of Pediatric Hematology and Oncology* **23**: 10–13.
75. Burchill SA, Kinsey SE, Picton S, Roberts P, Pinkerton CR, Selby P, et al. (2001) Minimal residual disease at the time of peripheral blood stem cell harvest in patients with advanced neuroblastoma. *Medical and Pediatric Oncology* **36**: 213–219.
76. Corrias MV, Haupt R, Carlini B, Parodi S, Rivabella L, Garaventa A, et al. (2006) Peripheral blood stem cell tumor cell contamination and survival of neuroblastoma patients. *Clinical Cancer Research* **12**: 5680–5685.
77. Avigad S, Feinberg-Gorenshtein G, Luria D, Jeison M, Stein J, Grunshpan A, et al. (2009) Minimal residual disease in peripheral blood stem cell harvests from high-risk neuroblastoma patients. *Journal of Pediatric Hematology/Oncology* **31**: 22–26.
78. Riley RD, Heney D, Jones DR, Sutton AJ, Lambert PC, Abrams KR, et al. (2003) A systematic review of molecular and biological tumor markers in neuroblastoma. *Clinical Cancer Research* **10**: 4–12.
79. Viprey VF, Corrias MV, Kagedal B, Oltra S, Swerts K, Vicha A, et al. (2007) Standardisation of operating procedures for the detection of minimal disease by QRT-PCR in children with neuroblastoma: quality assurance on behalf of SIOPEN-R-NET. *European Journal of Cancer* **43**: 341–350.
80. Beiske K, Burchill SA, Cheung IY, Hiyama E, Seeger RC, Cohn SL, et al. (2009) Consensus criteria for sensitive detection of minimal neuroblastoma cells in bone marrow, blood and stem cell preparations by immunocytology and QRT-PCR: recommendations by the International Neuroblastoma Risk Group Task Force. *British Journal of Cancer* **100**: 1627–1637.
81. Cheung IY, Feng Y, Gerald W, Cheung NK (2008) Exploiting gene expression profiling to identify novel minimal residual disease markers of neuroblastoma. *Clinical Cancer Research* **14**: 7020–7027.
82. Grasso YZ, Gupta MK, Levin HS, Zippe CD, Klein EA (1998) Combined nested RT-PCR assay for prostate-specific antigen and prostate-specific membrane antigen in prostate cancer patients: correlation with pathological stage. *Cancer Research* **58**: 1456–1459.
83. de la Taille A, Olsson CA, Buttyan R, Benson MC, Bagiella E, Cao Y, et al. (1999) Blood-based reverse transcriptase polymerase chain reaction assays for prostate specific antigen: long term follow-up confirms the potential utility of this assay in identifying patients more likely to have biochemical recurrence (rising PSA) following radical prostatectomy. *International Journal of Cancer* **84**: 360–364.
84. Okegawa T, Noda H, Kato M, Miyata A, Nutahara K, Higashihara E (2000) Value of reverse transcription polymerase chain reaction assay in pathological stage T3N0 prostate cancer. *Prostate* **44**: 210–218.
85. Mejean A, Vona G, Nalpas B, Damotte D, Brousse N, Chretien Y, et al. (2000) Detection of circulating prostate derived cells in patients with prostate adenocarcinoma is an independent risk factor for tumor recurrence. *Journal of Urology* **163**: 2022–2029.
86. Thiounn N, Saporta F, Flam TA, Pages F, Zerbib M, Vieillefond A, et al. (1997) Positive prostate-specific antigen circulating cells detected by reverse transcriptase-polymerase chain reaction does not imply the presence of prostatic micrometastases. *Urology* **50**: 245–250.
87. Ellis WJ, Vessella RL, Corey E, Arfman EW, Oswin MM, Melchior S, et al. (1998) The value of a reverse transcriptase polymerase chain reaction assay in preoperative staging and followup of patients with prostate cancer. *Journal of Urology* **159**: 1134–1138.

88. Gao C, Maheshwari S, Dean RC, Tatum L, Mooneyhan R, Connelly RR, et al. (1999) Blinded evaluation of reverse transcriptase-polymerase chain reaction prostate-specific antigen peripheral blood assay for molecular staging of prostate cancer. *Urology* **53**: 714–721.
89. Llanes L, Ferruelo A, Paez A, Gomez JM, Moreno A, Berenguer A (2000) The clinical utility of the prostate specific membrane antigen reverse-transcription/polymerase chain reaction to detect circulating prostate cells: an analysis in healthy men and women. *British Journal of Urology* **89**: 882–885.
90. Thomas J, Gupta M, Grasso Y, Reddy CA, Heston WD, Zippe C, et al. (2002) Preoperative combined nested reverse transcriptase polymerase chain reaction for prostate-specific antigen and prostate-specific membrane antigen does not correlate with pathologic stage or biochemical failure in patients with localised prostate cancer undergoing radical prostatectomy. *Journal of Clinical Oncology* **20**: 3213–3218.
91. Shariat SF, Gottenger E, Nguyen C, Song W, Kattan MW, Andenoro J, et al. (2002) Preoperative blood reverse transcriptase-PCR assays for prostate-specific antigen and human glandular kallikrein for prediction of prostate cancer progression after radical prostatectomy. *Cancer Research* **62**: 5974–5979.
92. Smith MR, Biggar S, Hussain M (1995) Prostate-specific antigen messenger RNA is expressed in non-prostate cells: implications for detection of micrometastases. *Cancer Research* **55**: 2640–2644.
93. Henke W, Jung M, Jung K, Lein M, Schlechte H, Berndt C, et al. (1996) Detection of PSA mRNA in blood by RT-PCR does not exclusively indicate prostatic tumor cells. *Clinical Chemistry* **42**: 1499–1500.
94. O'Hara SM, Veltri RW, Skipstunas P, et al. (1996) Basal PSA mRNA levels detected by quantitative reverse transcriptase polymerase chain reaction (Q-RT-PCR-PSA) in blood from subjects without prostate cancer. *Journal of Urology* **155**: 418 Abstract 430.
95. Gala J, Heusterspreute M, Loric S, Hanon F, Tombal B, Van Cangh P, et al. (1998) Expression of prostate-specific antigen and prostate-specific membrane antigen transcripts in blood cells: implications for the detection of hematogenous prostate cells and standardization. *Clinical Chemistry* **44**: 472–481.
96. Patel K, Whelan PJ, Prescott S, Brownhill SC, Johnston CF, Selby PJ, et al. (2004) The use of real-time reverse transcription-PCR for prostate-specific antigen mRNA to discriminate between blood samples from healthy volunteers and from patients with metastatic prostate cancer. *Clinical Cancer Research* **10**: 7511–7519.
97. Small EJ, Roach M III (2002) Prostate-specific antigen in prostate cancer: a case study in the development of a tumor marker to monitor recurrence and assess response. *Seminars in Oncology* **29**: 264–273.
98. Brossart P, Keilholz U, Willhauck M, Scheibenbogen C, Möhler T, Hunstein W (1993) Hematogenous spread of malignant melanoma cells in different stages of disease. *Journal of Investigative Dermatology* **10**: 887–889.
99. Tobal K, Sherman LS, Foss AJ, Lightman SL (1993) Detection of melanocytes from uveal melanoma in peripheral blood using the polymerase chain reaction. *Investigative Ophthalmology & Visual Science* **34**: 2622–2625.
100. Battayani Z, Grob JJ, Xerri L, Noe C, Zarour H, Houvaeneghel G, et al. (1995) Polymerase chain reaction detection of circulating melanocytes as a prognostic marker in patients with melanoma. *Archives of Dermatology* **131**: 443–447.
101. Hoon DS, Wang Y, Dale PS, Conrad AJ, Schmid P, Garrison D, et al. (1995) Detection of occult melanoma cells in blood with a multiple-marker polymerase chain reaction assay. *Journal of Clinical Oncology* **13**: 2109–2116.
102. Kunter U, Buer J, Probst M, Duensing S, Dallmann I, Grosse J, et al. (1996) Peripheral blood tyrosinase messenger RNA detection and survival in malignant melanoma. *Journal of the National Cancer Institute* **88**: 590–594.
103. Mellado B, Colomer D, Castel T, Muñoz M, Carballo E, Galán M, et al. (1996) Detection of circulating neoplastic cells by reverse-transcriptase polymerase chain reaction

in malignant melanoma: association with clinical stage and prognosis. *Journal of Clinical Oncology* **14**: 2091–2097.

104. Keilholz U, Willhauck M, Rimoldi D, Brasseur F, Dummer W, Rass K, et al. (for EORTC-MCG) (1998) Reliability of RT-PCR based assays for detection of circulating tumor cells. *European Journal of Cancer* **34**: 750–753.
105. Max N, Keilholz U (2001) Minimal residual disease in melanoma. *Seminars in Surgical Oncology* **20**: 319–328.
106. Brownbridge GG, Gold J, Edward M, MacKie RM (2001) Evaluation of the use of tyrosinase-specific and melanA/MART-1-specific reverse transcriptase-coupled–polymerase chain reaction to detect melanoma cells in peripheral blood samples from 299 patients with malignant melanoma. *British Journal of Dermatology* **144**: 279–287.
107. Strohal R, Mosser R, Kittler H, Wolff K, Jansen B, Brna C, et al. (2001) MART-1/Melan-A and tyrosinase transcripts in peripheral blood of melanoma patients: PCR analyses and follow-up testing in relation to clinical stage and disease progression. *Melanoma Research* **11**: 543–548.
108. de Vries TJ, Fourkour A, Wobbes T, Verkroost G, Ruiter DJ, van Muijen GN (1997) Heterogeneous expression of immunotherapy candidate proteins gp100, MART-1, and tyrosinase in human melanoma cell lines and in human melanocytic lesions. *Cancer Research* **57**: 3223–3229.
109. Sarantou T, Chi DD, Garrison DA, Conrad AJ, Schmid P, Morton DL, et al. (1997) Melanoma-associated antigens as messenger RNA detection markers for melanoma. *Cancer Research* **57**: 1371–1376.
110. Cormier JN, Hijazi YM, Abati A, Fetsch P, Bettinotti M, Steinberg SM, et al. (1998) Heterogeneous expression of melanoma-associated antigens and HLA-A2 in metastatic melanoma in vivo. *International Journal of Cancer* **75**: 517–524.
111. Dalerba P, Ricci A, Russo V, Rigatti D, Nicotra MR, Mottolese M, et al. (1998) High homogeneity of MAGE, BAGE, GAGE, tyrosinase and Melan-A/MART-1 gene expression in clusters of multiple simultaneous metastases of human melanoma: implications for protocol design of therapeutic antigen-specific vaccination strategies. *International Journal of Cancer* **77**: 200–204.
112. Kuo CT, Bostick PJ, Irie RF, Morton DL, Conrad AJ, Hoon DS (1998) Assessment of messenger RNA of beta 1–>4-N-acetylgalactosaminyl-transferase as a molecular marker for metastatic melanoma. *Clinical Cancer Research* **4**: 411–418.
113. Curry BJ, Myers K, Hersey P (1999) MART-1 is expressed less frequently on circulating melanoma cells in patients who develop distant compared with locoregional metastases. *Journal of Clinical Oncology* **17**: 2562–2571.
114. de Vries TJ, Fourkour A, Punt CJ, van de Locht LT, Wobbes T, Van Den Bosch S, et al. (1999) Reproducibility of detection of tyrosinase and MART-1 transcripts in the peripheral blood of melanoma patients: a quality control study using real-time quantitative RT-PCR. *British Journal of Cancer* **80**: 883–891.
115. Hoon DS, Bostick P, Kuo C, Okamoto T, Wang HJ, Elashoff R, et al. (2000) Molecular markers in blood as surrogate prognostic indicators of melanoma recurrence. *Cancer Research* **60**: 2253–2257.
116. Giese T, Engstner M, Mansmann U, Hartschuh W, Arden B (2005) Quantification of melanoma micrometastases in sentinel lymph nodes using real-time RT-PCR. *The Journal of Investigative Dermatology* **124**: 633–637.
117. Essner R (2006) Sentinel lymph node biopsy and melanoma biology. *Clinical Cancer Research* **12**: 2320–2325.
118. Martinez SR, Mori T, Hoon DS (2006) Molecular upstaging of sentinel lymph nodes in melanoma: where are we now? *Surgical Oncology Clinics of North America* **15**: 331–340.
119. Wang X, Heller R, VanVoorhis N, Cruse CW, Glass F, Fenske N, et al. (1994) Detection of submicroscopic lymph node metastases with polymerase chain reaction in patients with malignant melanoma. *Annals of Surgery* **220**: 768–774.

120. Mocellin S, Hoon DS, Pilati P, Rossi CR, Nitti D (2007) Sentinel lymph node molecular ultrastaging in patients with melanoma: a systematic review and meta-analysis of prognosis. *Journal of Clinical Oncology* **25**: 1588–1595.
121. Scoggins CR, Ross MI, Reintgen DS, Noyes RD, Goydos JS, Beitsch PD, et al. (2006) Prospective multi-institutional study of reverse transcriptase polymerase chain reaction for molecular staging of melanoma. *Journal of Clinical Oncology* **24**: 2849–2857.
122. Mangas C, Hilari JM, Paradelo C, Rex J, Fernández-Figueras MT, Fraile M, et al. (2006) Prognostic significance of molecular staging study of sentinel lymph nodes by reverse transcriptase-polymerase chain reaction for tyrosinase in melanoma patients. *Annals of Surgical Oncology* **13**: 910–918.
123. Tatlidil C, Parkhill WS, Giacomantonio CA, Greer WL, Morris SF, Walsh NM (2007) Detection of tyrosinase mRNA in the sentinel lymph nodes of melanoma patients is not a predictor of short-term disease recurrence. *Modern Pathology* **20**: 427–434.
124. Bénard J, Raguénez G, Kauffmann A, Valent A, Ripoche H, Joulin V, et al. (2008) MYCN-non-amplified metastatic neuroblastoma with good prognosis and spontaneous regression: a molecular portrait of stage 4S. *Molecular Oncology* **2**: 261–271.
125. Pantel K, Alix-Panabières C, Riethdorf S (2009) Cancer micrometastases. *Nature Reviews, Clinical Oncology* **6**: 339–351.
126. Lunt SJ, Chaudary N, Hill RP (2009) The tumor microenvironment and metastatic disease. *Clinical and Experimental Metastasis* **26**: 19–34.

11 Polymerase chain reaction and infectious diseases

Jim Huggett

As in numerous other areas, the comparatively simple technique of polymerase chain reaction (PCR) has revolutionized the field of infectious diseases. Whether this is through sequencing the genomes of key pathogens or developing vaccines by genetic manipulation, PCR-driven molecular biology has stamped its mark on infectious diseases. It is particularly fascinating to consider how PCR has influenced, and continues to influence, disease management and to realize how influential this research technology has become as a practical diagnostic tool.

Its role in this context can be broadly split into diagnosis, epidemiology, and prognostic monitoring. However, before considering the utility of the PCR, it is useful to discuss infectious diseases and the additional considerations required for using PCR for their management.

Infectious diseases can be broadly split into groups corresponding to the causative pathogen: bacterial (tuberculosis [TB], pseudomembranous colitis [PMC], sepsis) and viral (acquired immunodeficiency syndrome [AIDS], hepatitis C, influenza) are two simple groupings. Viral pathogens are the most common causes of infectious diseases worldwide (e.g., common cold with numerous viral causes[1]), with bacterial pathogens often being more serious when they strike. The remaining categories are more complex and include the eukaryotes. A major group are the Protozoa, including the causes of many classical tropical diseases (e.g., malaria and sleeping sickness). The fungal pathogens (*Pneumocystis pneumonia* [PCP]) are frequently opportunistic, causing infections worldwide in individuals with reduced or impaired immunity.

There are many other groups of eukaryotic pathogens as well as a small group of bacteria-like pathogens called mycoplasma and the prion that cause encephalopathy (Creutzfeldt–Jakob disease), but these will not be discussed further.

Appropriate disease management is made more difficult because many pathogens can cause illnesses with similar symptoms but have different management requirements. A patient with an infectious disease such as TB must, ideally, have the disease identified using methods that allow for the detection of other causes of respiratory illness (diagnostic differential). After a confirmed diagnosis is made, the physician needs to consider treatment. In the case of TB, specific guidelines exist from institutions such as the World Health Organization; however, there can be additional epidemiological factors (such as drug resistance) that can prevent the treatment from working. If the patient has had TB before, PCR can be used to answer important clinical questions such as whether this is a new infection or an old one that was never really cured.[2] Finally, PCR can be used to establish how the infective organism is responding to treatment and the likelihood of the patient having a good prognosis.[3] Although TB is a good example of how PCR should be used for management of a major disease, it also illustrates just how PCR is not being used in this context. This chapter provides examples of the use of PCR in infectious disease management, uses TB to discuss in detail where PCR could be used more extensively, and outlines reasons why it is not.

INFECTIOUS DISEASE DIAGNOSIS (JARGON)

Infectious disease diagnosis can be broadly described as the ability to determine what group of pathogens (or, occasionally, specific pathogen) is the cause of a patient's symptoms, while ruling out other possibilities. A good diagnostic test must be as able to tell physicians that a disease is *not* caused by pathogen X as it is able to tell them that it *is* caused by pathogen Y. The necessity for a test to both rule in and rule out infections is better illustrated by touching on the jargon that is used when discussing diagnosis. Unlike many other like subjects (especially molecular biology), the jargon to describe diagnostics is kept simple, focusing upon sensitivity and specificity. Sensitivity tells us how good the diagnostic test is at identifying the disease; a cough has pretty good diagnostic sensitivity for TB. Specificity focuses on how well the choice of diagnostic test rules out other diseases; for example, there are many things that cause cough, so although cough has high diagnostic sensitivity for TB, it has low specificity as the patient may also have another bacterial, fungal, or viral respiratory infection, allergic reaction, or environmental insult.

Infectious disease diagnosis can be split loosely into two categories: detecting a response to the pathogen (host marker) or detecting the pathogen itself. Fever is a host marker and is possibly the simplest and most common method for diagnosing an infection. Fever is sensitive; parents know that their child is ill when he or she has a temperature. The problem occurs because fever (like cough) is not specific; it tells you that something is wrong, but not what it is. If the child's high temperature persists, the parents will take him or her to a doctor who must diagnose from a battery of potential pathogens of varying severities. It is important that the doctor gets the diagnosis right and rules out other possible causes; for this we need a specific test.

If the child is from a more impoverished area, the doctor must consider potential confounding effects of poverty, such as impaired immune system through malnutrition; and if the child is from parts of the developing world, then the differential diagnosis must included malaria and/or infections that prey on individuals infected with human immunodeficiency virus (HIV).

The take-home message is that, for maximum clinical value, a diagnostic test must be both sensitive and specific. PCR has the potential to be both in the context of many infections, although to date PCR tests almost exclusively detect the pathogen rather than the host response.

DIAGNOSTIC TECHNIQUES

The most common method of diagnosing an infectious disease is empirical (i.e., a physician uses his or her knowledge combined with the clinical symptoms to make the diagnosis). It is empirical because frequently the infection is simple to diagnose (expensive diagnostic tools are not needed) but also because much of the world does not have access to the clinical tools (expensive diagnostic tools cost too much) and therefore is all that is available. Importantly, for treatment, the physician may not need to confirm the exact causative pathogen but just whether it is a bacterium (that will respond to antibiotics) or a virus (that will not).

When a more specific confirmed diagnosis is needed, microscopy is the most commonly used method available worldwide. With TB and malaria, two of the most common infectious causes of death worldwide, microscopy is frequently used (on sputum for TB and blood for malaria) and, as both diseases are major problems in the developing world, this can be (and often is) tailored to comparatively basic laboratories. Additional benefits include speed; a result is dependent on taking the sample, preparing the slide, and reading it. Microscopy also represents one of the oldest methods for investigating both of these infections, so the fact that it is established and thus accepted is a major factor in its use. Acceptance is important when considering new methods like PCR as we will discuss.

The main problem with microscopy is that it requires a considerable amount of human expertise to perform, which makes standardization difficult and automation almost impossible. An additional problem occurs with diseases, such as TB, that are not easily diagnosed by microscopy. If we use microscopy to investigate the sputum from 10 patients who have TB (and no other co-infections like HIV), using the best laboratories in the world, only 7 will be correctly diagnosed (a sensitivity of 70%). If we factor the reduction that will occur because of variation between laboratories and personnel along with the fact that co-infection with HIV (infecting one in three individuals in sub-Saharan Africa) dramatically reduces the efficacy of microscopy for TB, the result can easily become less than 5 of the 10 patients being correctly diagnosed. When things get this bad, then the physician will do just as well to flip a coin and call heads or tails to establish patient diagnosis.

Consequently, in the Western world and in advanced developing world laboratories, bacterial culture is considered as the best technique (gold standard) for diagnosing TB, but it can take as much as 6 weeks to get a result. Although it can be argued that TB constitutes an extreme example as it is so difficult to diagnose, it kills more than 1.6 million people every year and so provides a pressing example of where developing more rapid, sensitive, and specific diagnostic tests is needed. Other tests, for example those for culturing for methicillin-resistant *Staphylococcus aureus* (MRSA), will take 48 hrs. If a patient has septicemia caused by MRSA, even 48 hrs may be too long. The point here is that the physician has to decide whether to wait for the result or attempt to treat based on empirical information. This decision may be urgent because the patient may not only be unwell and at considerable risk if action is not taken, but he or she also may be infectious to others, representing considerable public health risk.

The benefit that PCR technology can provide is that it can be both rapid (like microscopy) while being sensitive (like culture). Yet PCR methods are not routinely used in the diagnosis of diseases such as TB or bacterial infections caused by MRSA. Why is this? Possible reasons are discussed in more detail (Why Is PCR Not Used More Routinely for Infectious Disease Diagnosis?) later in this chapter. Before these reasons are covered, however, we will discuss some areas where PCR is frequently used in infectious disease diagnosis.

VIRUSES

Viruses are the group of pathogen to which PCR-based diagnostic methods are most routinely applied. Viruses are units of infectivity that carry the genetic information sufficient for them to infect and corrupt a cell to enable it to reproduce and release their progeny, which must in turn find another cell to do the same.

Without a host cell no reproduction is possible, so a diagnostic method employing culture of the virus alone is not possible. Culture-based diagnostic tools have been developed in cases by infecting host cells, but such an approach poses a relatively high level of complexity and hence is not ideal.

Microscopy is occasionally employed in viral diagnosis but, unlike bacterial, parasitic, and fungal pathogens, viruses are tiny – ranging from 15 to 600 nm (from 0.00000015 to 0.000006 cm). The best light microscopes are capable of visualizing particles of approximately 180 nm, so many viruses like HIV (at approximately 130 nm) are simply not visible using this technique.

Immunological methods have been developed that recognize viral molecule (antigen) from the patient or the patients' immunological response to viral antigen (a good example of measuring a host marker used diagnostically). There are many examples of these types of diagnostic methods as they are easy to automate and highly robust. Furthermore, when the tests have been developed they are generally simple to perform.

The problem with immunological tests is that they are dependent on the immune system for both their development and their subsequent clinical application. Consequently, a simple change (different patient population, viral strain, etc.) can render the test obsolete. Although this change can also effect PCR-based methods, the fundamental difference with immunological methods is that they are easily modified so that they can quickly compensate for such a change.

Hepatitis C virus (HCV) is caused by a 50-nm single-stranded RNA virus that passes through blood and sexual contact.[4] Immunologically based methods using enzyme-linked immunosorbent assays (ELISA) to detect antibodies to HCV have been used but have low specificity (i.e., they result in the false-positive identification of healthy patients as HCV positive). Consequently, a PCR-based method that includes an initial reverse transcription step to convert the viral RNA to complementary DNA has been developed and now represents the gold standard diagnostic technique for HCV. Importantly, this method constitutes an essential prognostic test as well, as patients are required to be PCR negative before they are considered cured following treatment, which only occurs in approximately 50% of patients.

Although HCV infects approximately 170 million people worldwide and represents a serious health problem, its symptoms can be managed, even if patients do not get cured. PCR has become a valuable tool for the prognostic monitoring of the human immunodeficiency virus (HIV) – arguably the most serious viral pathogen of the last 50 years.

HIV remains a stigma-associated disease that leads to death without the treatment to keep the virus at bay. HIV is a retrovirus that infects and cripples the T cells of the immune system, making the patient more susceptible to diseases like TB and a range of infections caused by pathogens that are not commonly found in healthy individuals (opportunistic pathogens). HIV currently cannot be cured, but the infection can be held in check by preventing the virus from functioning by using a cocktail of drugs collectively called highly active antiretroviral therapy (HAART).

HIV viral load tests are essential in monitoring how patients are responding to current therapy that, unlike HCV, will never cure them. The monitoring assesses whether the virus has developed resistance to the therapy and is no longer being kept in check. If this occurs, the therapy must then be changed or the patient will be at risk of contracting acquired immune deficiency syndrome (AIDS) that leads to death.

PCR plays a major role in HIV viral load assessment, yet ironically HIV viral load assessment is possibly an example of a situation in which PCR is not the ideal method because HIV is constantly changing its genome sequence as part of its survival strategy.[5] This changing poses a major problem for PCR-based methods because although it is relatively easy to generate an assay that will detect a certain HIV sequence, there is always the danger that this sequence will change, rendering the PCR obsolete. Although PCR is simple to modify, there are so many potential variations in the HIV genome that modification to compensate for all sequences becomes impossible. The solution is to focus on regions that are more conserved[6]; however, despite this focus, tests must be carefully introduced into new areas prior to their clinical use.[7]

Influenza viruses are RNA viruses from the Orthomyxoviridae family. They are highly contagious viruses that typically infect the respiratory or gastric system and have the potential to kill far more people than do HIV and HCV combined. True influenza or flu is a serious condition and is often confused with less serious common colds.[8] Despite this severity and the fact that it frequently causes death in children and the elderly, flu is usually empirically diagnosed as a more severe set of symptoms than the common cold.[8] However, the threat of the highly virulent influenza type A (H5N1) has changed this.

UK residents have watched as reports of this potential pandemic have slowly made their way from the Far East across Europe until, on April 6, 2006, it was identified to have caused the death of a swan in Fife, Scotland. RNA PCR assays are now used to specifically diagnose H5N1, although (as with HIV) influenza represents a moving target due to genetic variability.[9] More recently the outbreak of H1N1 influenza (or swine flu) originating in Mexico has had the World Health organization on high alert with fears that a serious swine flu pandemic might occur or perhaps that the H1N1 and H5N1 strains might mix with even more serious medical consequences. PCR diagnostic methods represents the only feasible method for diagnosing and monitoring the spread of this infection, which has the potential of killing millions of people with huge economic consequences.[10]

Although viral infections represent the most common use of molecular diagnosis in the context of infectious disease diagnosis, there are some other routine uses of PCR-based molecular methods in bacterial diagnosis, monitoring, and epidemiology.

BACTERIA

Bacterial infections are not routinely diagnosed as frequently by PCR as viral infections; there are a few examples, however.

Chlamydia trachomatis and *Neisseria gonorrhoeae* are responsible for the sexually transmitted infections (STI) chlamydia and gonorrhea, respectively. Diagnosis of these infections is frequently conducted using molecular methods including PCR. The benefits of this over conventional culture are that preservation of the sample is less of a concern (as bacterial viability is not necessary for the test[11]), the additional sensitivity improves detection, and speed improves turnaround time.[12]

TB is a good example of an infection for which molecular diagnosis research is extensive, for which there are a number of commercial molecular

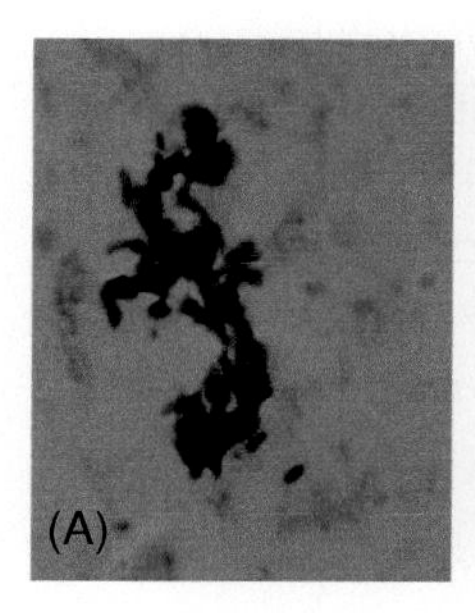

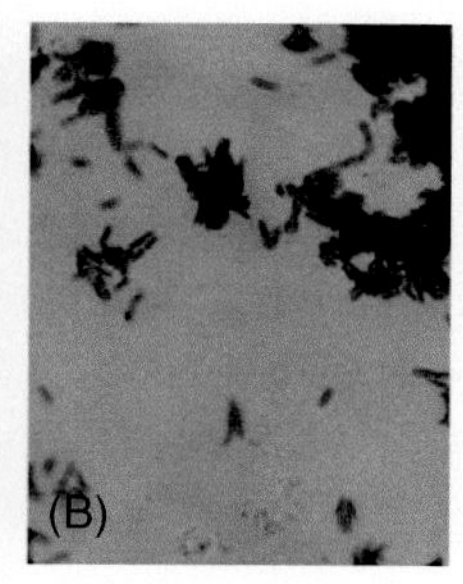

Figure 11–1. Sputum smear slides from two patients with tuberculosis (TB). *Mycobacterium tuberculosis* can be clearly seen, confirming the diagnosis. However, it is impossible to tell the drug-sensitive bacteria **(A)** from the strain resistant to two antibiotics **(B)**. *See Color Plates.*

diagnostic tests that use a variety of PCR-type methods,[2] but for which molecular methods are not generally used for independent diagnosis. Indeed, commercial methods are not recommended on their own for diagnosing TB.[13] In fact, with the exception of the STIs described, the majority of diagnostic laboratories do not use PCR for an initial diagnosis in bacterial disease. On the other hand, PCR has an essential role in assisting the disease management of bacterial infection.

MUTATION ANALYSIS

PCR-based methods play a crucial role in measuring mutations as surrogates for resistance to antibiotics. A patient with TB will start a regimen of four antibiotics. He or she must take these four antibiotics partly because *Mycobacterium tuberculosis* will be unlikely to develop resistance to all four at once. If a patient contracts TB caused by a drug-resistant *M. tuberculosis,* then the treatment will be less effective or not work at all. There is no way of distinguishing resistant from drug-sensitive *M. tuberculosis* by microscopy (Figure 11–1), and using culture can take up to 3 months, during which there is a risk that the patient will die and/or spread the drug-resistant disease.

Drug resistance is caused by mutations in key genes that reduce or stop the ability of the antibiotic to work. One of the key antibiotics of the quadruple regimen required for TB treatment is rifampicin, which is a bactericidal antibiotic that targets the bacterium DNA polymerase. Ninety percent of resistance to rifampicin is conferred through mutations in the DNA polymerase gene rpoB. A number of commercial PCR-based methods have been developed to target this gene; furthermore, these mutations usually occur in a specific region so PCR-based assays are able to very quickly establish if there is resistance.

Finally, as rifampicin resistance is a good marker for resistance to other drugs,[2] a physician can gain considerable information using PCR on how best to manage the patient, which, unlike culture, will be provided in a matter of hours. Recently commercially available PCR-based tests have been developed targeting rpoB and other key genes for assessing drug sensitivity that are suitable for more resource-poor settings.[14]

GENETIC RELATIONSHIPS

The bacterium *Clostridium difficile* causes PMC, another infection that has become a common occurrence in the news. It causes severe diarrhea and often death,

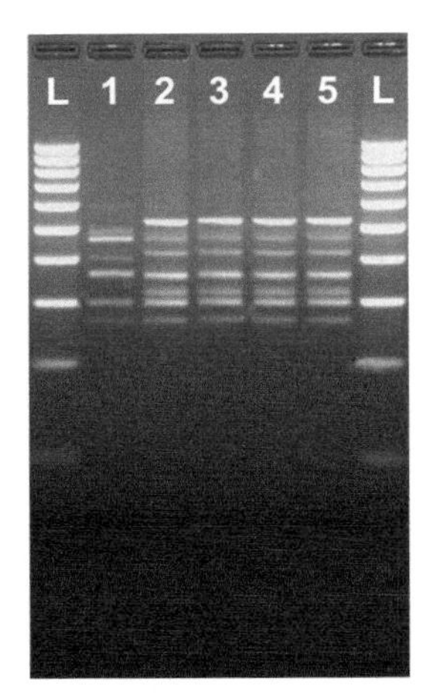

Figure 11–2. *Clostridium difficile* ribotype from 5 patients with pseudomembranous colitis. Patients 2, 3, 4, and 5 have the same ribotype, suggesting that they have contracted the bacteria from each other. Patient 1 has a distinct ribotype, confirming that the infection must have been contracted from elsewhere.

particularly in elderly patients. Diagnosis is particularly difficult as the variety of available tests work with low sensitivity and *C. difficile* diagnosis requires careful patient management, empirical experience, and repeat sampling. PCR is not used in clinical diagnosis of *C. difficile* as no acceptable test (i.e., one that is sensitive and inexpensive) has been developed yet; however, it has played a crucial role in patient management.

C. difficile poses a major problem as a nosocomial infection both for infection control and litigious reasons. Consequently, when there are two patients on a ward with PMC, it is very important to establish whether they constitute two independent infections or if one has contracted the infection from the other. The only way to establish quickly if the latter is true is genetically, and PCR-based ribotyping is the routine method currently.[15]

This technology uses PCR to amplify the genome of *C. difficile* between the 16 and 23 S ribosomal genes.[15] This sequence is not under tight selective pressure and varies between different strains. Consequently, the sequence differences can be observed as different sized molecules on a gel (Figure 11–2), and this pattern is termed a ribotype.

Although ribotyping is used clinically, it has its limitations. Because there appears to be no genetic link between many strains of the same ribotype, this technique has limited epidemiological value. Furthermore, to be clinically useful, this method of typing depends on an even distribution of a number of different ribotypes within the population. A predomination of any ribotype will increase the likelihood that two patients have the same ribotype by chance and not from one another.

A new method is being developed that may have clinical utility and supersede ribotyping.[16] This method targets variable number tandem repeats (VNTRs), which are sequences that vary between many different strains. This type of approach has been developed from an established method for genotyping higher organisms, including humans, and is used for genetic assessments such as paternity testing.

PCR has been used to genotype the causative strains of TB using VNTRs. This method allows rapid assessment of the genotype of a *M. tuberculosis* strain and is a useful tool for investigating strain epidemiology and evolution worldwide.[17] TB is capable of eliciting a destructive infection that has been associated with humans since before the use of medical records.[18] However, 90% of *M. tuberculosis* infection does not cause disease, with the bacterium causing a latent infection. The infected person, being completely unaware of his or her infection, can carry the latent bacteria and never contract TB disease. However, these individuals can sometimes contract TB disease when the latent infection reactivates. This is of

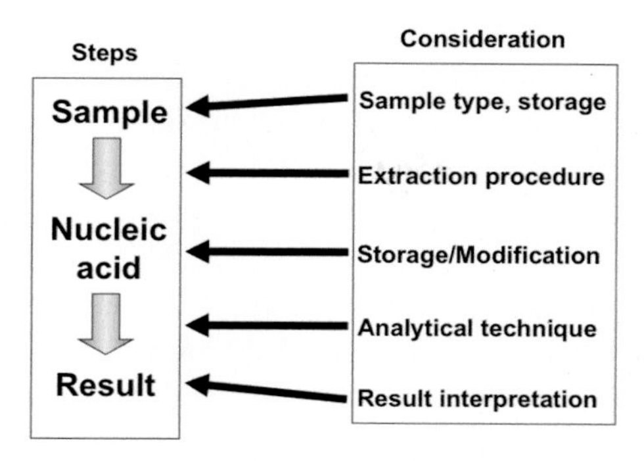

Figure 11–3. Flow chart illustrating the steps required to obtain a molecular diagnostic result from a clinical sample. Considerations that should be taken are outlined for each step.

increasing clinical concern with globalization and the associated migration of different populations who move from parts of the world where TB prevalence is high.

Genotyping of *M. tuberculosis* strains using VNTRs can provide fast epidemiological information as it is able to identify strains that have originated in different parts of the world.[19] Consequently, potentially more virulent strains can be tracked as they migrate with their human hosts. It is also of value to establish if TB disease is being caused by reactivation of existing infection or acquisition of new infections; the latter is of considerably more concern from an infection control point of view.

WHY IS PCR NOT USED MORE ROUTINELY FOR INFECTIOUS DISEASE DIAGNOSIS?

Theoretically, if a culture- or microscopically based approach is capable of diagnosing an infection by the presence of the pathogen, then a PCR-based method that detects DNA from that organism should be able to do the same job. Yet with the exception of viral and key bacterial diagnoses, PCR and other molecular methods are not generally used in routine diagnosis. Their clinical utility is frequently relegated to highly specialized clinical questions that simply cannot be answered in any other way. There are a number of potential reasons for this.

Molecular methods are generally expensive; a commercial diagnostic test for TB costs between about $40 and $80, whereas microscopy or culture costs about $1 or $5, respectively. These higher prices for molecular methods are frequently not considered cost effective in the West and completely rule out their use in the developing world. Another problem is the relative complexity of the molecular diagnostic protocol (Figure 11–3).

The PCR itself is simple to design and perform but it can be difficult to interpret and standardize between different laboratories. Furthermore, the process from clinical sample to PCR is multistep (Figure 11–3), which adds complexity and requires more demanding quality control. Finally, a major obstacle to the use of methods like PCR is the necessity for key components to be kept cold during storage and transport, also known as the cold chain.

In the developed world, this means additional transportation costs and sufficient laboratory cold storage space, and limits the time by which specific tests can be used. In the developing world, this problem becomes manifold, and molecular biological analysis is severely limited by the cold chain, which further adds to the complexity and cost. The use of PCR is simply too difficult for anything but highly focused research and is certainly not suitable for any routine developing world diagnostic laboratories below the reference level. Cost, complexity, and the cold chain are all fair reasons why PCR-based methods may not be currently used more in clinical diagnosis and management of infectious diseases, but they are not the only reasons.

The following describes two scenarios often occurring in London hospitals:

A physician asks a routine diagnostic laboratory to perform a PCR-based diagnostic test for TB in a patient who was negative by microscopy (remember in the best cases microscopy is only ~70% sensitive). The physician suspects empirically that the patient has TB. In the first example, the PCR comes back positive, confirming the physician's suspicions and the patient is treated for TB. In the second example, the PCR agrees with the microscopy and is negative; however, the physician decides to treat the patient for TB anyway.

The above scenario is not a criticism of physicians; it simply reflects the fact that PCR-based diagnostic methods are generally not trusted by clinicians and so are seldom used definitively, only supportively. Again this is not the fault of the medical profession; it is the scientists who develop and publish new molecular diagnostic tests who are at fault as there are two fundamental problems with their approach:

1) The processes that are required to measure a particular genetic sequence require a sample to be taken and nucleic acids to be extracted and then analyzed (Figure 11–3). To assess a PCR diagnostic protocol comprehensively, these steps must be considered individually. The first fundamental problem that has occurred in published diagnostic studies is that this is rarely done.

Diagnostic study reports need to ensure that the sample is not only appropriate but that it has been stored correctly. Furthermore, have the investigators selected the correct method for purifying the nucleic acids? There are a plethora of nucleic acid extraction methods that must not only recover the nucleic acids but must also remove all material that may affect that analysis, like assay inhibitors. Once purified, how are the nucleic acids stored prior to analysis? Will additional buffering components be added to the sample or will it be stored in water or dry? What temperature will be used – room temperature, 4°C, −20°C, −80°C? These are considerations that may affect the final result, but are almost never discussed in the literature.

The PCR (or molecular method) step must employ a sequence-specific component (called a primer) to detect the pathogen's DNA. This "priming" can be tailored to detect sequences that are not specific to a particular organism (e.g., all bacteria when investigating sepsis), or it can be tailored to detect sequences that differ from each other by a single base (as is used when detecting drug resistance). The rationale for choice of genetic target is all too often missing from publications and is too frequently simply copied from another article with no explanation as to why.

Table 11–1 Example of experimental considerations that should be addressed within a molecular diagnostic article

Step	Discuss in publication
Sampling	Sample type, time of sampling, additional clinical data (age, sex, etc.), ethics
Storage	Temperature, buffer, dry, etc.
Extraction	Extraction method, especially any variation on manufacturer's protocol. Rationale for choice of method
Storage	Temperature, buffer, dry, etc.
Analysis	Assay type, characteristics (PCR efficiency, reproducibility, coefficient of variation, detection sensitivity, dynamic range), genetic target, rationale for choice of target, description of positive/inhibition control
Interpretation	Positive and negative criteria, rationale

2) The second fundamental problem with many diagnostics publications is a manifestation of the first. Although studies that do not take a stepwise approach may remain useful for meta-analysis, this is often prevented by the rhetoric and frequent overinterpretation of the findings. A study that takes a single clinical sample, uses a single extraction technique, and measures this for pathogen DNA using a single reaction has not comprehensively evaluated PCR in the context of whatever infectious disease the researchers happen to be investigating.

Yet it will not take you long to find numerous studies that purport to have "evaluated PCR" and established the "usefulness of PCR" or "value of PCR" with just a single sample, single extraction, and single set of PCR primer. When this is combined with the absence of a detailed stepwise approach, as described earlier, it leads to a plethora of different findings in the context of serious infections like TB. It is not surprising that the clinical world is skeptical about the value of these methods.

A molecular method may be fantastic at diagnosing a disease, but without clinical acceptance and trust it will not be used. For molecular methods to be more comprehensively assessed, a set of guidelines that details how a study was approached would be favorable. An example of how this could be approached is outlined in Table 11–1. This could be similar to Minimum Information About a Microarray Experiment (MIAME) guidelines that have been established for gene expression analysis[20] and would considerably assist the experimental design, conclusions, as well as third-party interpretation. Importantly, by addressing these points, the readers (who may not have detailed molecular experience) will be able to compare different studies.

ARE MOLECULAR METHODS OF ANY VALUE FOR DIAGNOSING DISEASES LIKE TB?

Two interesting studies, both conducted by Noordhoek and colleagues,[21,22] clearly demonstrate that molecular methods are capable of diagnosing TB. These studies also demonstrate that different molecular methods do not always work well and that the same methods in different laboratories may also give different results. Therefore, molecular methods can diagnose TB, but the choice of methods and techniques is essential for this to work well.

In the absence of international standards that are routinely available for many viral diagnostic tests, direct comparison between laboratories will always be limited when assessing diseases like TB as laboratories will be forced to work in isolation. The need for such standards is great, and it is encouraging that standards are being put in place for many infections (e.g., malaria) that could arguably benefit from molecular diagnosis.[23]

Using a molecular method to identify a pathogen that is not usually present is theoretically easy. If you are HIV positive, then detection of HIV by PCR confirms this; a negative result suggests that the patient is HIV negative. PCR type methods are good at this as they are sensitive, and yes/no conclusions in this context can be made.

More difficult is an assessment of viral quantity as conventional PCR is not good at quantifying nucleic acids. PCR methods are used worldwide for assessing HIV viral load, but to do this the method must be modified to provide an idea of quantity. This is done by quantitative PCR or through limiting the sensitivity of conventional PCR to detect only higher viral load.

Another major challenge when diagnosing or monitoring an infectious disease occurs when the pathogen is opportunistic and may occur asymptomatically. A good example of this is *Pneumocystis jirovecii,* a fungal pathogen that causes PCP in immunocompromised patients. The epidemiology of *P. jirovecii* is not clearly understood, yet the fact that this fungus shows mammalian host specificity at the species level suggests that it co-evolved with humans. What is clear is that it is possible to detect in healthy individuals using both immunological and molecular methods. Consequently, a positive result tells us that a patient has the fungus "on board" but little about whether the patient has PCP. However, by employing quantitative molecular methods it is possible to establish that a particular amount of genetic material is more associated with disease and to use quantitative PCR to diagnose PCP.[24]

PCR FOR TROPICAL DISEASE DIAGNOSIS

Malaria is the archetypal tropical disease and is second only to TB as the most common infectious cause of death. Malaria is caused by a protozoan and is another good example of a disease in which a PCR-based test could potentially be a useful diagnostic tool.[25] Malaria is blood borne, so sampling is comparatively easy. The gold standard method of diagnosis, by looking for the pathogen in the patient's blood sample by using a microscope, is inexpensive, rapid, and effective.

There is a difference of opinion as to the utility of diagnosing malaria with PCR, as the costs will need to be considerably reduced for it to be considered and the existing microscopic methods are argued to be sufficient.[26] However, as malaria is often misdiagnosed empirically, there may be a niche for molecular methods as their costs come down.[27] Certainly where the two have been compared, PCR has been more effective[28] and is used for monitoring parasitemia (number of protozoa in the blood) in clinical trials.[29]

This is just one example; if you name a tropical disease you will find an example of a PCR diagnostic study arguing for its potential utility. As with malaria, this will often have the counterarguments of cost, cold chain, and that, in the developing world, where the tropical diseases often occur, these methodologies are simply not currently feasible. These two juxtaposed discussions apply to all infectious diseases in the context of the developing world, especially TB.

THE ROLE OF PCR IN THE FUTURE OF TB DIAGNOSIS

TB is a good example of a disease in which PCR can be used to diagnose, test drug sensitivity, and investigate epidemiology. Yet as a disease of poor populations, most people who die of TB are undiagnosed in the developing world (where the most basic diagnostic tools are not available). Even in the developed world, PCR tests are relegated to special assessments that are difficult or impossible to do using any other method.

This situation is likely to change as the technology advances and many associated patents run out (see Chapter 1); PCR becomes less expensive by the day. Consequently, PCR management of TB could replace microscopy in the developed world simply for economic reasons. The initiation of nongovernment organizations (like the Foundation for Innovative New Diagnostics) that aid the development of small diagnostics companies on the condition that they provide their products to the developing world on a royalty-free basis will ensure that research and development considers the developing world. Consequently, the use of molecular methods in poorer parts of the world is no longer being automatically disregarded for financial reasons. Research into infectious diseases is becoming ever more applied with increased collaboration between academia and industry to maximize patient impact.

A number of focused technical advances have been aimed at simplifying PCR and making it more tailored to the developing world. Experimentation using lyophilized reagents has been used for a while and may provide a solution to the cold chain problem.[30] The idea of drying samples has been applied, introducing the idea that, although the detection assay may need to be advanced, the sample storage can be highly simplified.[31] The utility of a TB diagnostic test using isothermal loop-mediated amplification (LAMP) has been investigated specifically in the developing world,[32] as has the PCR-based strip test that allows fast speciation and assessment of drug resistance.[33]

Despite these advances making PCR and other molecular methods simpler, if these techniques are to have a direct impact on developing world countries it will only be in their more advanced health facilities. A patient with a cough in a rural part of Africa will often only ever get to the most regional of health clinics, where

PCR type methodologies are unlikely to ever impact directly. However, PCR has a fundamental role in the advancement of techniques that could be used in these more rural parts of the world.

PCR is currently enabling comprehensive research into new approaches to diagnosing infections. It is allowing detailed investigations into novel methods like targeting *M. tuberculosis* DNA in urine,[34] enabling such a phenomenon to be characterized. As newer molecular methods are developed, which will better lend themselves to simple tests[35] more suitable in rural Africa, it will be findings from current PCR-based research that will guide how these tests should be used, considerably speeding up the impact of such methodologies on infectious disease diagnosis and management.

CONCLUSION

PCR has had a major impact on infectious disease diagnosis and management. However, it is not ironic that PCR has been most successful in situations where alternative methods are either impossible or severely lacking. In the example of TB diagnosis, alternatives do exist, and there are numerous reasons why PCR is not currently used more routinely. Despite this, PCR has a fundamental role in epidemiology and drug resistance testing for TB and many other infectious diseases. It is likely that, as the costs are reduced and the methods are perfected, molecular diagnosis for diseases like TB will become a more common occurrence in diagnostic laboratories.

ACKNOWLEDGMENTS

I thank Dr. Tim McHugh, Mr. Robert Shorten, and Professor Mike Wren for their assistance with the figures. I also thank Drs. Agnieszka Falinska and Weronika Szczecinska for critical review of the manuscript.

REFERENCES

1. D. Wat, The common cold: a review of the literature. *Eur J Intern Med* **15**(2), 79 (2004).
2. J. F. Huggett, T. D. McHugh, and A. Zumla, Tuberculosis: amplification-based clinical diagnostic techniques. *Int J Biochem Cell Biol* **35**(10), 1407 (2003).
3. F. M. Perrin, M. C. Lipman, T. D. McHugh et al., Biomarkers of treatment response in clinical trials of novel antituberculosis agents. *Lancet Infect Dis* **7**(7), 481 (2007).
4. J. N. Zuckerman and A. J. Zuckerman, Viral Hepatitis. In: G. C. Cook and A. I. Zumla (eds), Manson's Tropical Diseases. Elsevier Science, London, 2003.
5. J. D. Barbour and R. M. Grant, The role of viral fitness in HIV pathogenesis. *Curr HIV/AIDS Rep* **2**(1), 29 (2005).
6. C. Drosten, M. Panning, J. F. Drexler et al., Ultrasensitive monitoring of HIV-1 viral load by a low-cost real-time reverse transcription-PCR assay with internal control for the 5′ long terminal repeat domain. *Clin Chem* **52**(7), 1258 (2006).

7. B. S. Gottesman, Z. Grosman, M. Lorber et al., Measurement of HIV RNA in patients infected by subtype C by assays optimized for subtype B results in an underestimation of the viral load. *J Med Virol* **73**(2), 167 (2004).
8. R. Eccles, Understanding the symptoms of the common cold and influenza. *Lancet Infect Dis* **5**(11), 718 (2005).
9. Writing Committee of the Second World Health Organization Consultation on Clinical Aspects of Human Infection with Avian Influenza A (H5N1) Virus, A. N. Abdel-Ghafar, T. Chotpitayasunondh, Z. Gao et al., Update on avian influenza A (H5N1) virus infection in humans. *N Engl J Med* **358**(3), 261 (2008).
10. L. Mody and S. Cinti, Pandemic influenza planning in nursing homes: are we prepared? *J Am Geriatr Soc* **55**(9), 1431 (2007).
11. K. C. Chapin, Molecular tests for detection of the sexually-transmitted pathogens *Neisseria gonorrhoeae* and *Chlamydia trachomatis*. *Med Health R I* **89**(6), 202 (2006).
12. M. B. Miller, Molecular diagnosis of infectious diseases. *N C Med J* **68**(2), 115 (2007).
13. D. I. Ling, L. L. Flores, L. W. Riley et al., Commercial nucleic-acid amplification tests for diagnosis of pulmonary tuberculosis in respiratory specimens: meta-analysis and meta-regression. *PLoS ONE* **3**(2), e1536 (2008).
14. D. I. Ling, A. A. Zwerling, and M. Pai, GenoType MTBDR assays for the diagnosis of multidrug-resistant tuberculosis: a meta-analysis. *Eur Respir J* **32**(5), 1165 (2008).
15. S. L. Stubbs, J. S. Brazier, G. L. O'Neill et al., PCR targeted to the 16S-23S rRNA gene intergenic spacer region of *Clostridium difficile* and construction of a library consisting of 116 different PCR ribotypes. *J Clin Microbiol* **37**(2), 461 (1999).
16. R. J. Van Den Berg, I. Schaap, K. E. Templeton et al., Typing and subtyping of *Clostridium difficile* isolates by using multiple-locus variable-number tandem-repeat analysis. *J Clin Microbiol* **45**(3), 1024 (2007).
17. P. Supply, S. Lesjean, E. Savine et al., Automated high-throughput genotyping for study of global epidemiology of Mycobacterium tuberculosis based on mycobacterial interspersed repetitive units. *J Clin Microbiol* **39**(10), 3563 (2001).
18. E. Crubezy, B. Ludes, J. D. Poveda et al., Identification of Mycobacterium DNA in an Egyptian Pott's disease of 5,400 years old. *C R Acad Sci III* **321**(11), 941 (1998).
19. J. T. Crawford, Genotyping in contact investigations: a CDC perspective. *Int J Tuberc Lung Dis* **7** (12 Suppl 3), S453 (2003).
20. A. Brazma, A. Hingamp, J. Quackenbush et al., Minimum information about a microarray experiment (MIAME) – toward standards for microarray data. *Nat Genet* **29**(4), 365 (2001).
21. G. T. Noordhoek, S. Mulder, P. Wallace et al., Multicentre quality control study for detection of Mycobacterium tuberculosis in clinical samples by nucleic amplification methods. *Clin Microbiol Infect* **10**(4), 295 (2004).
22. G. T. Noordhoek, J. D. van Embden, and A. H. Kolk, Reliability of nucleic acid amplification for detection of Mycobacterium tuberculosis: an international collaborative quality control study among 30 laboratories. *J Clin Microbiol* **34**(10), 2522 (1996).
23. D. J. Padley, A. B. Heath, C. Sutherland et al., Establishment of the 1st World Health Organization International Standard for *Plasmodium falciparum* DNA for nucleic acid amplification technique (NAT)-based assays. *Malar J* **7**, 139 (2008).
24. J. F. Huggett, M. S. Taylor, G. Kocjan et al., Development and evaluation of a real-time PCR assay for detection of *Pneumocystis jirovecii* DNA in bronchoalveolar lavage fluid of HIV-infected patients. *Thorax* (2008) Feb; **63**(2):154–9.
25. P. F. Mens, G. J. Schoone, P. A. Kager et al., Detection and identification of human Plasmodium species with real-time quantitative nucleic acid sequence-based amplification. *Malar J* **5**, 80 (2006).
26. T. Hänscheid and M. P. Grobusch, How useful is PCR in the diagnosis of malaria? *Trends Parasitol* **18**(9), 395 (2002).

27. P. Mens, N. Spieker, S. Omar et al., Is molecular biology the best alternative for diagnosis of malaria to microscopy? A comparison between microscopy, antigen detection and molecular tests in rural Kenya and urban Tanzania. *Trop Med Int Health* **12**(2), 238 (2007).
28. P. Bejon, L. Andrews, A. Hunt-Cooke et al., Thick blood film examination for *Plasmodium falciparum* malaria has reduced sensitivity and underestimates parasite density. *Malar J* **5**, 104 (2006).
29. L. Andrews, R. F. Andersen, D. Webster et al., Quantitative real-time polymerase chain reaction for malaria diagnosis and its use in malaria vaccine clinical trials. *Am J Trop Med Hyg* **73**(1), 191 (2005).
30. P. R. Klatser, S. Kuijper, C. W. van Ingen et al., Stabilized, freeze-dried PCR mix for detection of mycobacteria. *J Clin Microbiol* **36**(6), 1798 (1998).
31. U. Tansuphasiri, B. Chinrat, and S. Rienthong, Evaluation of culture and PCR-based assay for detection of Mycobacterium tuberculosis from sputum collected and stored on filter paper. *Southeast Asian J Trop Med Public Health* **32**(4), 844 (2001).
32. C. C. Boehme, P. Nabeta, G. Henostroza et al., Operational feasibility of using loop-mediated isothermal amplification for diagnosis of pulmonary tuberculosis in microscopy centers of developing countries. *J Clin Microbiol* **45**(6), 1936 (2007).
33. C. M. Quezada, E. Kamanzi, J. Mukamutara et al., Implementation validation performed in Rwanda to determine whether the INNO-LiPA Rif.TB line probe assay can be used for detection of multidrug-resistant Mycobacterium tuberculosis in low-resource countries. *J Clin Microbiol* **45**(9), 3111 (2007).
34. A. Cannas, D. Goletti, E. Girardi et al., Mycobacterium tuberculosis DNA detection in soluble fraction of urine from pulmonary tuberculosis patients. *Int J Tuberc Lung Dis* **12**(2), 146 (2008).
35. H. A. Ho, K. Dore, M. Boissinot et al., Direct molecular detection of nucleic acids by fluorescence signal amplification. *J Am Chem Soc* **127**(36), 12673 (2005).

12 Polymerase chain reaction and respiratory viruses

Ian M. Mackay

A BRIEF INTRODUCTION TO THE FEATURES OF RESPIRATORY VIRUSES AND THE ROLE OF POLYMERASE CHAIN REACTION IN DETECTING THEM

The principal detection of a virus in respiratory secretions tends to bestow upon it the colloquial title of a "respiratory virus." Human respiratory viruses are the most numerous and most highly diverse, organ-defined group of viruses that we currently know of; not surprising, considering their efficient method of transmission. They infect with greater frequency (infections per person per year) and with a broader coverage (annual number of infected people worldwide) than does any other infectious agent. To date, most acute respiratory tract infectious entities are known to be viruses, and most of these have a monopartite ribonucleic acid (RNA) genome. Viruses, being what they are, intimately associate with human cells and secretions, from which they must be discriminated. This can make them difficult to detect, let alone characterize. Most of the modern focus on respiratory virus detection is now drawn by molecular methods. At the forefront of these, both in research and routine diagnostic laboratories, is the polymerase chain reaction (PCR), a technique lauded for its ability to detect a target from among a far superior number of nontarget sequences.[1]

Data from the reinvigorated arena of respiratory virus research are pushing many new and old infectious disease issues to the forefront of microbiology, none more so than the question of what is required of our experimental design to determine whether a virus detected from the respiratory tract is the cause of respiratory symptoms. A similar question prompted Robert Koch, in 1890, to postulate ways to associate infection with disease.[2] Koch's postulates suggested that a true and nonfortuitous pathogen isolated and purified from a diseased host would induce disease anew upon infection of a healthy body, a moot point for most of the recently identified respiratory viruses that have yet to be isolated and propagated using in vitro culture. However, if the putative pathogen is known to cause severe disease, then it may be ethically unconscionable to reproduce that state in humans by intentional infection. There would be even less justification to replicate that disease in the major at-risk populations of neonates (0 to 1 month old), infants (1 to 12 months), toddlers (12 to 24 months), and children (2 to 14 years; Figure 12–1). We can use in vitro primary cell culture models for preliminary investigations of the molecular response to respiratory virus infection[3] and employ animal models to try and duplicate the disease process – a slow and artificial approach in which there may be only a tenuous link between data from the animal model and disease progression in a human host. Nonetheless, Koch's postulates were not meant to be absolutes, rather they were intended to prompt topical discussion about the best way to implement contemporary methods to identify and define associations between infection and clinical outcome. Our own studies recently led us to tweak already revised versions of the postulates[4] to better suit the needs of this field of virus research.[5]

Recent respiratory virus discoveries made through molecular studies (and there have been many in the past seven years from 2001 through 2007) have been even more difficult to definitively link with disease than were similar discoveries made

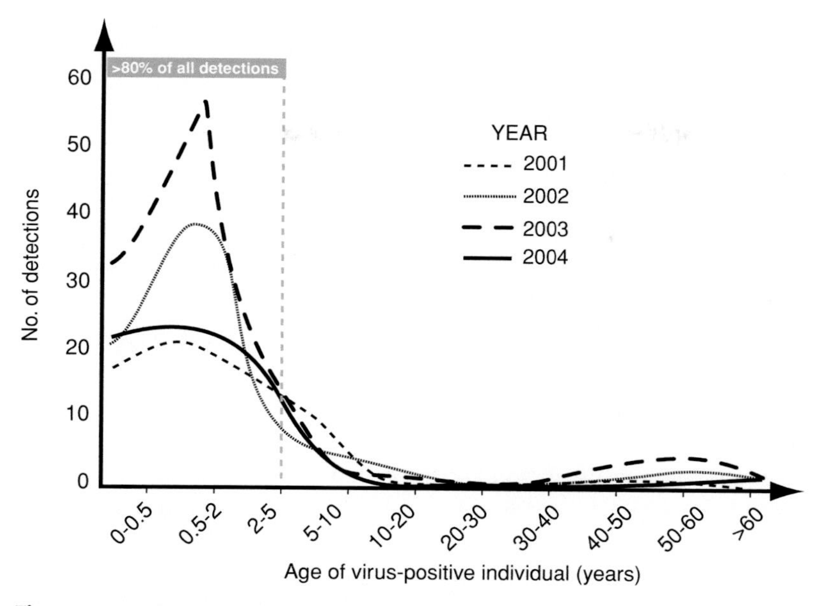

Figure 12–1. Reverse transcriptase–polymerase chain reaction (RT–PCR) detection of human metapneumovirus (hMPV) in a hospital-based population. Data were derived from a four-year, all-season epidemiological study of hMPV. More than 700 detections were achieved using RT–PCR to determine the circulation characteristics of the respiratory newly identified virus. Figure indicates a feature shared by many respiratory virus–positive populations – a prominent peak of detection among young children (>80% of positives occurred in persons younger than five years). Our studies indicated that nearly every variable that we examined (including predominant viral strain, peak age of positivity, month of peak infection, and frequency of detection) drifted from year to year.

during the 1930s when diagnostic screening meant infecting an animal[6] and in vitro cell and organ cultures later became the tools du jour. Unfortunately, PCR has also generated confounding data simply because the technique only detects nucleic acids. Such data do not inform us of the infectivity of their viral source. Additionally, PCR can only find that about which we already know something. Gene-specific primers need a specific target sequence and oligo-(dT) needs a poly(A) tract. Random hexamers, pentadecamers, or even highly degenerate primers work best in the presence of copious amounts of template bolstered by some luck – components not always available from the study of respiratory tract specimen extracts. Antigen-based detection systems and, at least for some viruses, in vitro culture methods are not limited by these same issues, although they do have their own problems. Approaches to antigen-based detection are stymied by an absence of high-quality reagents for all the targets of interest, and undamaged protein is required for detection by antibody. Culture requires inoculation of a suitable cell type with sufficient intact, viable, and cultivable virus, which can be difficult to achieve routinely. There is no dispute that PCR-based research tools have identified and contributed to teaching us much about many respiratory viruses. Innovative molecular methods have greatly improved the identification of fastidious viruses. Microarrays, high-throughout sequencing, and discovery methods such as virus discovery complementary deoxyribonucleic acid (cDNA)–amplified fragment length polymorphism (cDNA–AFLP; VIDISCA7); sequence-independent, single-primer amplification (SISPA[8]); and primer extension enrichment reaction (PEER[7]) have contributed to significant discoveries; but the cornerstone of these and many emerging molecular techniques is PCR. In its simplest, plain vanilla form, PCR has been the tool most responsible for a vast improvement in the rate of virus detection from individuals with symptoms of acute respiratory tract infections (ARTI). Despite these improvements and even when all likely respiratory viruses are sought by PCR (a rare event), more than a third of all (presumably correctly) diagnosed ARTI cases cannot be accounted for by infection with a known virus, and thus ill individuals, mostly children, return home or are admitted to hospital, without laboratory confirmation of a clinical diagnosis.

The remaining pages of this chapter are devoted to discussing beneficial aspects of using PCR for respiratory virus detection and characterization but also dwell on the disappointments and myths associated with the main features of its use. Here, PCR has been challenged by the targets and sampling environment like no other application of the technique to date. PCR will be the technique of choice for screening specimens should a true respiratory virus pandemic befall us in the near future, and rightly so. It then falls to us to ensure that we expeditiously address some of the serious issues raised by recent viral discoveries and outbreaks so that we can be better prepared to deal with, and even understand, future developments in an extremely large and recurring source of global morbidity.

WHY RESPIRATORY VIRUS PCR IS NOTHING TO SNEEZE AT

Sensitivity, specificity, and speed – these three words are the central tenet of using PCR in any field of microbiology. In particular, PCR has been most helpful in unveiling many new issues for the study of ARTIs. ARTIs are associated with a wide range of virus concentrations. Human bocavirus (HBoV) can be present in the respiratory tract in low numbers, requiring a sensitive diagnostic method for reliable detection. ARTIs are associated with many different clinical entities manifesting in both the upper and lower respiratory tract, including viruses that also have been detected in extrarespiratory tissues, such as HBoV in feces,[9–11] respiratory syncytial virus (RSV; normally respiratory) in blood,[12] and enteroviruses (normally enteric) in the respiratory tract.[13]

HBoV provides an example of an increasingly frequent problem for traditional diagnostic methods – an inability to isolate virus using in vitro culture methods. The most common reasons for failed isolation are the poor quality of specimen collection (suitability of the site and quantity sampled from it) and the lack of care used in specimen transport and handling during processing (poor temperature control and the absence of preprocessing specimen stabilization). These issues are not nearly as important for PCR because it only needs intact, amplicon-sized nucleic acid template, not infectious or even intact virus. Among all those respiratory viruses we can isolate, there remains the problem that cytopathic effects (due to replication of the desired virus in vitro) cannot always be distinguished from those caused by any other virus that also may be present in the same specimen, so an additional layer of diagnostic specificity is usually necessary. Such methods traditionally have relied on antibodies of which there are few compared to the number of distinctly circulating respiratory viruses. Additionally, the detection of two or more viruses in the one specimen is not easy to achieve using in vitro culture. Multiplex PCR, defined here to discriminate it from duplex PCR, is "the simultaneous amplification of three or more targets (including an internal amplification control) in the same reaction vessel." This approach has considerable theoretical promise for simplifying the task of respiratory virus screening. Reminded that PCR is our favored diagnostic tool, we will take a journey of necessary introspection to discuss more specifically some aspects of using PCR to detect respiratory viruses.

PCR IS THE BEST TOOL WE HAVE . . . BUT NOT THE PERFECT TOOL

This statement then begs the question, "What are the criteria for a perfect respiratory virus test?" The following list addresses some of the requirements of an improved method:

- Direct specimen testing without a need for additional handling needs to be used to minimize exposure of the specimen to the environment by reducing aliquoting, sampling, and streamlining nucleic acid extraction (if nucleic acids are required).
- The test must not be influenced by inhibitory substances coextracted with the nucleic acids.
- The virus identification system should have a reporting signal that is stronger than any current system and increases in proportion to any amplified target.
- The reporter should be coupled with an exceptionally sensitive signal detection system.
- The test should have the capacity to detect all relevant targets and be able to accept additional identification modules as new viral discoveries are made.
- No cross-reactivity should be seen with closely related targets or human genomic DNA.
- The reagents used should be nonradioactive and nontoxic and have long shelf lives.

In the broad field of respiratory virus research and clinical diagnostics, the success of PCR has raised issues of concern that have yet to be resolved. As with any technological advance, the pace of diagnostic change has left some in its wake, leaving little time for problem solving. These problems relate both directly and indirectly to the technique itself but also to the nature of the respiratory tract (Figure 12–2A), particularly the difficulties encountered when trying to standardize

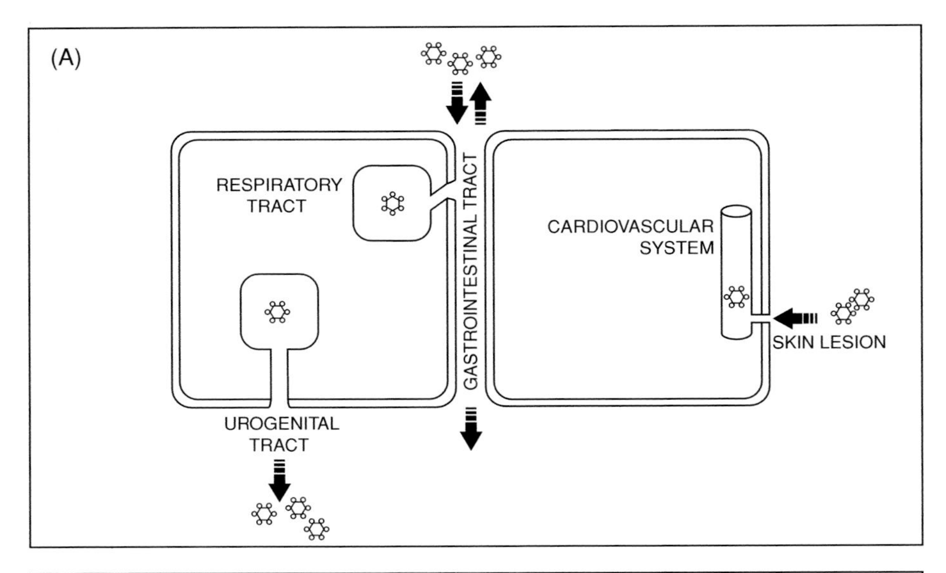

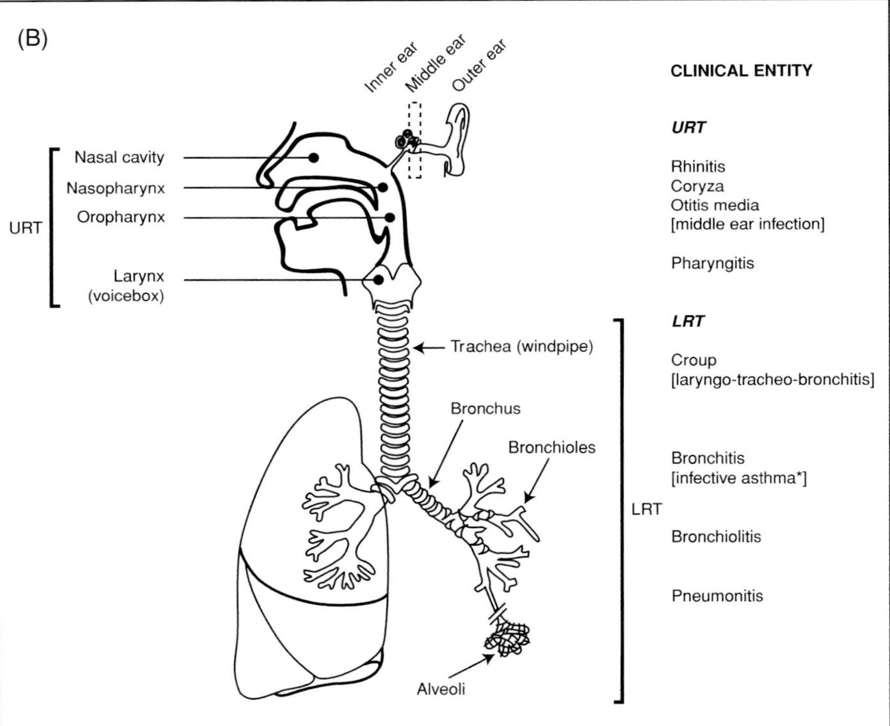

Figure 12–2. Schematic representation of human systems from which respiratory viruses have been detected. **(A)** Extremely simplified scheme of the human body and the various invaginations afflicted by viral replication that can be associated with symptoms of "respiratory virus" infection. Each site can be sampled and nucleic acids extracted for polymerase chain reaction (PCR) screening. **(B)** Upper and lower respiratory tract (URT/LRT) and components of the ear are indicated as are major anatomical sites of interest. Beside the schematic are the approximate locations of URT and LRT diseases associated with infection by respiratory viruses. *Recurrent attacks of shortness of breath and wheezing caused by spasmodic contraction of the bronchi attributed to infection. Part **(B)** reproduced with permission of the publisher. From Mackay IM, et al. (2007), in Real-Time PCR, pages 269–318. Caister Academic Press.

the collection of respiratory secretions for PCR quantification. Problems with the use of PCR and raised as a result of its use include:

- how best to sensitively detect the daunting number of human respiratory viruses;
- paucity of comprehensive sequences representing all likely viral strains of a given virus;
- the inherent and uncontrollable variability associated with the collection of respiratory specimens that has so far precluded true quantification (be it absolute or relative);
- poor understanding of the implications of subtle sequence variation on oligonucleotide hybridization and of the extent to which it occurs;
- incomplete knowledge of all the viruses circulating (there is no one to blame here; we just need to expend more resources on finding all the agents before we can hope to conduct comprehensive analyses seeking to link the detection of a virus to active infection and to clinical outcome);
- uncertainty due to the inability of multiplex PCR to detect all desirable targets with equivalent sensitivity (missed detections are not acceptable if we wish to answer complicated questions about infection-related respiratory tract conditions such as asthma);
- disappointing lack of standardization for diagnostic PCR, particularly quantitative PCRs (qPCRs), and the rarity of standardized commercial real-time PCR kits (both have restricted the generation of comparable data);
- confusion over which fluorogenic chemistries, if any, are best suited to diagnostic real-time PCR and whether some are more relevant to microbiology than to human gene studies;
- the scarcity and poor planning of clinical studies (frequently, studies are weakened by incomplete virus-screening panels, further limiting their power to define links between clinical outcome and viral load and between single and multiple detections, and inaccurately defining the impact of newly identified viruses (NIVs) on clinically "well" populations;
- the paucity of therapeutic options to employ when infection is suggested by a positive PCR; and
- our failure, in at least a third of inpatients and outpatients, to identify any microbial contributor to a clinically suspected infectious respiratory illness, even if we use PCR to test for every currently known respiratory virus.

It is interesting to note that issues of improved analytical and clinical sensitivity and speed are as often the focus of today's respiratory virus PCR publications as they were in the early 1990s when PCR was first used to detect RSV,[14,15] adenoviruses,[16] rhinoviruses,[17] influenza viruses,[18] parvoviruses,[19] and papillomaviruses.[20] In many respects, little has changed. Many of the answers to that daunting list of issues will rely on the generation of new data from additional research and by altering the way we think about some of the questions. Closer global collaboration is needed, together with long-term surveillance studies to best address these issues. One thing is certain – without PCR we would have no idea of just how little we know.

GOOD PCR ASSAYS NEED GOOD PRIMERS, AND GOOD PRIMERS NEED CHARACTERIZED SEQUENCE TARGETS

During the past few decades researchers have contributed numerous gene sequences and sequence fragments as well as an increasing number of complete respiratory virus genome sequences to public databases such as Entrez, the largest such repository (including GenBank and accessed through http://www.ncbi.nlm.nih.gov/entrez/query.fcgi?db=Nucleotide). These contributions have made the design of highly specific PCR primers much simpler and in turn have permitted rapid and global diagnostic responses to the appearance of newly identified but endemic (e.g., human metapneumovirus [HMPV], HBoV, human coronaviruses [HCoVs] NL63 and HKU1, and the Karolinska Institute [KI] and Washington University [WU] polyomaviruses) or newly emerged viruses (e.g., severe acute respiratory syndrome–coronavirus [SARS-CoV] or influenza virus strains). Accurate and rapid identification of emerging respiratory viruses is especially important because they have the potential to rapidly impart serious clinical consequences to a nonimmune population. Because PCR is largely our front-line diagnostic screening technique, it will be expected to perform quickly and reliably under the pressures of extremely high numbers of specimens accompanied by the enormous scientific, public, personal, financial, laboratory management, and political pressures that future infectious pandemics will bring to bear on every microbiology laboratory; a lot of pressure for two little pieces of DNA. Interestingly an unintended benefit from the completion of the Human Genome Project has been the potential for enhanced primer design by improving computer-based predictions of cross-reaction, a molecular diagnostic assay designer's archnemesis. Although an imperfect approach, such in silico studies are a useful guide to what might occur during the attempted amplification of viral gene sequences, and the results are usually more beneficial than misleading.

How do we make oligonucleotides into good primers?

In microbiology there are many answers to this question, not all of them well informed. Nonetheless, few would argue that familiarity with the chosen virus helps one select the target sequence because awareness of any notable sequence features can only fine-tune the choice of an amplification site. Paramyxoviruses transcribe their genes such that a concentration gradient forms with more transcripts from the genes at the 3′ genomic terminus than from those at the 5′ terminus of the RNA genome so one can target a PCR assay to the genes likely to be in greatest abundance. It is best to target conserved sequences for diagnostic assays. Choose a region not currently or ever likely to be under selective pressure that additionally exhibits good sequence stability at the nucleotide level, such as one encoding a polymerase that is often conserved to high levels in the taxonomic hierarchy. Some structural formations are essential to the replication of viruses and are conserved at the level of the nucleotide sequence (e.g., the 5′ untranslated region of picornaviruses that form stem-loop structures). In these instances, assay sensitivity and reliability can be enhanced by targeting primers to genes closer to the 5′ end of the genome as long as the secondary structures do not interfere with hybridization. Next, an alignment of all sequences in the target region of every known strain or variant permits preliminary identification of the location and extent of nucleotide variation and determines whether there exist certain intragenic regions that should be avoided completely or alternatively bracketed by primers to permit later genotyping using sequencing and phylogeny. Now the practical work can commence. First, use a positive control to optimize the reaction's yield. Next, screen as many more positives as possible and confirm these using another discreetly targeted PCR assay. Sequence variation occurring over distance can be identified during the in silico alignment process unless the target is an NIV, when few or no other sequences may be available. In this case, the assay designer must weigh the benefits of rolling out the assay quickly against waiting until more sequences are available upon which to base a more robust design. In the interim, new lineages and sublineages (genetic variants that have not yet been subjected to any phenotypic testing) are likely to be identified as additional laboratories bring their own quirks of assay design to bear on finding the virus in question. Data from the early adopter also will contribute to the global knowledge base – benefiting the slow and steady designer, but probably not his or her publication record.

One additional factor is just as important for a good primer: exclusive hybridization with the intended template. It is frequently difficult to find virus-specific sequence without some partial homology to human genomic DNA. In our experience, this can be more of a problem for certain viruses, such as the coronaviruses, than for others.

THERAPIES DRIVE BETTER DIAGNOSTICS, BUT RESEARCH PRIORITIZES THE TARGETS FOR DRUG DEVELOPMENT

The existence of a therapeutic option is a driving force for the development and commercialization of qualitative and qPCR assays in microbiology; virus detection can then have a tangible clinical benefit because a treatment can be given. Morbidity, and perhaps mortality, can be reduced, and the response of infection to treatment may be monitored, so dosage can be adjusted or treatment changed if the virus demonstrates an ability to develop resistance. Drugs provide a tangible reason to have the fastest and most sensitive and discriminatory molecular assays possible – and then to package them with standardized controls, and even with instruments, so that results can be compared between laboratories and countries. Unfortunately, many respiratory virus infections do not have specific therapeutics available, and there is both an equally poor range of commercial kits available and a limited variety of competitive options among the kits that do exist for a given viral target.

Another significant issue hindering kit developers is the extent of the expense borne by the company. This expense includes the cost of setting the assay up and the extensive sourcing of specific reagents as well as the lengthy validations, patenting costs, and licensing fees and the time and cost of obtaining approvals from the relevant regulatory bodies. These expenses ultimately must be borne by the kit purchaser, and the expense of a single kit reaction can far exceed that which can be achieved in-house. This may in turn mean that volume sales are low, resulting in slow investment return. There is clearly room for closer international collaborations and for enhancing existing quality assurance programs of the sort commonly conducted in Europe (e.g., the Quality Control for Molecular Diagnostics program [http://www.qcmd.org/]).

THE PROMISE OF PERFECT PCR: THE REAL-TIME REVOLUTION

The detection of PCR products as they accumulate has reduced many a scientist, from technician to professor, to square-eyed, slack-jawed, screen gazers. The almost instant gratification of seeing one's amplicon accumulate as it happens can be mind-numbingly fascinating; perhaps our parents were right and watching the "idiot box" really does rot our brains (or, at least, make our eyes turn square). For some, this panacea to the problems of conventional PCR in microbiology has seemingly brought their common sense to a similarly stupefying halt as we shall further explore in the following sections, which investigate this latest iteration of PCR.

Self-contained amplification means never having to say you contaminated your laboratory

Real-time PCR is described as a homogenous PCR method, meaning that after the reaction mix and template are added, the reaction vessel need not be opened again because the detection is performed through the wall of the closed tube by measuring fluorescent emissions; the completed amplification reaction simply can be discarded. There are exceptions to this process, which mostly apply during assay development or if a strange fluorescent curve appears, when it may be important to examine amplicon size to ensure that only the correct band is present. Although the presence of multiple bands is not the end of a primer set's usefulness, it does, at least by conventional wisdom, herald its failure for quantitative applications because inefficient amplification has occurred; more than one target has been draining the resources of the reaction, possibly with greater efficiency than the intended target. The number of assay designers who do not check the quantity and nature of their amplicons, instead relying solely upon threshold cycle (C_T) values, is both surprising and shocking. Perhaps real-time PCR has made life so easy for us that we have forgotten why some of the potential avenues for faster turnaround times are better left unexplored.

PCR was fast; real-time PCR is speedy

Real-time PCR does away with agarose gel electrophoresis. For many, that headline was sufficient to cause us to inter our old electrophoresis tanks. No more ethidium bromide disposal nightmares; no more tip packing, agarose and molecular-weight-ladder purchases; taping of gel trays; pouring of superheated agarose; and loading of endless wells. These tasks are akin to doing the cooking and the dishwashing; they must be done but it's the eating that is the good bit. For PCR, it is the result of the amplification, not the setup or gel running that is the good bit, and, with the advent of self-reporting amplicon detection systems, every night has become takeaway night. Here is where most of the time savings occur; real-time PCR assays do not even need to be run to completion before a strong virus-positive specimen can be identified. Further savings can be made using fan-forced air-heated thermal cyclers such as the LightCycler® (Roche, Indianapolis, IN); PCR performance is even faster, at rates approaching the native performance characteristics of the enzymes rather than the restricted pace of heating a slab of metal. Faster instrument designs permit sub–60-minute assays in the presence of suitably high template loads, and this is ideally suited to special-request testing – the sort of testing that can enhance hospital infection control, help differential diagnosis during epidemic or pandemic situations, and prevent both unnecessary invasive sample collection and the use of antibiotics.

Conventional PCR was exceptionally sensitive; real-time PCR... is too

A commonly discussed belief is that the addition of a fluorogenic oligoprobe will automatically improve assay sensitivity without regard for the primers or reaction conditions. Astonishingly, such a claim is sometimes made when employing two entirely different primer sets for the comparison (if indeed a comparison is performed at all), which may each yield different amplicon sizes and have quite different predicted melting temperature (T_m) values. Furthermore, there is usually no examination of the existence of primer, template, or amplicon secondary structures that might cause any difference in amplification efficiency. So, again, well-controlled data examining this issue are needed to substantiate claims of improved performance.

A more real problem introduced by real-time PCR is that of the aberrant kinetic curve. Once upon a time, the finding that a single band of DNA occurred at the expected size was enough to call the result, but the extra data produced by today's real-time PCR assays have changed that. A curve may have an early plateau, a late C_T cycle, a "funny" slope, or a poor linear phase. What do these all mean? The answers can vary and are often specific to the user, laboratory, assay, and perhaps even the phase of the moon. There is no doubt, however, that real-time

PCR is much more than just an enhanced amplicon detection system; it can quickly identify levels of inhibition and error previously unknown to PCR without considerable additional experimental investigation. This most useful feature of real-time PCR is related to analytical and clinical sensitivity but probably should be distinctly identified as "assay performance" sensitivity.

Making our own standards, or "why molecular diagnostic laboratories should not clone"

Cloning, or in this context, the addition of a PCR target and perhaps an additional sequence, to a plasmid vector, resulting in a renewable resource of DNA that can be more easily quantified than culture-derived material (particularly for NIVs that cannot be cultured) and used for positive control, internal amplification control, or as a template for the in silico production of RNA. However, cloning methods produce massive amounts of DNA ($>10^{10}$ copies/μL) compared to the amount of target in a patient's specimen. It will take very little time before that DNA starts to appear as a false positive or two, especially in a high-throughput laboratory environment. It is then a short step to total assay failure resulting from consistent contamination. Although they may not be commonly undertaken in close proximity, cloning endeavors can work in harmony with the high-throughput laboratory next door, but the time it takes to fix and track down contamination, should it occur, can bring a clinical molecular laboratory to its knees. Every hour of delay causes specimen backlogs, increased workloads, and frayed tempers. Time also can be lost if new oligonucleotides need to be purchased. In the worst case, an entirely new assay may have to be designed, optimized, and validated, and the laboratory will have to withdraw the current test from its diagnostic menu. An idealized laboratory design is presented in Figure 12–3. It is just one component of an approach that minimizes amplicon carryover contamination while still permitting the manufacture of synthetic PCR templates to help improve result quality. In effect, an off-site facility is used to perform the high-titer work, and only suitably dilute reagents emerge from that facility. Such an approach may be useful to overcome an absence of commercial controls and for live virus templates.

Cost is absolutely relative

Yes, a real-time thermal cycler is more expensive than a conventional thermal cycler. Yes, a fluorogenic oligoprobe is more expensive than no oligoprobe or than the intercalating, nonspecific fluorogenic dyes, but these issues can be raised out of context. The instrument is an investment by the laboratory and one that is often repaid quickly in a pay-per-test laboratory environment where it is used to accurately obtain more results, in a shorter period of time, with less handling. Even in a research laboratory, the savings in time can offset the cost outlay. Additionally, the extra costs of gloves, agarose, ethidium bromide disposal, gel-loading tips, buffers, and molecular-weight ladders for conventional amplicon

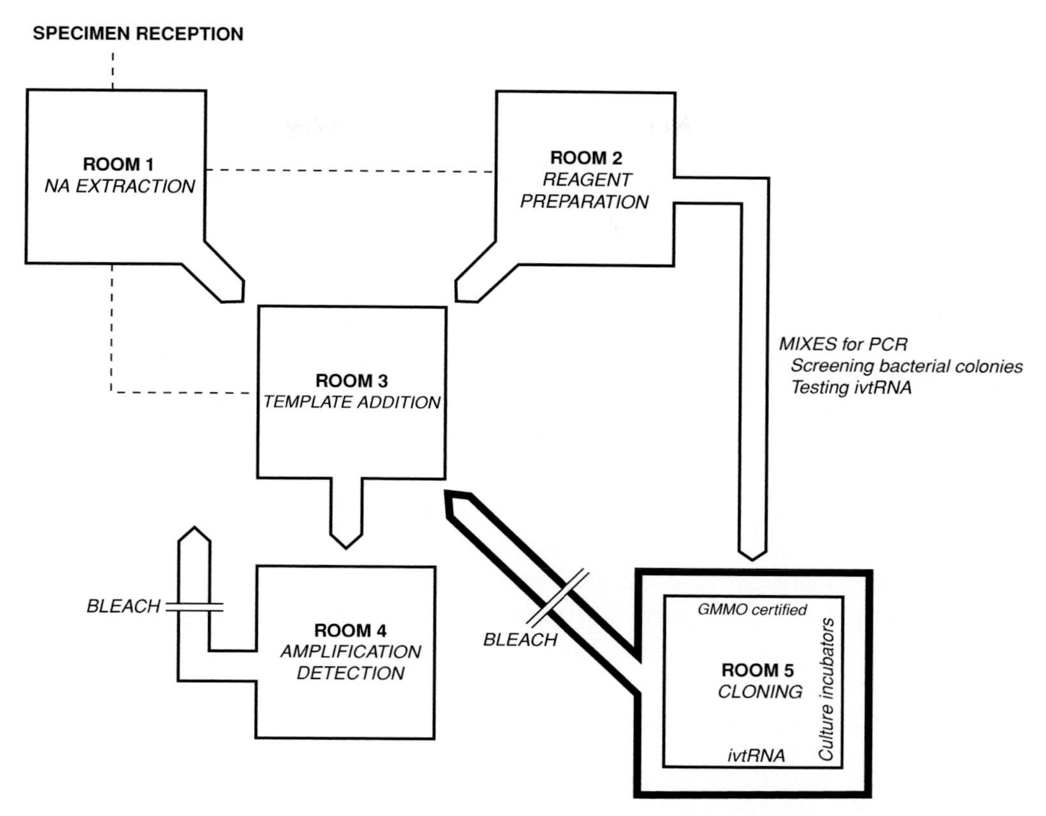

Figure 12–3. A proposed five-unit idealized polymerase chain reaction (PCR) suite. The suite incorporates a physically separate room (indicated by bold outline surrounding Room 5) for molecular cloning of amplicons and other genetically modified microorganisms (GMMO) and for the production of synthetic control materials including in vitro transcribed RNA (ivtRNA). The extraction of nucleic acid (NA) templates (Room 1) is also separated from the reagents and reaction mix preparation area (Room 2) and from the template addition area (Room 3). Solid arrows indicate unidirectional workflow (during any person's shift) and airflow to prevent amplicon being transported back into Rooms 1, 2, or 3. Bleach baths are used to remove amplicon from returning racks. Dashed lines indicate communication paths essential for notifying different rooms when tasks that require action will arrive, for example, how many specimens are coming, what reaction mixes they will require (after the lodgment processes have been completed), which specimens will require urgent testing, and which may be batched.

detection more than outweigh the cost of a single aliquot of a common oligoprobe chemistry. Glass capillaries and more complex fluorogenic oligoprobe formats alter that balance slightly, but not significantly.

Chemistry: will this be on the exam?

Florescence has proven itself to be the guiding light in a wilderness of diagnostic clinical PCR. To detect respiratory viruses, a specific chemistry is the best choice (i.e., one or more sequence-specific oligonucleotides labeled with a fluorescent moiety that will produce a signal only after hybridization to its specific target). The choice of chemistry can be daunting, however. The best rule of thumb is to keep things simple and choose something with a good history of functionality in the intended role. That narrows the choice considerably: hydrolysis probes (also known as TaqMan® [Applied Biosystems, Foster City, CA] or 5′ nuclease probes). These chemistries have a dominant presence in the relevant literature (validated by extensive commercial support and in-house expertise) and have been kept fresh by incorporating developments in the field such as the marriage of a minor grove binding (MGB) moiety to the TaqMan® oligoprobe to create a chemistry well suited to mutation detection. In microbiology, these TaqMan–MGB® probes are useful for maintaining suitably high hybridization temperatures when used in regions of low guanine–cytosine (GC) content, but if unknowingly subjected to sequence variation, these new probes are easily disturbed from their hybrid state reducing their diagnostic effectiveness.

The addition of "dark" quenchers has also freed up some spectral space, most commonly described as being of benefit for the next generation of real-time multiplex PCRs. This small benefit arose because dark quenchers do not emit captured reporter emissions as fluorescence of another wavelength (thus freeing up a portion of the spectrum for an additional reporter dye), but rather as heat. Sadly, other issues have held up the success of multiplex assays. However, the dark quenchers do reduce "leaky" fluorescence emitted by earlier quenchers, and this improves the signal-to-noise ratio, permitting more sensitive (earlier) detection of reporter fluorescence.

qPCRs – Why most are not

For a number of reasons, PCR data, especially those touted as quantitative, can be difficult to replicate. Real-time qPCR assays must include the capacity to correct for variations caused by a range of factors. Without this capacity, qPCR data are only semiquantitative, and at worst they produce biologically irrelevant results.[21,22] The continued production of data that cannot be compared between laboratories, except in the broadest sense, should sound alarms because poor reproducibility can damage research collaborations and diagnostic quality assurance programs and ultimately confound our attempts to better understand the interaction between microorganism and host.[23] Unfortunately, it remains rare to find well-controlled real-time qPCR assays among those in the clinical microbiology literature. Why is this so? Controls have been commonly used for human genetic applications of real-time qPCR; can their use to detect viruses in the human respiratory tract be so different? Indeed it can, it is, and so far these differences have meant that accurate virus quantification from samples of the respiratory tract, as is commonly performed for human genes or blood-borne viruses from samples of blood or solid tissue is currently unachievable. This disparity can most simply be attributed to the lack of a single target against which we can normalize our input nucleic acid quantity and is probably a contributor to the limited number of commercial kits.

We already include a nonparticipating (or "passive") internal reference to overcome fluorescence changes caused by reaction mix composition, mix volume variation, or nonspecific quenching, but the passive reference neither determines template quality and/or enzyme inhibition nor accounts for variation in nucleic acid amount. These require a preanalytical process, but even such a process will not address the biggest hurdle: How do we interpret our results if we have no idea of the amount of nucleic acid we started with? For example, Patient 1 has 10^8 copies of virus X in her bronchoalveolar lavage and Patient 2 has 10^3 copies, but Patient 1 may have had more cells collected than Patient 2 had (because of inflammation unrelated to the target virus, because of a disease process, or due to chance), or less fluid retrieved (effectively increasing the cellular concentration). Without a suitable normalizing control, we're stymied. Viruses further complicate matters by residing both within and external to cells, making a cell count or the use of an endogenous "housekeeping" gene target of little value as a normalizer.

The concern here is that we generally apply real-time qPCR methods better suited to the study of human transcription instead of considering the variables that set clinical microbiology apart. The major implication of inaccurate qPCR assays is a slowing of our understanding of infectious disease etiology. Because we live in a time when the discovery of newly emergent or previously unknown endemic pathogens is increasing in frequency, we must strive harder than ever before to expand our understanding of infectious diseases, and for that we need trustworthy, organism-specific results from our virus-specific qPCR tools before we can conduct suitably comprehensive clinical studies.

Size may matter

The theory goes that an amplicon about the size of two primers and an oligoprobe is the best length to aim for when using real-time PCR. In reality, the differences in amplification efficiency for longer amplicons, even containing some secondary structure, are likely to be so small as to be of little significance for qualitative real-time PCR applications. Size may even be a negligible concern for quantitative applications; data addressing the issue are far from extensive or convincing.

An important size constraint, especially for respiratory virus quantification, is the need to place primers so as to minimize cross-reaction with ever-present human genomic DNA. The size of an amplicon also must be influenced by the need for primers to bind to sites of minimal secondary structure and suitable GC content and to hybridization sites that are suitably conserved (i.e., not the subject of inter-strain sequence variation). All these reasons can mean that primers have to be placed much farther apart than the constraints of "best practice" oligonucleotide design would advocate; ultimately it is impractical to apply stringent rules to amplicon length for virus-detection assays.

IDENTIFYING THE EMERGING CHALLENGES OF NIVs

Many of us lucky enough to work in the gray area between routine virus diagnostics and pure virology research have in recent years been hunting for and then characterizing the swag of agents we feel are yet to be described, but which have probably been endemically contributing to the global infectious respiratory disease burden for many, many years. An example is HCoV-NL63, first described in 2004 and proposed to have first diverged from an ancestor virus approximately 900 years ago. Our group has categorized such findings as NIVs so as to differentiate them from emerging viruses the appearance of which is more closely tied to a zoonotic event.[24,25] NIVs are found to circulate endemically but require modern molecular techniques for their identification due to pernickety or as yet unknown in vitro propagation requirements. How long a virus remains categorized as an NIV is a matter of debate. Also, we must strive to better understand what impact subtle sequence variation will have on the reliability of molecular diagnostic techniques because this is unknown for the RNA and DNA NIVs. Here we refer to both quasispecies variation and inter-host strain variation, currently considered to be two different beasts. Generally speaking, quasispecies constitute a swarm of subtly different sequence variants within a single specimen that may act as a reservoir of evolutionary potential. They contain the capacity for future development of antiviral resistance and can act as a pool of phenotypes containing the capacity to quickly escape host-induced immune pressures. As discussed previously, such sequence variation can be avoided during assay design if we have some idea of what gene(s) are subject to variation, but for the NIVs, at least during the early research, it is the nature of the virus, its epidemiology, and its clinical impact that are of interest to investigators rather than which genes vary under selective pressures.

It is disappointing that the gaps in our understanding also extend to some of the classical respiratory viruses, even decades after their description. For example, we have yet to contribute in a significant way to the understanding and molecular characterization of the intra-strain variation, epidemiology, seasonal recurrence, and epidemic peaks of the human rhinoviruses. These viruses constitute a dauntingly large pack of distinct viruses still considered by most to be a single entity usually referred to as "the rhinoviruses" and frequently dismissed as unimportant players in infectious disease.

Raging infection or just gene detection?

An interesting question raised by the recent flurry of NIV data is what impact, if any, does the infrequent culture isolation of NIVs have on their characterization? The literature is increasingly citing the detection of viruses from an individual's specimen extract and too frequently labeling that as an infection. Pedantically, an infection describes, in this case, viral invasion of an individual who then hosts the replication of that virus without regard to the host's clinical status. All we can describe using PCR is the *detection* of nucleic acids from the virus target. The sensitivity of PCR is theoretically as low as a single input copy, and this has caused some researchers to claim that qualitative PCR is too sensitive, detecting virus when it is unlikely to be the cause of symptoms. Some examples of possible causes of such false-positive results that are not due to amplicon contamination

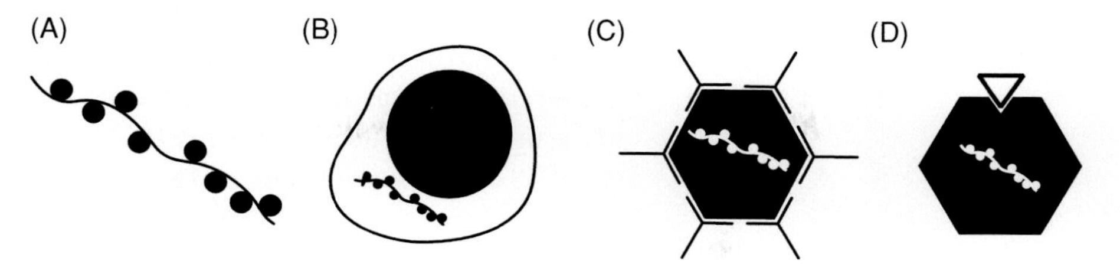

Figure 12–4. False-positive virus polymerase chain reaction (PCR). Four examples in which viral nucleic acids may still produce a positive PCR signal, despite being noninfectious and therefore unable to replicate in tissue culture. **(A)** Extracellular viral nucleic acids protected from degradation by viral proteins. **(B)** Virus nucleic acids phagocytosed, endocytosed, or left over within a cell as a result of abortive replication. **(C)** Opsonized (antibody-coated) virus particles that are unable to enter cells via specific receptors and therefore would be negative by culture. **(D)** Virus that has been inactivated by small-molecule antiviral drugs designed to interfere with virus binding or entry into target cells.

are presented in Figure 12–4. Nonetheless, it is this sensitivity that has caused an overall increase in detection frequency for any given respiratory virus because PCR does not rely on infectious particles to be present in the specimen, nor does it require a specific cell line to accommodate fastidious tropism; in vitro cell culture systems do, which is largely why PCR outperforms culture. This factor is especially applicable to the NIVs because they cannot be grown using in vitro culture (or at least not in a timely and efficient manner). By breaking the link between detection and infection, we also lose an important clinical predictor of pathogenicity: that the virus is actively replicating in the ill host and, simply due to this foreign activity, is more than likely causing some form of disruption to the host's clinical state. More and more viruses are being identified, but our ability to define their role in disease causation is quite limited. Thankfully, another tool has begun to show potential to fill in some blanks. Microarrays may be used to categorize the host's response patterns upon and during infection, serving to identify the clinical effect of the presence of certain microorganisms.[26]

You say co-infection, I say codetection

If we cannot be certain of the impact of a positive respiratory virus PCR, then we are entirely bamboozled by the role of two, three, or four positive virus detections in a single patient's specimen extract, an increasingly common occurrence now that nonbiological detection systems are in play. The study of NIVs has been intimately associated with PCR, and three viruses in particular (KI and WU polyomaviruses and HBoV) provide some good examples of the importance of PCR for identifying multiple agents in single human specimens.

It is important to clarify that codetections occur among all respiratory viruses; the NIVs are not unique in this regard. Surprisingly, this fact is not commonly relayed by studies (our own included) citing increased rates of codetection. In fact, one could be forgiven for understanding that each new virus accounts for, or "clears up," a significant portion of the respiratory specimens previously lacking a laboratory diagnosis. The NIVs we know of to date have provided only a

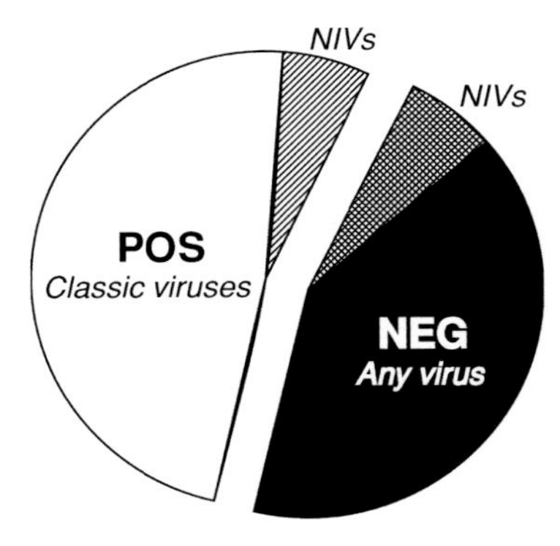

Figure 12–5. Data derived from the Brisbane respiratory virus research study.[27] The fraction of all polymerase chain reaction (PCR)-negative and -positive specimen extracts attributed to sole detection of newly identified viruses (NIVs). Only one sixth of all previously negative extracts were accounted for by screening for four newly identified respiratory viruses. Many more were cleared up by including more classic viral targets.

fractional contribution to the clear-up rate (Figure 12–5). This fact also provides some indirect data to bolster what some researchers in the field already take for granted; there *must* be more respiratory viruses to find because the current rate of clear-up described in our example would require the identification of many more viruses before every suspected ARTI specimen had at least one virus detected in it. (For the purposes of this review, I am not mentioning bacteria, which is a deliberate simplification; however, the potential for future bacterial discovery is likely to be equally rosy). Some ill populations, such as expiratory wheezers, already have high clear-up rates, at least in terms of single detections, whereas others, such as hospital-based populations comprising in- and outpatients, have quite poor clear-up rates.

Accumulating data from more comprehensive respiratory virus studies that employ larger PCR panels indicate that multiple detections are more common than we used to think; their occurrence may eventually become the norm rather than the exception if these findings continue. How many more viruses could we have missed? If current detection formats ever prove to be reliable markers of active infection, then one can foresee a time when it is found that the sheer burden of the microbial mix, and perhaps the order of infection, tips the balance from healthy host to ill patient. If this is the case, then the number of currently unknown viruses likely to be involved is extremely large; we will not be looking at single detections (but at multiple detections) to improve clear-up rates. The role of PCR in these continuing studies will be essential to the value and completeness of their conclusions.

The KI and WU polyomaviruses

Two newly identified, discrete polyomaviruses (KI and WU) were independently described in 2007. PCR-based techniques, augmented in the case of WU virus by microarray technology, were used to discover, characterize, and describe the preliminary epidemiology of both. To date, neither virus has been isolated in culture nor has any recombinant protein been described. Proteins are used to identify the seroprevalence of a virus within ill or healthy populations. Among other things, such data can indicate at what age humans first acquire a virus infection. Using PCR, we can extrapolate the most at-risk population from the highest rate of positivity.

Both polyomaviruses have relatively low detection frequencies (no more than 1% of tested specimens to date), which may downplay their importance in the eyes of some researchers. Nonetheless, PCR also has quickly identified additional instances or variants of these and other NIV genomes during molecular epidemiology studies, serving to confirm that the viruses circulate among the population. This further reassures those scientists who have concerns about our reliance upon PCR-based methods for virus discovery. In particular, concerns have been raised about sequences that may have been derived from PCR contaminants or that could have arisen from mixed amplicons so that the newly identified "virus" is nothing more than a molecular artifact. Whether conducted on prospectively collected or retrospectively stored specimen extracts, these NIV epidemiology studies also suggest the at-risk populations, the peak season of circulation (useful for studies in geographically removed and temporally shifted populations, i.e., other laboratories), and whether the virus recurs each year at the same testing site. When combined with phylogeny studies employing sequences obtained from multiple complete coding regions of local strains, complete genome sequence, and in silico–derived structural predictions, researchers can, without isolating the virus in vitro, build a convincing picture of a novel entity. However, for some researchers, the extent of the convincing will be directly proportional to their willingness to see beyond current or previous paradigms.

HBoV

At the time of this writing, HBoV had been detected by PCR in 0% to 19% of respiratory specimens (averaging 7%) from an equally broad variation in study populations[27,28]; the highest values generally occur during studies of hospitalized young children, from specimens collected during months of peak infectious respiratory disease. The lowest values occur among adults. When we heard of HboV, we employed PCR methods and rapidly detected HBoV DNA in 5% to 6% of specimen extracts from Australia.[29,30] These positive specimens also contained other viral nucleic acids in approximately two thirds of all detections, considered to be an unusually large fraction. Perhaps obviously, these codetections mostly occurred with viruses displaying overlapping peak seasons or affected age groups. Our studies exemplified, at least to us, how respiratory NIV identification, sequencing, phylogeny, and epidemiology studies could be quickly assembled using PCR. The original description of HBoV reported a codetection frequency of 18%, and, after our study, reports appeared describing rates as high as 90% (averaging 42%).[31] Higher values occurred from studies using more inclusive diagnostic screening panels and reliable PCR assays rather than classical techniques. Intriguingly, recent studies have identified HBoV in acute- and convalescent-phase sera and in fecal extracts. This identification has strengthened the case for systemic and perhaps persistent infection by this virus.[28,31] Today's thinking, for the most part, associates disease with infection by one virus (e.g., HRSV is often described as the most common cause of bronchiolitis and pneumonia among young children).[32] Past studies (on which such thinking is based), however, relied on biological detection systems (less sensitive) and were conducted without knowledge of, and thus testing for, the NIVs (less comprehensive). The relatively extreme rates of codetection and the detection of HBoV in multiple, anatomically distinct sites understandably cloud the link between HBoV detection and disease causation. Nonetheless, there is precedent suggesting that suspected co-infection is associated with more serious clinical outcomes[33–35] and also that "respiratory viruses" can be detected beyond their principal realm.[36,37]

Reports associating serious outcomes with the detection of any NIV, including HBoV, are relatively common. Unfortunately, they frequently do not describe PCR testing for all likely respiratory pathogens and potentially render many conclusions meaningless. Clinical features in HboV-positive patients cannot be differentiated from those features attributed to infection by many other respiratory viruses, so laboratory diagnoses using PCR play a critical role in any comprehensive diagnostic process.

Control studies including asymptomatic or differently symptomatic populations report no or few HBoV detections from asymptomatic populations, but clinical follow-up is not described, so subsequent development of symptoms is missed.[9,28,38,39] We should be cautious of overinterpreting these PCR-positive results; detection of HBoV (or any respiratory virus for that matter) in the presence of subclinical illness does not, alone, assign the target to an insignificant role in disease. It is possible that true infection causes only transient and/or very mild illness in an immune host. Such an illness may have gone undetected in past clinical studies either because the criteria for asymptomatic cases were biased against the collection of mild illness data[13] (e.g., a headache or allergy-like symptoms such as watery eyes or a stinging nostril) or because symptoms developing after sample collection or before study enrolment were not identified because of a lack of clinical follow-up or collection of a pre-enrolment clinical history. These infections will be PCR positive but may be clinically categorized as false positives, enforcing the belief that PCR has a low positive predictive power. This is conjecture of course, but conjecture that it is hoped will be examined in the future. Because further applied virus research (e.g., vaccine or antiviral developments) is often instigated on the basis of associated symptom severity, we have to ensure that clinical study conditions are such that we make those judgments using the most comprehensive data and do not leap to conclusions because of poorly designed or enacted studies that cast doubts on both the performance of molecular assays and the role of a given virus.

CONCLUSIONS

Future developments in respiratory virology (from better virus detection to improved microarrays, enhanced differential diagnostic capacity [benefitting future pandemic preparedness], to sturdier clinical study design) are contingent upon us quickly identifying and characterizing NIVs and each and every emergent virus. PCR is clearly our foremost tool for analyzing viruses, but the highly specific hybridization that is the hallmark of PCR is proving both a blessing and a curse. For PCR to work at its best in a clinical microbiology capacity, oligonucleotides must be designed to hybridize efficiently to their targets, so we need some advance knowledge of these targets. This approach clearly has diminishing returns in the current era of virus discovery. The field is large and growing, and the issues of concern for determining the etiology of infectious respiratory disease, not just related to PCR, are numerous. This may be why, after more than three quarters of a century of concerted study, the most widespread and frequent causes of human infections – the respiratory viruses – are proving to be as confounding as ever. The latest discoveries are posing questions faster than we can answer them. If not for PCR, we certainly would have less of an understanding of the scope of the tasks that lie ahead of us. PCR has single-handedly helped us shake up perhaps the largest field of human infectious disease and in the process given us new viruses to study and an ongoing capacity to look at all new issues. Viva la PCR revolution!

ACKNOWLEDGMENTS

As ever, the tireless patience of my wife, Dr. Katherine Arden, is gratefully acknowledged, in particular for her contribution to reviewing this work but also, along with Peter McErlean, for countless discussions of the issues herein. John F. Mackay is thanked for access to his fathomless knowledge, particularly on PCR-related topics, and Dr. Michael Nissen for imparting pivotal clinical concepts during my short time studying the viruses of ARTI. Additional thanks go to Corin Mackay for two years of joy and many, many personal experiences with ARTIs. Many thanks also to the creative forces behind my current vices (caffeine, chocolate, and SG-1) and to the Royal Children's Hospital Foundation for continued funding, support, and vision.

REFERENCES

1. Persing DH, Landry ML (1989) In vitro amplification techniques for the detection of nucleic acids: new tools for the diagnostic laboratory. The *Yale Journal of Biology and Medicine* **62**: 159–171.
2. Falkow S (2004) Molecular Koch's postulates applied to bacterial pathogenicity – a personal recollection 15 years later. *Nature Reviews. Microbiology* **2**(1): 67–72.
3. Wark PAB, Johnston SL, Bucchieri F, et al. (2005) Asthmatic bronchial epithelial cells have a deficient innate immune response to infection with rhinovirus. *Journal of Experimental Medicine* **201**(6): 937–947.
4. Fredricks DN, Relman DA (1996) Sequence-based identification of microbial pathogens: a reconsideration of Koch's postulates. *Clinical Microbiology Reviews* **9**(1): 18–33.
5. Mackay IM, Arden KE, Nissen MD, Sloots TP (2007) Challenges facing real-time PCR characterization of acute respiratory tract infections. In: Real-Time PCR in Microbiology: From Diagnosis to Characterization, pages 269–318. Norfolk: Caister Academic Press.
6. Smith W, Andrewes CH, Laidlaw PP (1933) A virus obtained from influenza patients. *Lancet* ii: 66–68.

7. Biagini P, de Lamballerie X, de Micco P (2007) Effective detection of highly divergent viral genomes in infected cell lines using a new subtraction strategy (primer extension enrichment reaction-PEER). *Journal of Virological Methods* **19**: 106–110.
8. Ambrose HE, Clewley JP (2006) Virus discovery by sequence-independent genome amplification. *Reviews in Medical Virology* **16**: 365–383.
9. Maggi F, Andreoli E, Pifferi M, Meschi S, Rocchi J, Bendinelli M (2007) Human bocavirus in Italian patients with respiratory diseases. *Journal of Clinical Virology* **38**(4): 321–325.
10. Lau SKP, Yip CCY, Que T-L, et al. (2007) Clinical and molecular epidemiology of human bocavirus in respiratory and fecal samples from children in Hong Kong. *Journal of Infectious Diseases* **196**(7): 986–993.
11. Vicente D, Cilla G, Montes M, Pérez-Yarza EG, Pérez-Trallero E (2007) Human bocavirus, a respiratory and enteric virus. *Emerging Infectious Diseases* **13**(4): 636–637.
12. Rohwedder A, Keminer O, Forster J, Schneider K, Schneider E, Werchau H (1998) Detection of respiratory syncytial virus RNA in blood of neonates by polymerase chain reaction. *Journal of Medical Virology* **54**: 320–327.
13. Parody R, Rabella N, Martino R, et al. (2007) Upper and lower respiratory tract infections by human enterovirus and rhinovirus in adult patients with hematological malignancies. *American Journal of Hematology* **82**(9): 807–811.
14. Okamoto Y, Shirotori K, Kudo K, et al. (1991) Genomic sequences of respiratory syncytial virus in otitis media with effusion. *Lancet* **338**: 1025.
15. Paton AW, Paton JC, Lawrence AJ, Goldwater PN, Harris RJ (1992) Rapid detection of respiratory syncytial virus in nasopharyngeal aspirates by reverse transcription and polymerase chain reaction amplification. *Journal of Clinical Microbiology* **30**(4): 901–904.
16. Allard A, Girones R, Juto P, Wadell G (1990) Polymerase chain reaction for detection of adenoviruses in stool samples. *Journal of Clinical Microbiology* **28**(12): 2659–2667.
17. Gama RE, Hughes PJ, Bruce CB, Stanway G (1988) Polymerase chain reaction amplification of rhinovirus nucleic acids from clinical material. *Nucleic Acids Research* **16**(19): 9346.
18. Zhang W, Evans DH (1991) Detection and identification of human influenza viruses by the polymerase chain reaction. *Journal of Virological Methods* **33**: 165–189.
19. Salimans MMM, Holsappel S, van de Rijke FM, Jiwa NM, Raap AK, Weiland HT (1989) Rapid detection of human parvovirus B19 DNA by dot-hybridization and the polymerase chain reaction. *Journal of Virological Methods* **23**: 19–28.
20. (1989) Human papillomaviruses and the polymerase chain reaction. *Lancet* (8646): 1051–1052.
21. Dheda K, Huggett JF, Chang JS, et al. (2005) The implications of using an inappropriate reference gene for real-time reverse transcription PCR data normalization. *Analytical Biochemistry* **344**(1): 141–143.
22. Tricarico C, Pinzani P, Bianchi S, et al. (2002) Quantitative real-time reverse transcription polymerase chain reaction: normalization to rRNA or single housekeeping genes is inappropriate for human tissue biopsies. **309**(2): 293–300.
23. Hoorfar J, Cook N, Malorny B, et al. (2003) Making an internal amplification control mandatory for diagnostic PCR. *Journal of Clinical Microbiology* **41**(12): 5835.
24. Taylor LH, Latham SM, Woolhouse MEJ (2001) Risk factors for human disease emergence. *Philosophical Transactions of the Royal Society of London Series B, Biological Sciences* **356**(1411): 983–989.
25. Woolhouse ME, Gowtage-Sequeria S (2005) Host range and emerging and reemerging pathogens. *Emerging Infectious Diseases* **11**(12): 1842–1847.
26. Relman DA, Falkow S (2001) The meaning and impact of the human genome sequence for microbiology. *Trends in Microbiology* **9**(5): 206–208.
27. Lin F, Zeng A, Yang N, et al. (2007) Quantification of human bocavirus in lower respiratory tract infections in China. *Infectious Agents and Cancer* **2**: 3.

28. Allander T, Jartti T, Gupta P, et al. (2007) Human bocavirus and acute wheezing in children. *Clinical Infectious Diseases* **44**(7): 904–910.
29. Arden KE, McErlean P, Nissen MD, Sloots TP, Mackay IM (2006) Frequent detection of human rhinoviruses, paramyxoviruses, coronaviruses, and bocavirus during acute respiratory tract infections. *Journal of Medical Virology* **78**(9): 1232–1240.
30. Sloots TP, McErlean P, Speicher DJ, Arden KE, Nissen MD, Mackay IM (2006) Evidence of human coronavirus HKU1 and human bocavirus in Australian children. *Journal of Clinical Virology* **35**(1): 99–102.
31. Fry AM, Lu X, Chittaganpitch M, et al. (2007) Human bocavirus: a novel parvovirus epidemiologically associated with pneumonia requiring hospitalization in Thailand. *Journal of Infectious Diseases* **195**(7): 1038–1045.
32. Hall CB. (2001) Respiratory syncytial virus and parainfluenza virus. *New England Journal of Medicine* **344**(25): 1917–1928.
33. Semple MG, Cowell A, Dove W, et al. (2005) Dual infection of infants by human metapneumovirus and human respiratory syncytial virus is strongly associated with severe bronchiolitis. *Journal of Infectious Diseases* **191**(3): 382–386.
34. Templeton KE, Scheltinga SA, van den Eeden WC, Graffelman AW, van den Broek PJ, Claas EC (2005) Improved diagnosis of the etiology of community-acquired pneumonia with real-time polymerase chain reaction. *Clinical Infectious Diseases* **41**(3): 345–351.
35. Papadopoulos NG, Moustaki M, Tsolia M, et al. (2002) Association of rhinovirus infection with increased disease severity in acute bronchiolitis. *American Journal of Respiratory and Critical Care Medicine* **165**(9): 1285–1289.
36. Vabret A, Dina J, Gouarin S, Petitjean J, Corbet S, Freymuth F (2006) Detection of the new human coronavirus HKU1: a report of 6 cases. *Clinical Infectious Diseases* **42**(5): 634–639.
37. Schildgen O, Glatzel T, Geikowski T, et al. (2005) Human metapneumovirus RNA in encephalitis patient. *Emerging Infectious Diseases* **11**(3): 467–470.
38. Kesebir D, Vazquez M, Weibel C, et al. (2006) Human bocavirus infection in young children in the United States: molecular epidemiological profile and clinical characteristics of a newly emerging respiratory virus. *Journal of Infectious Diseases* **194**(9): 1276–1282.
39. Manning A, Russell V, Eastick K, et al. (2006) Epidemiological profile and clinical associations of human bocavirus and other human parvoviruses. *Journal of Infectious Diseases* **194**(9): 1283–1290.

13 Polymerase chain reaction and severe acute respiratory syndrome

Weijun Chen and Yang Huanming

Severe acute respiratory syndrome (SARS) first emerged in Guangdong Province, China, in November 2002, and presented as an outbreak of a typical pneumonia that was soon recognized as a global threat.[1] In Mainland China, it infected 5,327 people and caused 349 deaths within the first seven months of its recognition. In Hong Kong, SARS caused considerable disruption as this area faced the largest outbreak outside of Mainland China. It infected 1,755 people and caused 299 deaths (a fatality rate of 17.04%).[2] Among the infected, 405 people (23.08%) were health care workers and medical students in hospitals and clinics.[3,4] Within a month of recognition of this as a new type of infection, but before the disease pathogen was identified, it had spread to thirty-three countries and regions over the world, largely as a result of international air travel.[2,5]

Although the number of worldwide cases remained relatively low (8,098 cases), the mortality rate (774 deaths) remained relatively high until July 7, 2003.[2] This

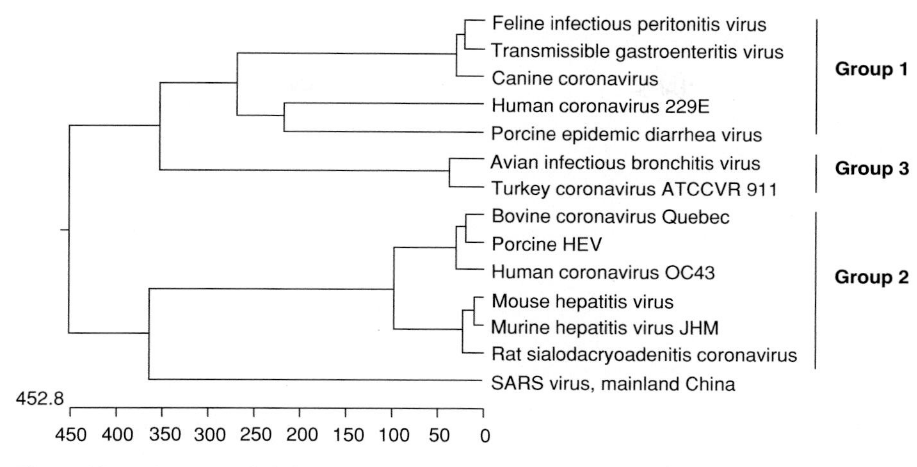

Figure 13–1. A proposed phylogenetic tree of the severe acute respiratory syndrome–associated virus based on the genome sequences.[13]

rate resulted in widespread concern, sometimes to the point of panic, in both affected and nonaffected populations. It was viewed in the same category as acquired immunodeficiency syndrome (AIDS) – a severe and readily transmissible new disease to emerge in the twenty-first century.[6]

The SARS epidemic highlighted the need for a rapid international response to disease control. The recent outbreak of H5N1 influenza in birds in Southeast Asia has only reinforced the potential for a pandemic spread of newly emerging or evolving infectious agents.

SARS CORONAVIRUS

The taxonomy of the SARS coronavirus

SARS is characterized by a fever, a nonproductive cough, dyspnea, chest pain, lung infiltrates and fibrosis, and a decreased lymphocyte count.[7] The causative agent of SARS was isolated from SARS patients and identified as a novel coronavirus, now known as the SARS coronavirus (SARS-CoV).[8–11] The complete genome sequence of SARS-CoV has been determined, and the virus is classified within the order *Nidovirales*, family *Coronaviridae*, genus *Coronavirus*.[12,13] Coronaviruses include viruses that cause human diseases, as well as viruses that cause mammal- and bird-species–specific infections.

As measured by serological analysis response, coronaviruses are made up of three different antigen groups: Group 1 and Group 2 are mainly mammal and human infective coronaviruses, and Group 3 is primarily infectious for poultry. However, phylogenetic trees based on the genome and protein coding sequence indicated that SARS-CoV did not belong to any of the three known coronavirus groups and was closer to the second coronavirus group (Figures 13–1, 13–2).[14,15]

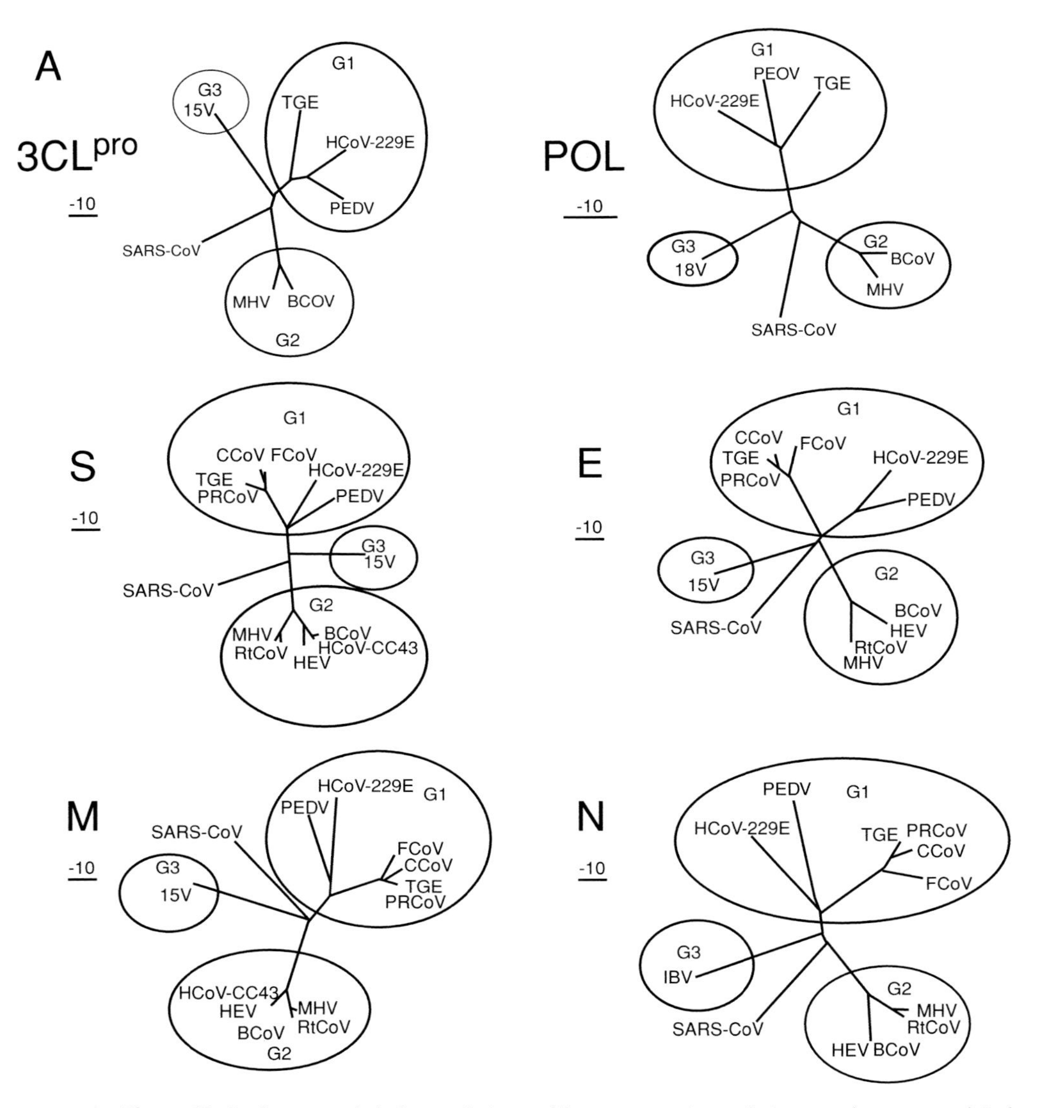

Figure 13–2. A proposed phylogenetic trees of the severe acute respiratory syndrome–associated viruses based on the complete amino acid sequences of the main coding protein.[13]

According to these results, it is reasonable to place SARS-CoV into a new evolutionary group.

Structure and morphography of SARS-CoV

SARS-CoV's molecular weight is 4 × 108 Da. Its buoyancy density in a sucrose gradient is 1.15 to 1.19 g·cm^{-3}, and its settling ratio is 300 to 500 S. SARS-CoV has the same basic shape as that of other known coronaviruses – that is, the viral capsid can take on a circular, ellipsoid, or pleomorphic shape with a diameter of 60 to 220 nm. The surface has several clavate vesicles that can be released and are approximately 20 nm in length. The viral capsid core is composed of viral

Plate 5–1. A fluorophore is slowly forged under the supervision of the dye chemist (Matt Lyttle, PhD).

Plate 5–2. David Seebach, oligo technician, operating the SuperSAM^{TM} high-throughput DNA synthesizer.

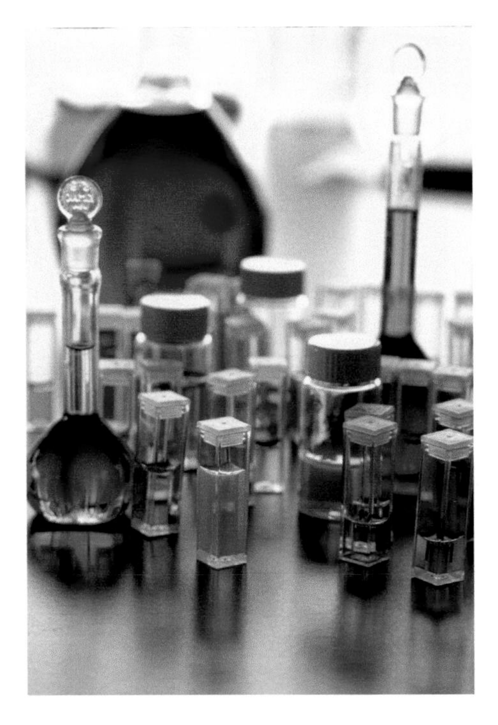

Plate 5–3. Flasks and cuvettes display an assortment of fluorophores and other dye labels intended to modify oligos.

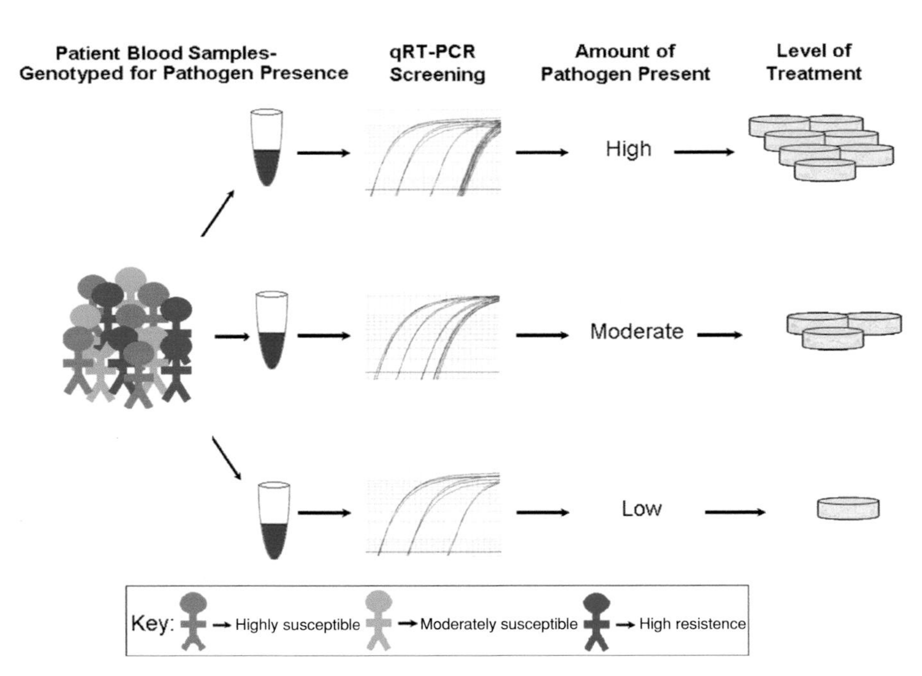

Plate 7–1. Course of clinical treatment against an infectious agent is determined by both patient genotype and quantitative polymerase chain reaction (qPCR) of the pathogenic agent. qRT–PCR, quantitative reverse transcriptase–PCR.

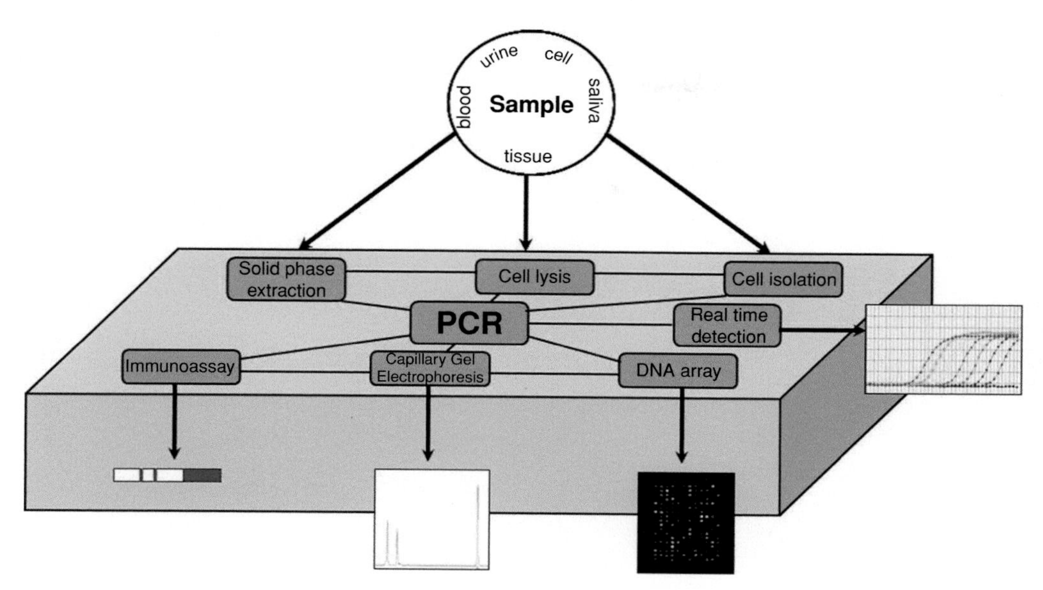

Plate 7–3. Micro total analysis system platform integrating pre-polymerase chain reaction (PCR), PCR, and post-PCR stages onto one microfluidic chip. DNA, deoxyribonucleic acid.

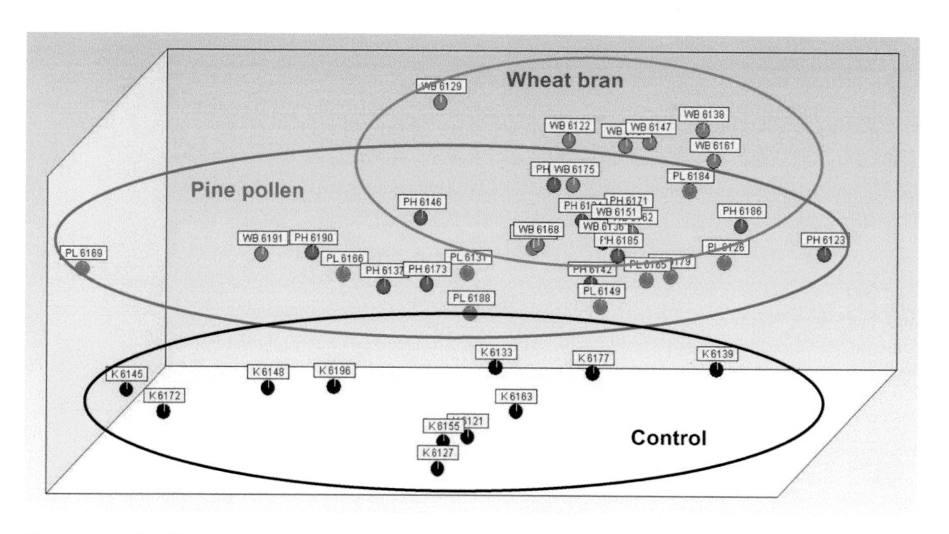

Plate 8–2. Multidimensional regression analysis via a three-dimensional scatter plot of daily intake, daily gain, and area of follicles in lymph node. Three different feeding regimens were investigated: wheat bran, pine pollen, and untreated control group.

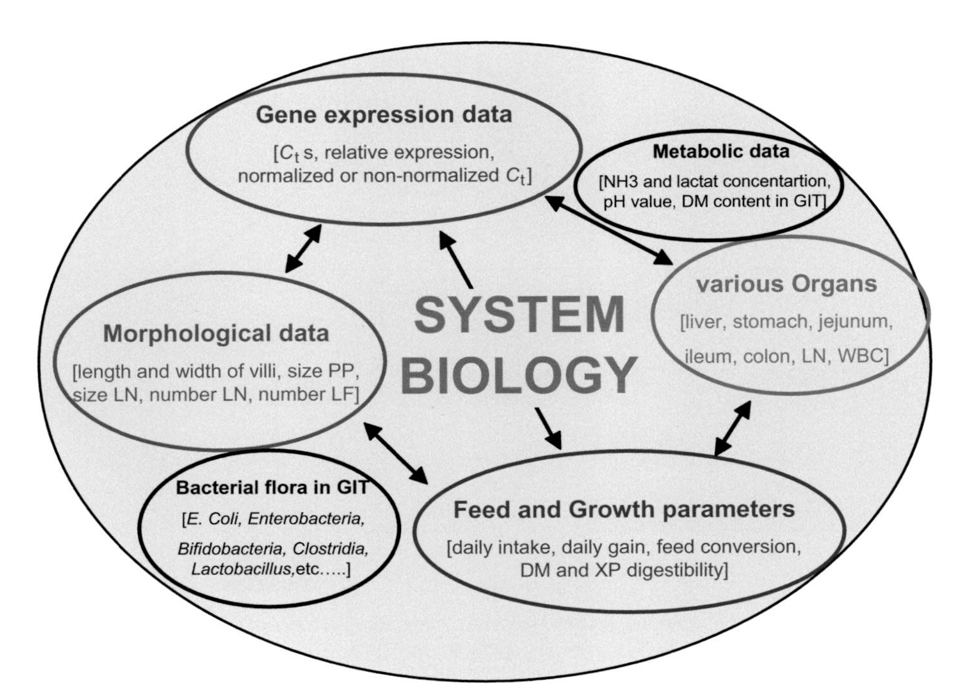

Plate 8–3. Multiple comparison of gene expression data in various organs, feed parameters, growth parameters, and morphological data (metabolic and bacterial) for a piglet feeding study for development of a system biology approach.

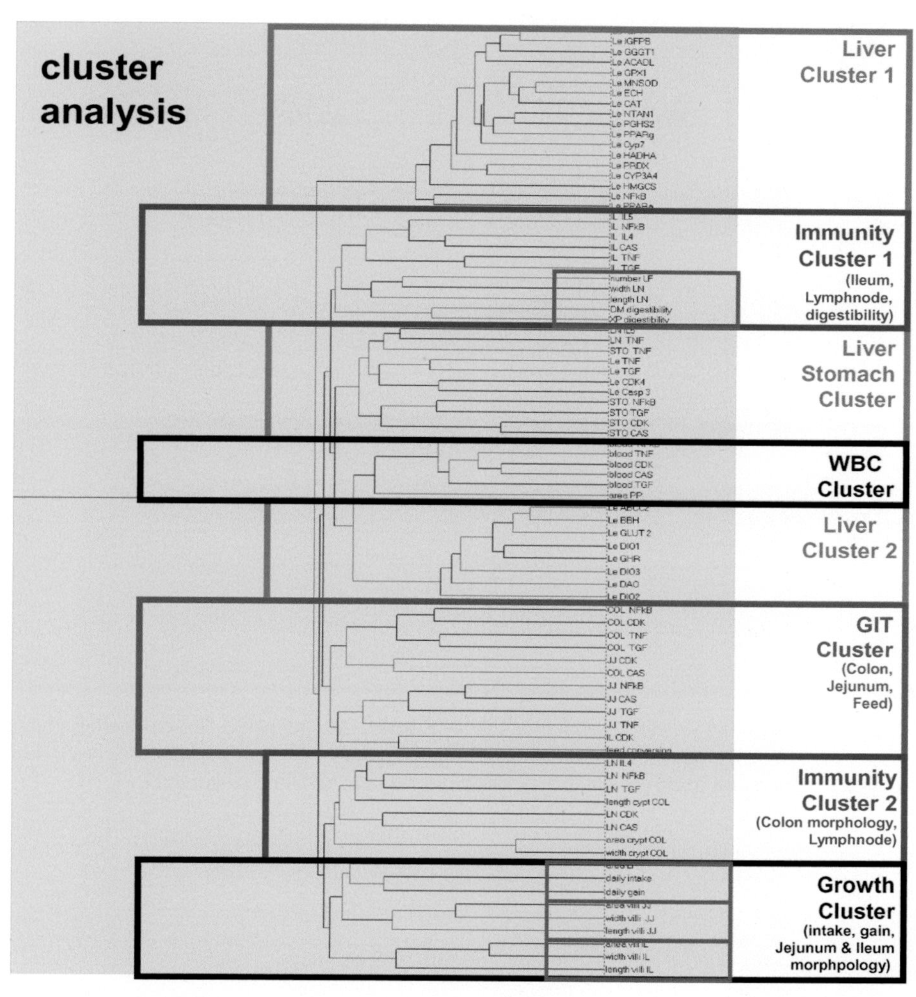

Plate 8–5. Dendrogram as result of cluster analysis. Various data sets cluster in main cluster and subcluster. WBC, white blood cell.

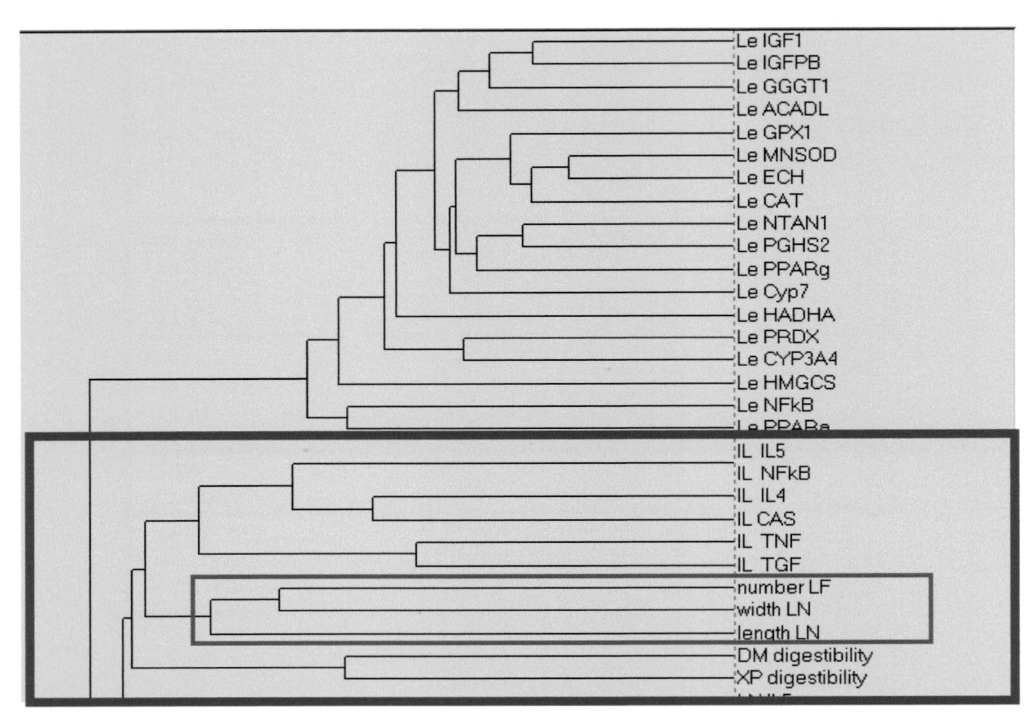

Plate 8–6. Dendrogram of liver and immunity subclusters. All liver genes cluster together in the upper part. Gene expression of immunological marker genes in the ileum, lymph node morphology, dry matter, and crude protein digestibility cluster clearly together (blue frame).

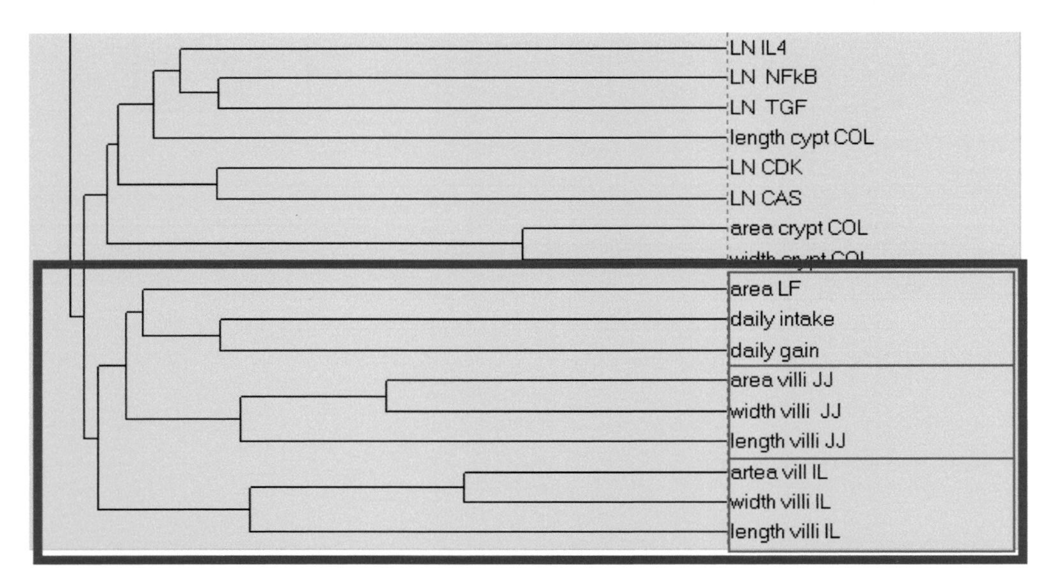

Plate 8–7. Dendrogram of growth cluster and immunity subclusters.

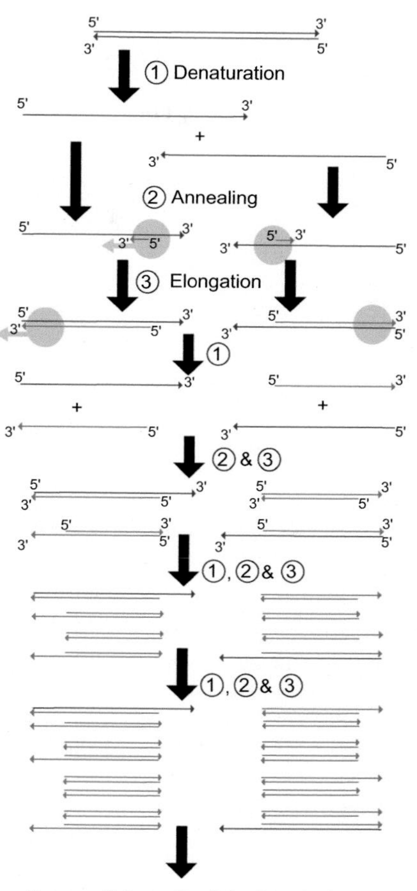

Plate 9–1. The polymerase chain reaction (PCR). The first step of the PCR is to separate the strands of target deoxyribonucleic acid (DNA). Short PCR primers are then annealed to specific sequences and elongated using a DNA polymerase to incorporate deoxynucleotide triphosphates (dNTPs). In theory each target strand is duplicated with each cycle, resulting in a logarithmic increase in the region encompassed by the primers.

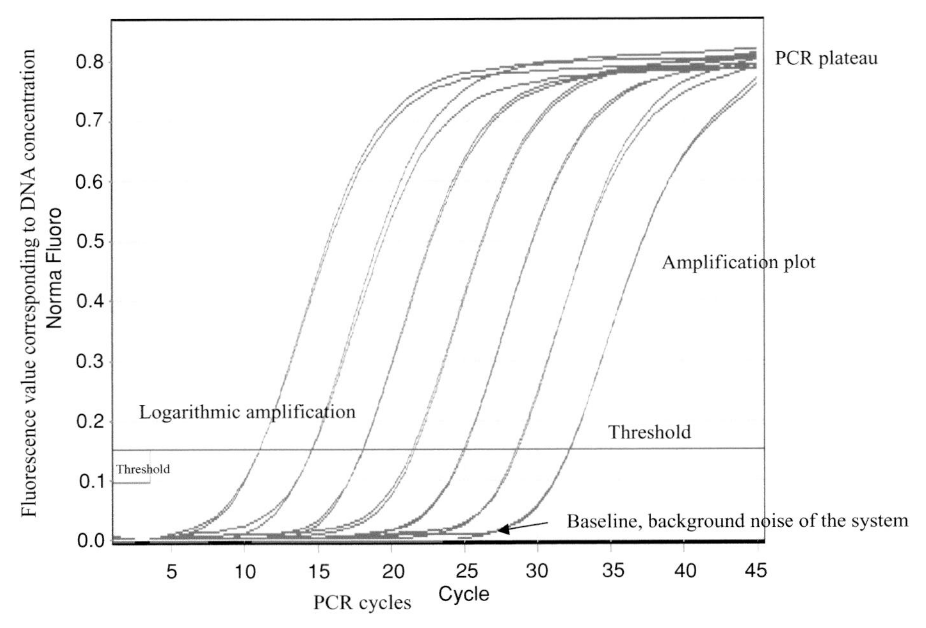

Plate 9–3. Phases of polymerase chain reaction (PCR). The PCR is a logarithmic reaction. When using real-time quantitative PCR (qPCR), a fluorescent label is added to the PCR that associates with the amplicon at each cycle. This addition allows the increase in deoxyribonucleic acid (DNA) concentration to be monitored. Also, the phases of PCR can be visualized. During early cycles the changes in DNA concentration cannot be determined because the signals are below the detection limit (baseline) of the qPCR system. When sufficient amplicon molecules have been produced, the PCR logarithmic amplification is observed until the PCR is limited and reaches the plateau phase. In the experiment shown, a fold serial dilution of template DNA was amplified. With each dilution there is a corresponding delay in the amplification plot indicating that more cycles are required to produce enough DNA (fluorescent signal) to be detected. Measurements are conventionally taken relative to the threshold.

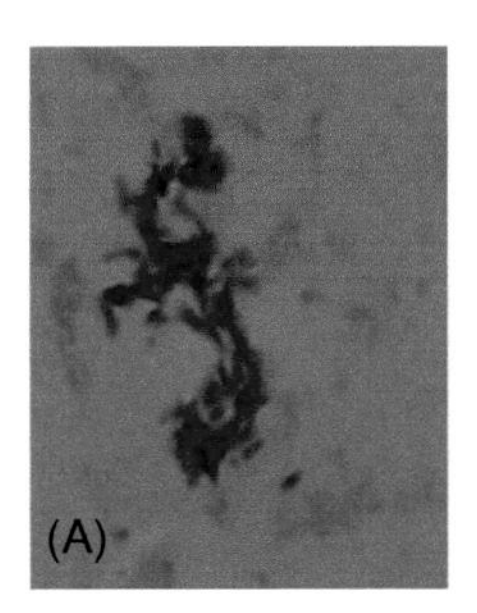

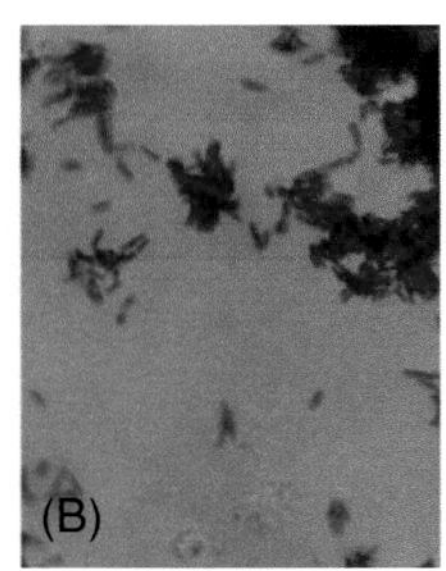

Plate 11–1. Sputum smear slides from two patients with tuberculosis (TB). *Mycobacterium tuberculosis* can be clearly seen, confirming the diagnosis. However, it is impossible to tell the drug-sensitive bacteria **(A)** from the strain resistant to two antibiotics **(B)**.

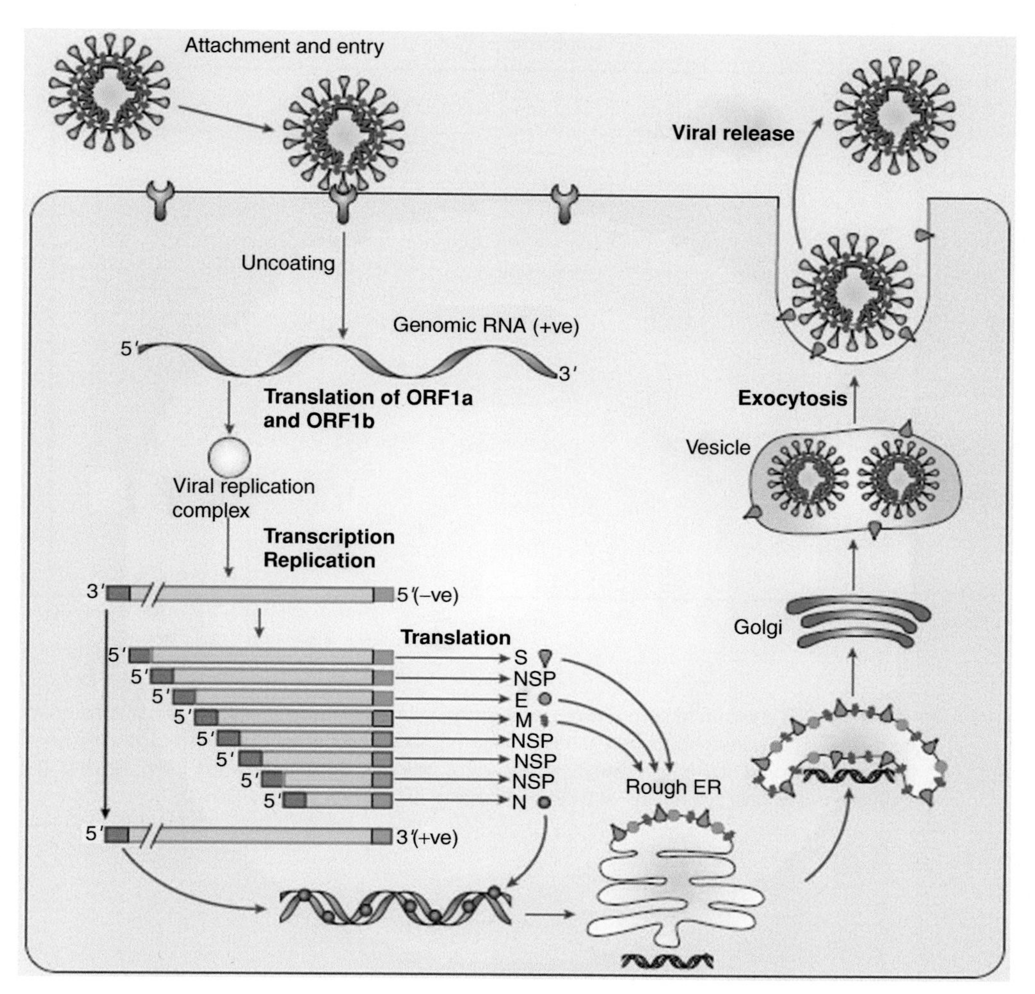

Plate 13–4. Life cycle of coronaviruses. +ve, positive; ORF, open reading frame; S, spike, NSP, non-structure protein; E, envelope; M, matrix; ER, endoplasmic reticulum. Plate from Stadler K, Masignani V, Eickmann M, Becker S, Abrignani S, Klenk HD, et al. (2003), *Nature Reviews (Microbiology)* **1**: 209–218.

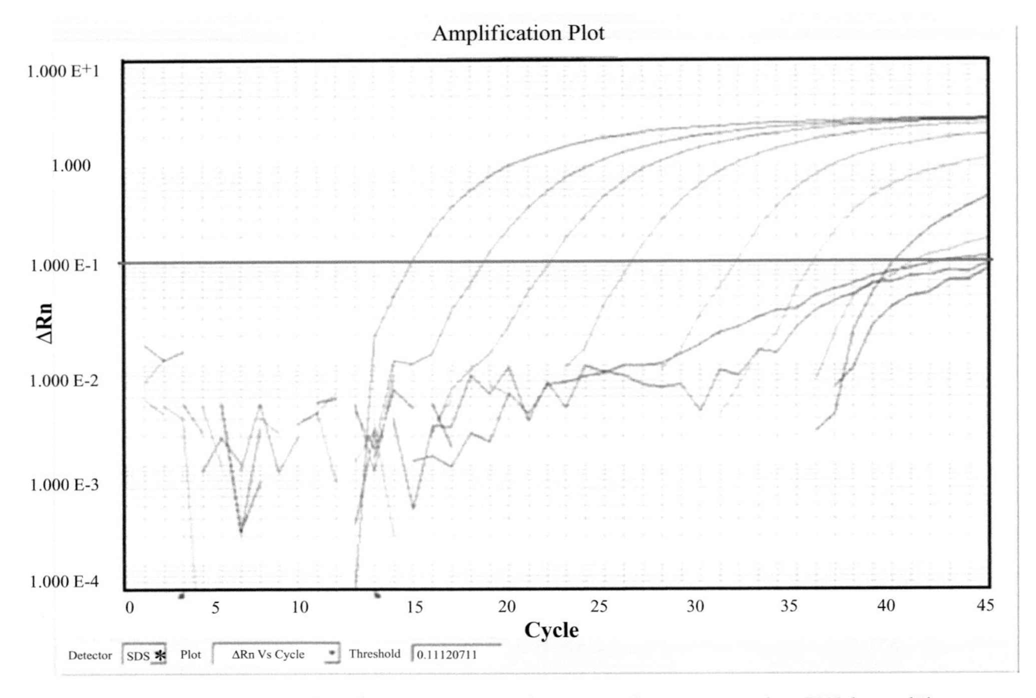

Plate 13–7. Detection of severe acute respiratory syndrome-coronavirus RNA by real-time quantitative reverse transcription–polymerase chain reaction (RT–PCR) for the nucleocapsid region of the viral genome. An amplification plot of △Rn, which is the fluorescence intensity over the background (*y*-axis) against the PCR cycle number (*x*-axis) (Chen W. et al, 2004).

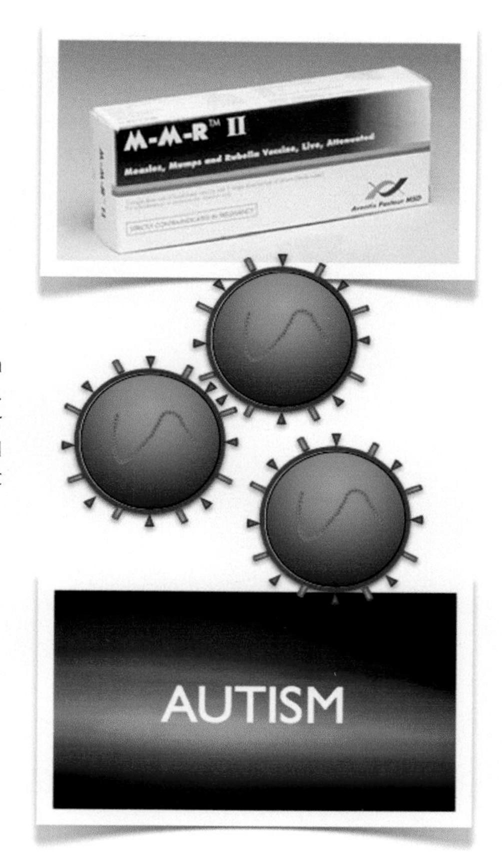

Figure 14–1. Measles, mumps, and rubella (MMR), measles virus (MV), and autism. Detection of MV (*middle*) is a prerequisite for a link between the triple MMR vaccine (l) and the development of a disorder on the autistic spectrum (lm).

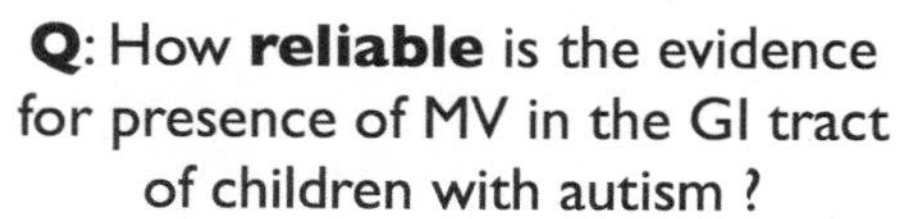

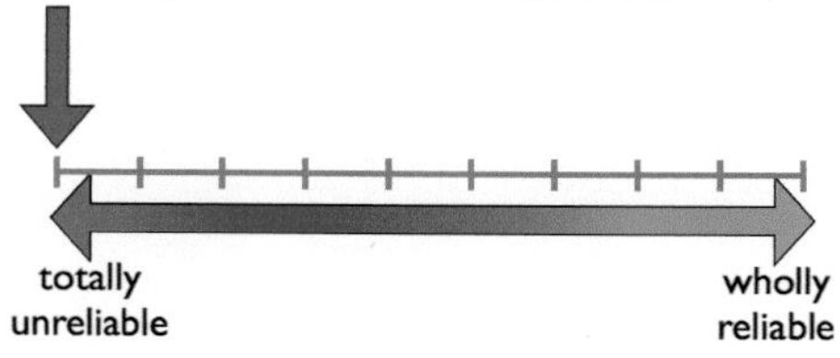

Plate 14–2. Reliability of quantitative polymerase chain reaction evidence. MV, measles virus; GI, gastrointestinal.

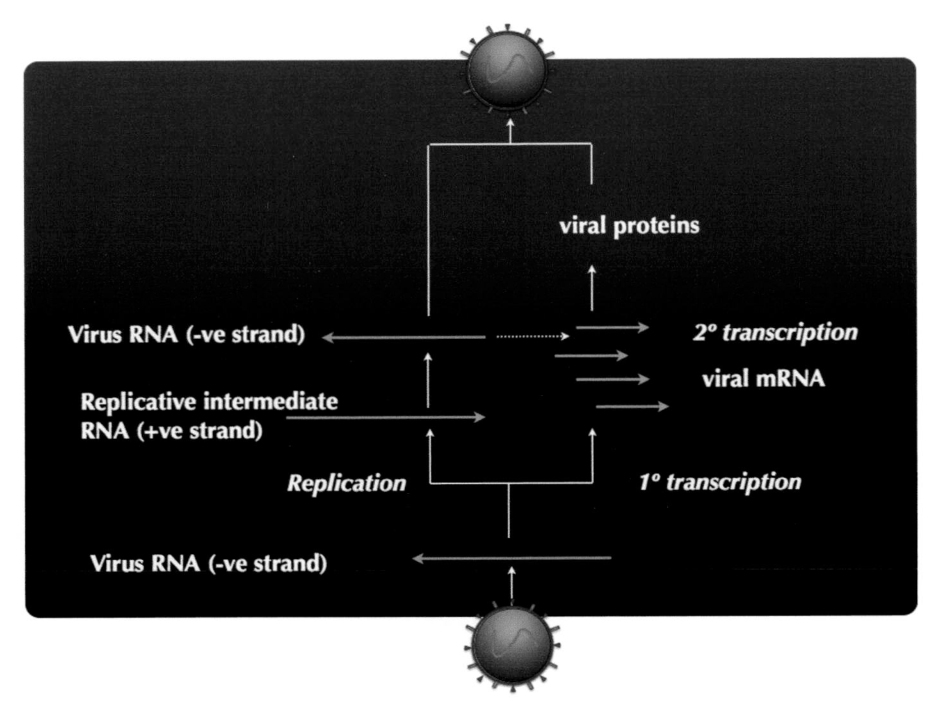

Plate 14–3. Measles virus life cycle. mRNA, messenger RNA, +ve, positive; −ve, negative.

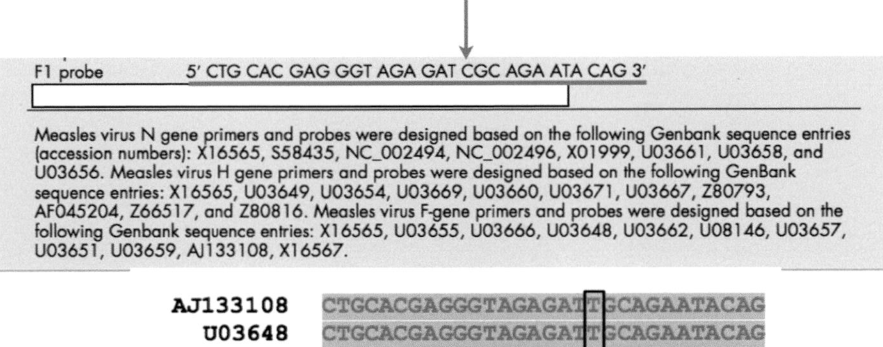

AJ133108	CTGCACGAGGGTAGAGATTGCAGAATACAG
U03648	CTGCACGAGGGTAGAGATTGCAGAATACAG
U03651	CTGCACGAGGGTAGAGATTGCAGAATACAG
U03655	CTGCACGAGGGTAGAGATTGCAGAATACAG
U03657	CTGCACGAGGGTAGAGATTGCAGAATACAG
U03659	CTGCACGAGGGTAGAGATTGCAGAATACAG
U03662	CTGCACGAGGGTAGAGATTGCAGAATACAG
U03666	CTGCACGAGGGTAGAGATTGCAGAATACAG
x16567	CTGCACGAGGGTAGAGATTGCAGAATACAG
x16565	CTGCACGAGGGTAGAGATTGCAGAATACAG
Consensus	**CTGCACGAGGGTAGAGATTGCAGAATACAG**

Plate 14–4. Identification of single base mismatch in the hydrolysis probe used for the detection of the measles virus (MV) F gene. Arrow points to the mistaken "C," which should clearly be a "T," if the authors had used the consensus sequences from the Genbank database, as claimed.

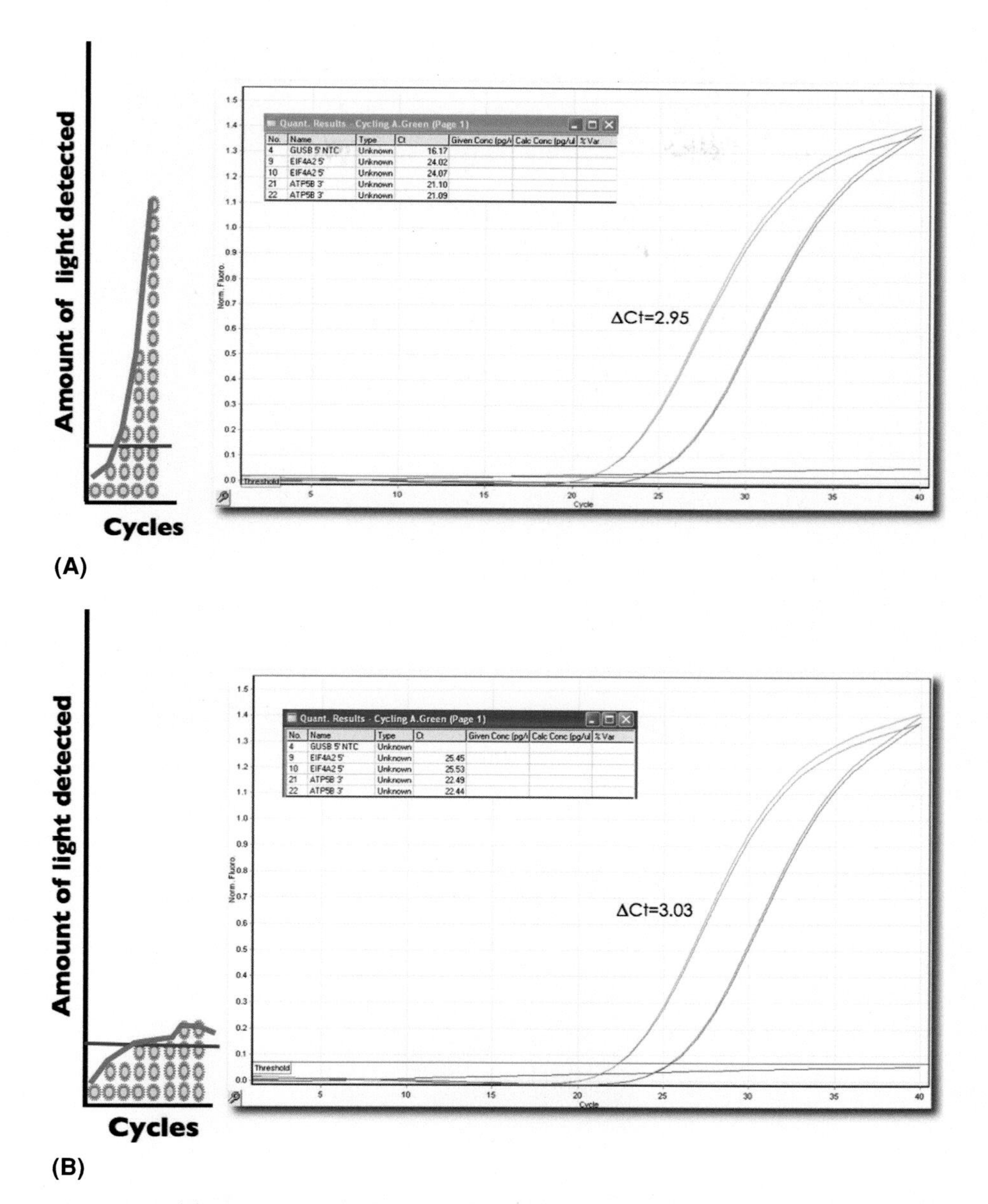

Plate 14–5. (A) Exponential amplification results in amplification plots that clearly cross the detection threshold of the real-time polymerase chain reaction (PCR) detection instrument. This positive amplification is evident from the two samples run in duplicate, which differ in their C_qs by 2.95. A third sample, shown in purple and barely crossing the threshold, does not show this pattern and is clearly spurious. **(B)** Adjusting the threshold more or less retains the ΔC_q value, but now shows a negative result for the spuriously amplifying sample.

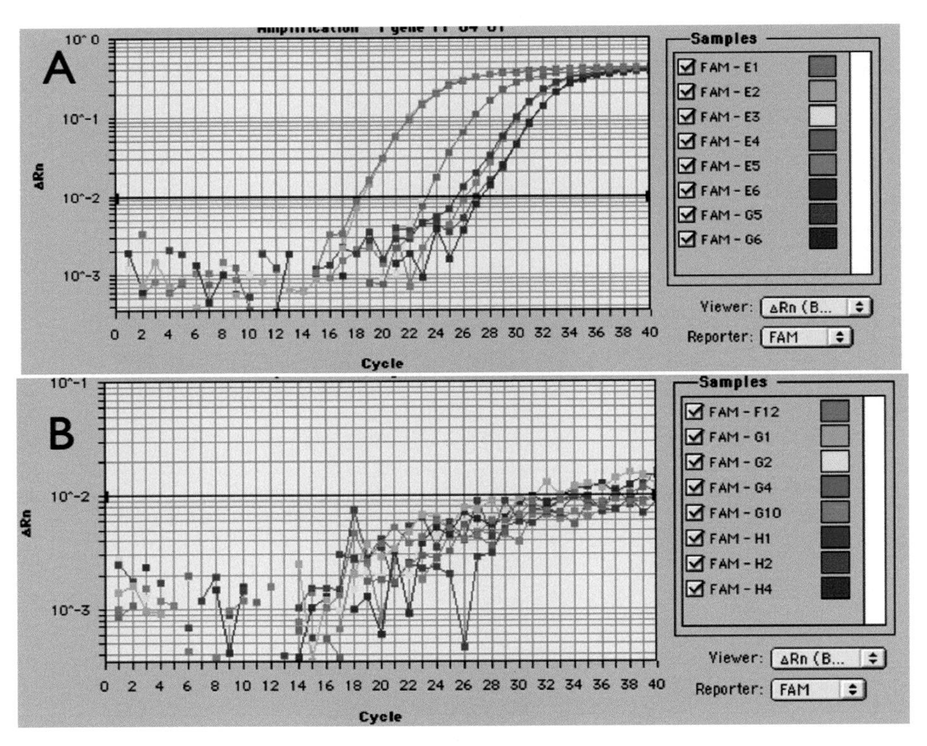

Plate 14–6. **(A)** Genuine amplification products. **(B)** Positive samples that should have been analyzed differently (i.e., by adjusting the threshold as suggested in Figure 14–5).

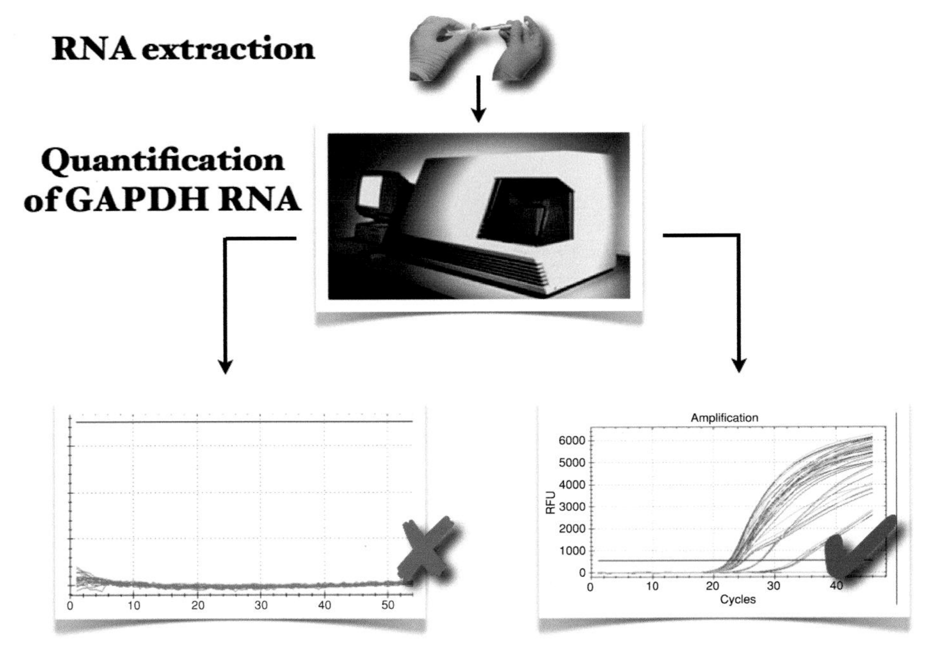

Plate 14–9. Schematic description of standard operating procedure requiring discarding of samples negative for glyceraldehyde 3-phosphate dehydrogenase (GAPDH; *red cross*) and further analysis for samples positive for GAPDH (*green check mark*).

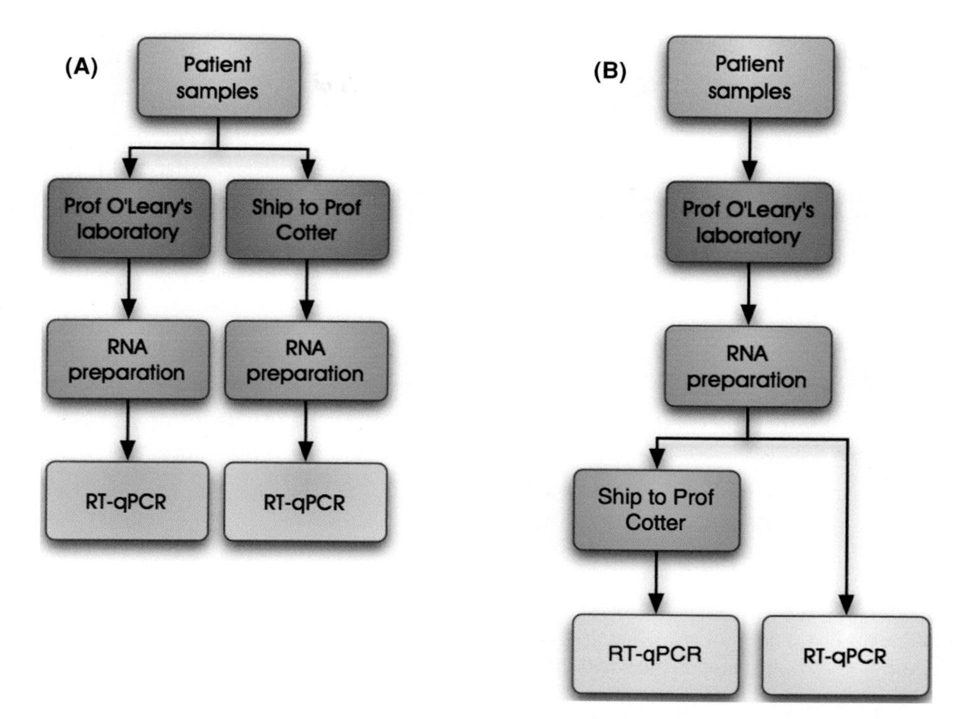

Plate 14–7. (A) Experimental setup 1: Both laboratories extract RNA from duplicate samples. **(B)** Experimental setup 2: O'Leary's laboratory extracts RNA from samples. Both laboratories analyze the same RNA samples. RT–qPCR, real-time reverse transcription–polymerase chain reaction.

	RNA extracted by Cotter		RNA extracted by O'Leary			
Sample	Cotter 1A	Cotter 1B	Cotter 2A	Cotter 2B	O'Leary A	O'Leary B
1						
2						
3						
4						
5						
6						
7						
8						
9						
10						
11						
12						
13						

negative for MV
positive for MV

Plate 14–8. RNA extracted by the independent laboratory gives negative results for duplicate runs (Cotter 1A and 1B). RNA extracted in O'Leary's laboratory is positive, albeit discordantly, in the independent laboratory (Cotter 2A and 2B) as well as in O'Leary's laboratory (O'Leary A and B). MV, measles virus.

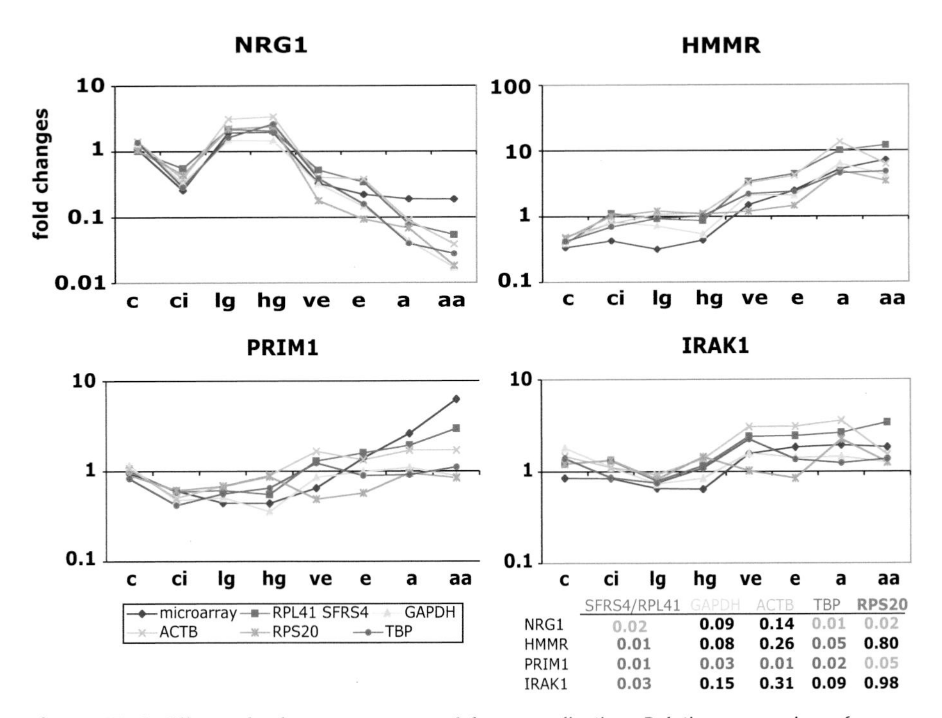

	SFRS4/RPL41	GAPDH	ACTB	TBP	RPS20
NRG1	0.02	0.09	0.14	0.01	0.02
HMMR	0.01	0.08	0.26	0.05	0.80
PRIM1	0.01	0.03	0.01	0.02	0.05
IRAK1	0.03	0.15	0.31	0.09	0.98

Figure 17–4. Effects of reference genes used for normalization: Relative expression of neuregulin 1 (NRG1), hyaluronan-mediated motility receptor (HMMR), primase polypeptide 1 (PRIM1), and interleukin-1 receptor-associated kinase 1 (IRAK1) for all stages of hepatitis C virus (HCV)-induced hepatocellular carcinoma (HCC). Quantitative real-time reverse transcriptase–polymerase chain reaction (qPCR) data were normalized to ribosomal protein L41 (RPL41) and splicing factor, arginine/serine-rich 4 (SFRS4; *pink*), to glyceraldehyde 3-phosphate dehydrogenase (GAPDH; *yellow*), to actin, beta (ACTB *light blue*), to ribosomal protein S20 (RPS20; *green*), and to TATA binding protein (TBP; *brown*). Microarray data are shown in *dark blue*. Fold changes are indicated on the *y* axis. Stages of hepatocarcinogenesis: c = control; ci = cirrhosis; lg = low-grade dysplasia; hg = high-grade dysplasia; ve = very early HCC; e = early HCC; a = advanced HCC; aa = very advanced HCC. Table shows *p* values for the change in gene expression from high-grade dysplasia to very early HCC for NRG1, HMMR, PRIM1, and IRAK1 (*rows*) when normalized to the genes indicated (*columns*). Significant ($p < 0.05$) upregulation between these stages is indicated in *red*; downregulation in *green*. Figure and table are from Waxman S, Wurmbach E (2007), *BMC Genomics* **8**: 243.

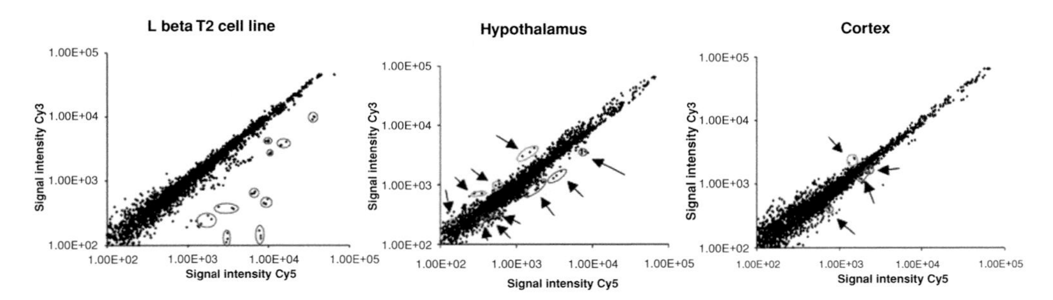

Plate 17–5. Scatter plots of microarray assays from tissues with varying complexity: Plotted are the signal intensities of the LβT2 cell line (Cy3: vehicle, Cy5: gonadotropin releasing hormone [GnRH]), the hypothalamus (Cy3: euglycemic, Cy5: hypoglycemic), and the cortex (Cy3: saline, Cy5: 2,5-dimethoxy 4-iodoamphetamine [DOI]) experiments. The triplicates confirmed by quantitative real-time reverse transcriptase–polymerase chain reaction (qPCR) are marked with *red circles* indicate upregulation, and those marked with *green circles* indicate downregulation. Figures adapted from Wurmbach E, et al. (2002), *Neurochemical Research* **27**: 1027–1033.

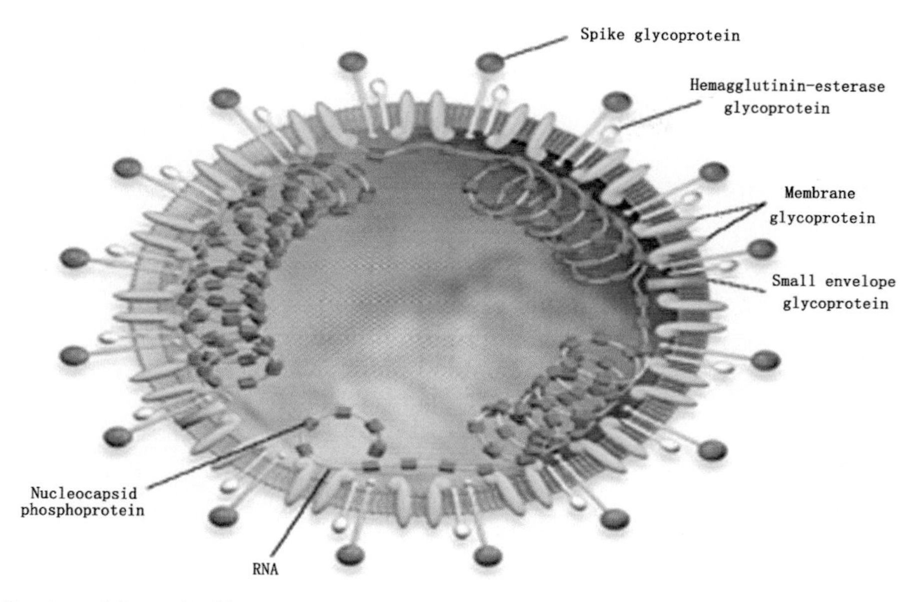

Figure 13–3. Illustration of the severe acute respiratory syndrome-coronavirus.[16]

RNA, and the external membrane is made up of a protein-embedded lipid bilayer (Figure 13–3).[16] The viral capsid structure is relatively stable in the presence of Mg2+ but is sensitive to the presence of salt, liposolvents, non-ionic detergents, formaldehyde, and oxidants.

Propagation of SARS-CoV

SARS-CoV could culture in many cell lines including Vero and Vero-E6 cells, MDCK, Hep-2, Hela, BHK-21, and more. Vero-E6 cells infected with SARS-CoV would shrink in size and gradually fall off. After 24 to 48 hours all cells fall off. Coronavirus's reproduction occurs as follows: The virion binds to cell surface receptors using its projecting spikes. It then enters the cell by membrane fusion with the plasma membrane or by receptor-mediated endocytosis. The viral positive, single-stranded RNA genome enters the cytoplasm and is reverse transcribed by a virus-encoded RNA-dependent RNA polymerase, forming a replication complex attached to membrane. These replication complexes then produce new genomic RNAs and (subgenomic) messenger RNAs (mRNAs) coding for new viral proteins. The N protein joins the new genomic RNA to form new ribonucleoproteins (RNPs). These RNPs attach to the membrane where S proteins and M proteins have previously assembled. The RNPs bud into the lumen of the vesicle and remain there as immature virions. These particles progress up the periphery of the Golgi apparatus, maturing as they do so into a denser and more icosahedral form. The new virus particles collect in large vesicles and are finally released onto the cell surface to start the cycle again (Figure 13–4).[17,18]

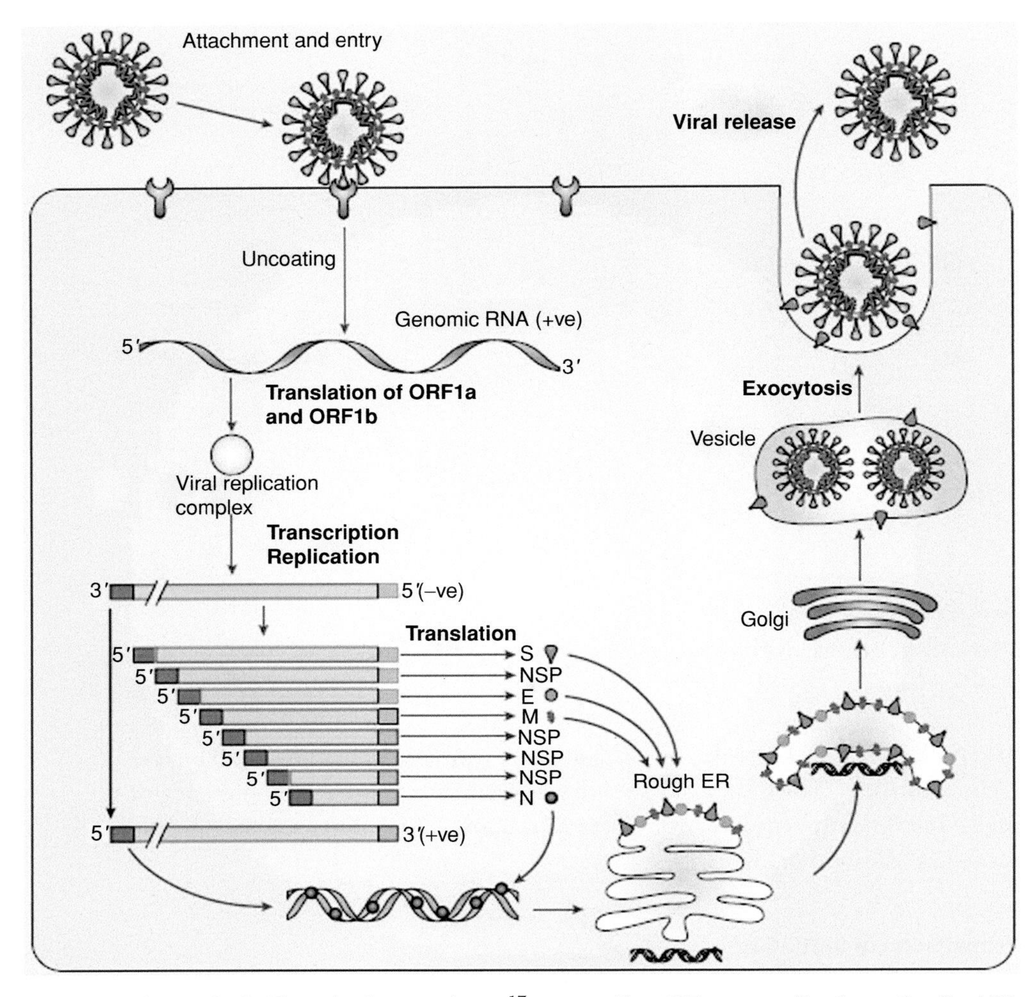

Figure 13–4. Life cycle of coronaviruses.[17] +ve, positive; ORF, open reading frame; S, spike; NSP, non-structure protein; E, envelope; M, matrix; ER, endoplasmic reticulum. *See Color Plates.*

SARS-CoV's genome

The SARS-CoV genome is approximately 29.7 kb in length, and the percentage of guanine–cytosine residues is 41% (the range for published, complete coronavirus genome sequences is between 37% and 42%). It is a positive, single-stranded RNA that corresponds to a polycistronic mRNA, consisting of 5′ and 3′ untranslated regions (UTRs), thirteen to fifteen open reading frames (ORFs), and approximately ten intergenic regions (Figure 13–5).[14,19,20] Its genes encode two replication polyproteins: the SARS-CoV *rep* gene (pp 1a) and the RNA-dependent-RNA polymerase (pp 1ab), which undergo cotranslational proteolytic processing. The *rep* gene comprises approximately two thirds of the genome. A set of ORFs at the 3′ end code for four structural proteins: surface S glycoprotein (1,256 aa.), envelope (E, 77 aa.), matrix (M, 222 aa.), and nucleocapsid (N, 423 aa.) proteins, which are present in all known coronaviruses. The gene encoding

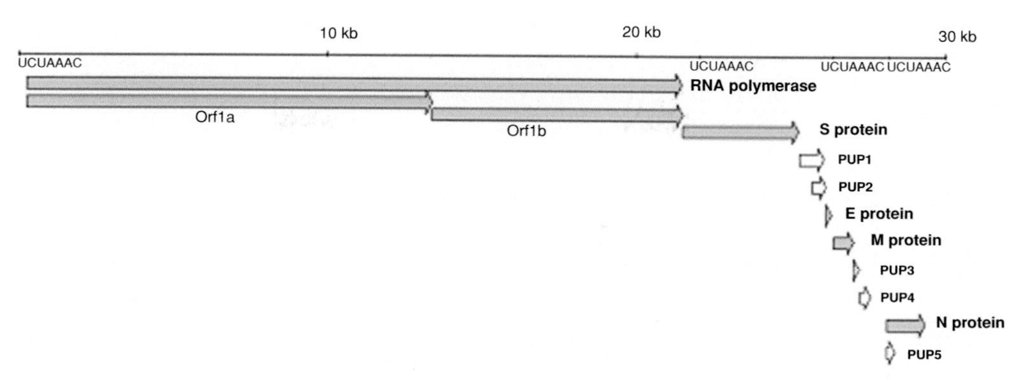

Figure 13–5. Genome structure of severe acute respiratory syndrome-coronavirus, Beijing 01 strain.[14]

hemagglutinin-esterase, which is present between ORF1b and S in Group 2 and some Group 3 coronaviruses, was not found.[21] There are an additional eight to nine predicted ORFs in the SARS-CoV genome that are not present in other known coronaviruses; however, the function of their protein products is still under investigation.[22] Coronaviruses, including SARS-CoV, also encode a number of nonstructural proteins (Figure 13–4), which are located between S and E, between M and N, and downstream of N. These nonstructural proteins, which vary widely among the different coronavirus species, are of unknown function and seem dispensable for virus replication.

Characterization of the 5′-UTR of the genome using a 5′-rapid-amplification of cDNA ends (RACE) assay indicated that the predicted AUG initiation codon of ORF1 of the Beijing 01 strain was 264 bp upstream. Alignment of the seven 5′-RACE sequences showed a consensus of 72 bp, which is composed of a leader sequence of 61 bp and an intergenic sequence (IGS) at the last 9 bp. The IGS 5′-UAAACGAAC-3′ was identical for all of the ORFs, except X2. Sequencing of the 3′-RACE products showed that there is a unique discontinuous transcription system in CoVs, which generates a nested set of transcripts that have common 3′ ends and common leader sequences on the 5′ ends. Evidence indicates that the 3′-UTR (the sequence downstream of the N protein) is crucial in CoV transcription regulation.[23]

Sequence alignment of the predicted amino acid sequence of ORF1 revealed recognizable ORFs including the replicase 1a and 1b translation products, the S glycoprotein, the E protein, the M protein, and the N protein, with a strong match to the transcriptional regulatory site consensus sequence upstream of the potential initiating methionine residue. In addition, unique features of SARS-CoV are a number of ORFs larger than 40 aminoacids with no matching database sequences. Preliminary analysis of the ORFs identified the following: The replicase 1a ORF (base pairs 265–13,398) and replicase 1b ORF (base pairs 13,398–21,485) occupy 21.2 kb of the SARS virus genome. A frame shift interrupts the protein-coding region and separates the 1a and 1b reading frames.

ORF3 (base pairs 25,268–26,092) encodes a predicted protein of 274 amino acids that lacks significant similarity to any known proteins. The most likely structural model of this protein from analysis is that the C terminus and a large 149–amino acid N-terminal domain would be embedded in the viral or cellular membrane. The C-terminal (interior) region of the protein may encode an adenosine triphosphate (ATP)-binding domain.

ORF4 (base pairs 25,689–26,153) encodes a predicted protein of 154 amino acids. This ORF completely overlaps ORF3 and the E protein.

The gene encoding the small E protein (base pairs 26,117–26,347) yields a predicted protein of 76 amino acids. Amino acid sequence comparisons indicate that the predicted protein exhibits significant matches to many E (alternatively known as small membrane) proteins found in several coronaviruses.

The S glycoprotein (base pairs 21,492–25,259) encodes a surface projection glycoprotein precursor predicted to be 1,255 amino acids in length. Mutations in the gene encoding the S protein have been correlated with altered pathogenesis and virulence in other coronaviruses.[18]

The gene encoding the M glycoprotein (base pairs 26,398–27,063) yields a predicted protein of 221 amino acids. BLAST and FASTA analyses of this protein reveal significant similarity to a large number of coronaviral M glycoproteins. The association of the S glycoprotein with the M glycoprotein is an essential step in the formation of the viral envelope and in the accumulation of both proteins at the site of virus assembly.[18]

ORF7 (base pairs 27,074 to 27,265) encodes a predicted protein of 63 amino acids, TMpred analysis predicts the presence of a transmembrane helix at residues 3–22, with the N terminus located outside the viral particle.

ORF8 (base pairs 27,273–27,641), encoding a predicted protein of 122 amino acids, is likely to be a type I membrane protein, with the major hydrophilic domain of the protein (residues 16–98) and with the N terminus oriented inside the lumen of the endoplasmic reticulum/Golgi apparatus or on the surface of the cell membrane or virus particle, depending on the membrane localization of the protein.

ORF9 (base pairs 27,638–27,772) encodes a predicted protein of 44 amino acids. The TMpred program predicts the existence of a single strong transmembrane helix. Currently, studies do not indicate if the N terminus of this helix is located inside or outside the particle.

ORF10 (base pairs 27,779–27,898), encoding a predicted protein of 39 amino acids, maybe encode a transmembrane helix.

ORF11 (base pairs 27,864–28,118) encodes a predicted protein of 84 amino acids. It exhibits only very short (nine to ten residues) matches with a region of the human coronavirus S glycoprotein precursor (starting at residue 801).

The gene encoding the N protein (base pairs 28,120–29,388) yields a predicted protein of 422 amino acids. It contains a domain that may contain a nuclear localization signal, which could play a role in pathogenesis. In addition, the basic nature of this peptide suggests that it may assist in RNA binding.

ORF13 (base pairs 28,130–28,426) encodes a predicted protein of 98 amino acids, and no transmembrane helices are predicted.

ORF14 (base pairs 28,583–28,795) encodes a predicted protein of 70 amino acids with a single transmembrane helix.[12]

EPIDEMIOLOGICAL MULTIDIRECTIONAL TRANSMISSION OF SARS-CoV

Epidemiological studies of index SARS cases in Guangdong Province provided initial evidence that the agent responsible for the outbreak was zoonotic in origin. Between November 2002 and February 2003, the most important goal was to determine the route and direction of transmission as this would lead to an understanding of zoonotic disease emergence and allow the development of strategies to control future outbreaks. For SARS-CoV, during the outbreaks of SARS in 2002–2003 and 2003–2004, there was evidence to suggest four possible routes of transmission: animal-to-human, animal-to-animal, human-to-human, and human-to-animal. Here we consider each.

Animal-to-human transmission

During the sporadic outbreaks of 2003–2004, a total of four patients were independently infected with SARS-CoV.[24,25] No direct link could be found between any of these four cases, nor had any of these patients had a direct or indirect contact history with previously documented SARS cases. All of them, however, had a history of contact with animals. Epidemiologic studies revealed that animal handlers and people working in the food industry had a higher representation than did any other group among early SARS patients. Retrospective serologic studies indicated that there were no antibodies to SARS-CoV in the human population prior to the SARS outbreak, indicating that SARS-CoV had not been an already existent human coronavirus.[9]

Molecular epidemiologic studies confirmed that the earliest genotypes of human SARS-CoV from the 2002–2003 outbreaks were most closely related to animal SARS-CoV isolates.[26] Furthermore, genome sequences of SARS-CoVs from human patients in 2003–2004 were nearly identical to SARS-CoVs isolated from civets present in marketplaces during this same time period, but that they were more divergent from human SARS-CoVs obtained during the 2002–2003 outbreaks.

Taken together, these results demonstrated that animal-to-human transmission was responsible for the introduction of SARS-CoV into the human population.

Animal-to-animal transmission

Sampling of six masked palm civets from a marketplace in China revealed that all had been exposed to SARS-CoV. These animals were sampled at the same time in the same market, but had each originated from different regions of southern China. Thus it appeared that most, if not all, of them were infected in the market through animal-to-animal transmission.[26]

Animal-to-animal transmission has also been demonstrated in experimental situations. Martina et al. showed that ferrets (*Mustela furo*) and domestic cats

(*Felis domesticus*) are susceptible to infection by SARS-CoV and that they can efficiently transmit the virus to previously uninfected animals that are housed with them.[27] This observation strongly indicated the occurrence of inter-species transmission among the animals.

Human-to-human transmission

Numerous epidemiological studies have demonstrated the rapid human-to-human transmission of SARS-CoV, which spread the virus to more than thirty countries in less than five months.[2] Countries that had a moderate-to-large number of cases played a pivotal role in large-scale transmission of the virus through superspreading events (SSEs). In such circumstances, a small number of infected individuals caused a much higher number of secondary infections.

Hong Kong in particular experienced a series of SSEs: in a Hong Kong hotel and in a Hong Kong hospital setting, then on a flight from Hong Kong to Beijing, which ultimately led to SSEs in health care settings in Beijing, Singapore, and Toronto.[28] In the SSE in a Beijing hospital, one patient infected 33 of 74 persons who had close contact with that patient. These secondary cases resulted in a further 43 cases before this chain of transmission subsided.[29]

Human-to-animal transmission

There have been reports suggesting spread of the virus through bathroom-plumbing U-traps that had been contaminated with SARS-CoV. Other studies indicate a possible role for environmental spread by city-dwelling or domestic animals, such as rats and cats.[27,30] Domestic cats living in an apartment complex were found to be infected with SARS-CoV, which indicated possible human-to-animal transmission. This notion was subsequently supported by experimental infection of domestic cats with a human SARS-CoV isolated from a Hong Kong patient.[27]

In another potential example of human-to-animal transmission, SARS-CoV was isolated from a pig during a surveillance study in farming villages outside of Tianjin, where a SARS outbreak occurred in the spring of 2003.[31] The genome sequence of the pig isolate (designated TJF) revealed it to be closely related to the human isolate BJ01 obtained from a patient in Beijing. More importantly, the TJF genome contained a 29-nt deletion, which was the genetic feature characterizing the SARS-CoV that circulated among human patients during the latter phases of the 2002–2003 outbreaks, but that was never observed in any of the animal SARS-CoV isolates.

DIAGNOSTIC METHODS

SARS is caused by a novel coronavirus in human beings and poses a continuing global human public health risk. The most important measures that need to be taken to create an ongoing strategy for preventing further national or international spread are to develop the means for rapid diagnosis, to create mechanisms for global surveillance, and to report every incidence of SARS.

Clinical symptoms

The prodrome of the illness is generally the presence of fever (>38°C), chills/rigor, headache, myalgia, malaise, and mild respiratory symptoms. After 3 to 7 days, the respiratory symptoms progress to a dry, nonproductive cough and dyspnea. In approximately 10% to 20% of cases, the dyspnea is severe enough to require intubation and mechanical ventilation. Chest radiographs show that some patients have early focal infiltrates, which could progress to more generalized, patchy, interstitial infiltrates, and areas of consolidation in the late stages of SARS.[32]

Early in the course of disease, there can be a reduced lymphocyte count. In general, however, white cell counts are normal or only slightly decreased. A rise in creatine phosphokinase levels and hepatic transaminases also has been seen. At the peak of the respiratory illness, up to half of patients have leukopenia and thrombocytopenia or below-normal platelet counts. A rapid clinical response (during which patients receive broad-spectrum antibiotics and antiviral agents, such as oseltamivir or ribavirin, in combination with steroids) could be effective for disease treatment; however, this requires a faster means of recognizing the infection.

Laboratory methods for SARS diagnosis

In the absence of effective drugs or a vaccine for SARS, control of this disease relies on the rapid identification of cases and the appropriate management of their close contacts. However, SARS-CoV infections are symptomatically similar to other acute febrile illnesses (such as influenza and atypical pneumonias); it is difficult to differentiate SARS from other respiratory infections through disease presentation. So rapid laboratory methods for SARS diagnosis are very important. Currently, laboratory identification for SARS-CoV infections is carried out by detection of SARS-CoV–specific RNA by reverse transcription–polymerase chain reaction (RT–PCR), seroconversion by enzyme linked immunosorbent assay (ELISA) or immunofluorescence test (IFA), or isolation of viral strains in cell culture.[33]

Confirmation of a SARS virus infection using PCR technology requires multiple specimens or tests. This can be done in various ways: obtaining two different types of clinical specimens (e.g., nasopharyngeal and stool), collecting the same type of clinical specimen from two or more days during the course of the illness (e.g., two or more nasopharyngeal aspirates), or carrying out two different assays or repeated PCRs using a new sample from the original clinical sample for each test.

To confirm a SARS infection by seroconversion by ELISA or IFA requires a negative antibody test on acute serum followed by a positive antibody test on convalescent serum or a fourfold or greater rise in antibody titer between the acute and convalescent phase sera tested in parallel.

To confirm a SARS infection via virus isolation, virus isolation was carried out through the examination of typical cyto-pathogenic effect (CPE) on the culture cells along with PCR confirmation using a validated method.

Although virus isolation and antibody-based diagnosis of SARS have been demonstrated to be a reliable proof of SARS infection, neither of these two methods are sensitive enough for detection during the early phase of the disease.

Virus isolation of is a traditional test for detection of pathogen. The advantage of virus isolation is that it demonstrates the presence of infectious virus and thus proves active and potentially infectious SARS-CoV infection in the patient. However, the success rate for virus isolation is low, and results from the Chinese University of Hong Kong indicate that PCR testing is superior to attempting to isolate virus.[34] Virus isolation can detect only live virus; consequently, virus that may have been inactivated during long-term shipping or storage would not be detected, whereas their RNA may potentially be detected by RT-PCR. Furthermore, virus isolation is hazardous for routine clinical laboratories because of the risk of laboratory-acquired infections with this virus. Biosafety level 3 laboratory facilities are required for cell culture recovery and identification of this virus.[35] Inactivation of SARS-CoV by autoclaving prior to testing by PCR may provide the potential for the safe processing of the specimen by laboratory personnel.[36] Finally, virus isolation is time-consuming and often takes 3 to 5 days.

Because specific antibodies to SARS-CoV could be generated in SARS patients, serological tests were also used for SARS diagnosis. Four-fold or greater rise in antibody titer between acute- and convalescent-phase sera of SARS patients is also one of the criteria for SARS-CoV infection. Peiris et al. found seroconversion in 93% of 75 patients by day 30 after onset of symptoms.[10] None had antibodies prior to day 10, and the mean time period to seroconversion was 20 days.[37] However, some cases were emergency admissions, had no time to recover, and the patient died.[38] Some well-advertised "rapid tests" have caused much excitement in the media, as they seemed to offer a quick answer; the speed with which an antibody test result is available, however, does not help at all with the commonest and most urgent practical problem (i.e., to establish a reliable diagnosis in suspect cases), for which antibody testing is unsuitable due to the appearance of antibodies 8 to 14 days after onset of illness. Therefore, it is unsuitable for early diagnosis of the disease. In contrast, the finding of SARS-CoV antibodies in animal handlers without a history of clinical disease compatible with SARS and presumably exposed to closely related but possibly nonhuman-pathogenic coronaviruses isolated from different species of animals in southern China,[26,39] may point to a possible explanation for such phenomena. However, most studies found no background seroprevalence against SARS-CoV in the control populations screened so far.

Antibody testing is most suited to retrospectively confirming the diagnosis of SARS and may help to further elucidate the epidemiology of this novel disease. Although unsuitable during the acute phase of illness when a reliable diagnosis is needed most urgently, it has the advantage of requiring only a blood specimen and probably being little time-consuming after patients are beyond the first few weeks of their illness.

Compared to serology, the use of PCR technology is critical because target nucleic acid of the virus can be detected in specimens from patients in the early stages of infection. WHO recommends nested PCR for the detection of the SARS-associated coronavirus. Poon et al. found that of 50 nasopharyngeal aspirates collected 1 to 3 days after onset of disease, 40 (80%) were positive for SARS-CoV target nucleic acid.[40]

PCR METHODS FOR SARS-CoV EARLY DIAGNOSIS

Considering that the timely recognition of SARS-CoV infection would prevent the spread of the disease to populations at large in huge geographic areas, an early warning system must include a rapid and specific detection of the causative agent, which at present is a major task. Thus, developing early, rapid diagnostic methods is urgent and should be of the highest priority for monitoring and controlling SARS. PCR is a powerful technique for the identification of SARS-CoV and has the potential for confirming infection type within several hours or, ultimately, minutes.

When SARS first emerged, WHO recommended nested PCR for the detection of the SARS-associated coronavirus, but it is easy to contaminate PCR products and cause false-positive results with general PCR. In addition, the ethidium bromide used for staining agarose gels is a strong mutagen and is harmful for laboratory personnel. Real-time fluorescent quantitative PCR (qPCR), in which a fluorescent labeled probe was used to quantitate the copy number of a target gene, has been developed (Figure 13–6).[41–43] Because all steps are performed in a closed tube, the possibility of contamination with PCR products is decreased, and the use of a target-specific probe increases the specificity of detection.

Legacy PCR assays

As noted, the SARS-CoV genome is single-stranded RNA, and the complementary DNA must be synthesized first by using an RT polymerase. The procedure for amplifying the RNA genome (RT–PCR) requires a pair of oligonucleotide primers. The target primer sequences must be unique to identify a specific organism or an organism group. These primer pairs are designed on the basis of the known RNA polymerase or nucleocapsid sequence of SARS-CoV and can specifically amplify RNA. DNAs generated by using these specific primers can be further analyzed via molecular genetic techniques such as sequencing. The sensitivity of PCR tests for SARS depends on the specimen and the time of testing during the course of the illness. Thus, depending on the state of the specimen or the time of illness during sampling, a false negative by PCR testing in real cases of SARS can occur. Sensitivity can be increased if multiple specimens and/or multiple body sites are tested.

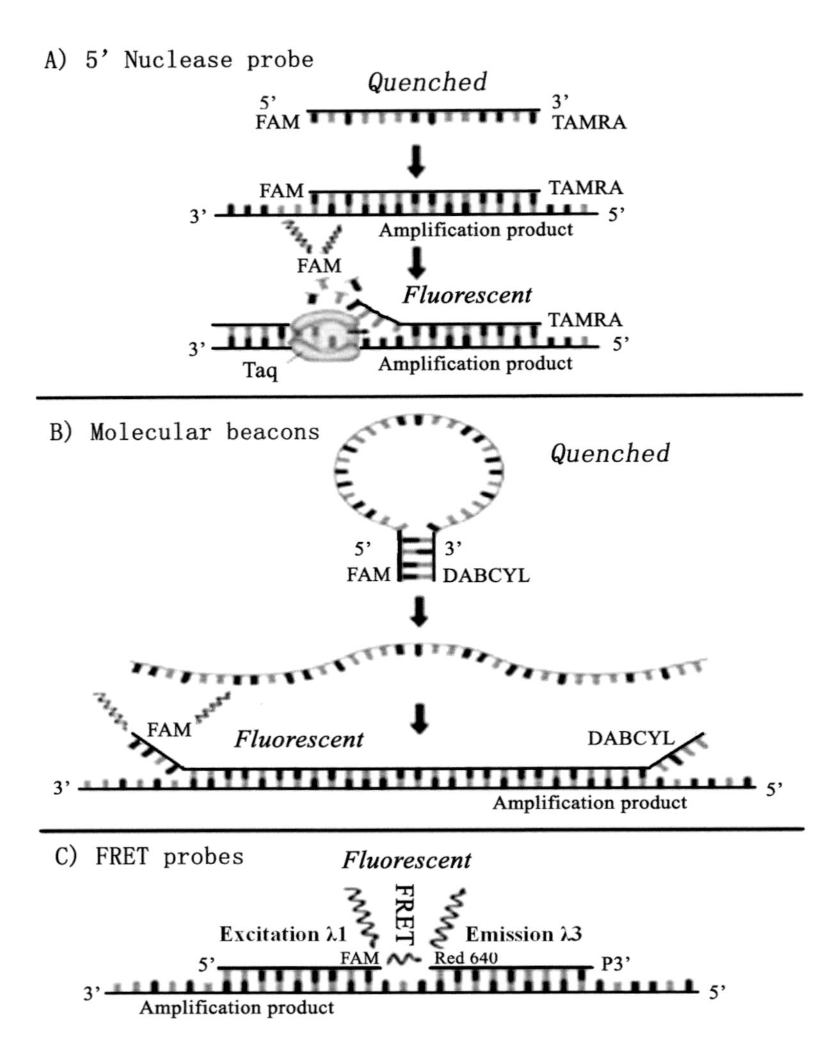

Figure 13–6. Different probes used in real-time quantitative reverse-transcription-polymerase chain reaction.

During the SARS outbreak, the PCR-based testing for SARS focused mainly on the analysis of nasopharyngeal aspirates, urine, and stool; these different samples can provide different results. Data from an early study indicated that PCR testing indicated positive for SARS in 32% of nasopharyngeal aspirates from SARS patients obtained at a mean time of 3.2 days after the onset of illness; the detection rate increased to 68% at day 14. In the same study, SARS-CoV RNA was detected in 97% of stool samples collected at a mean of 14.2 days after symptom onset.[44,45] Otherwise, SARS-CoV RNA was detected in 75% of blood samples collected at a mean of 14 days after symptom onset.[46]

Real-time PCR assays (qPCR)

qPCR assays can provide a novel, rapid means of virus detection in diagnostic laboratories: the kind needed to obtain quick and accurate confirmation of

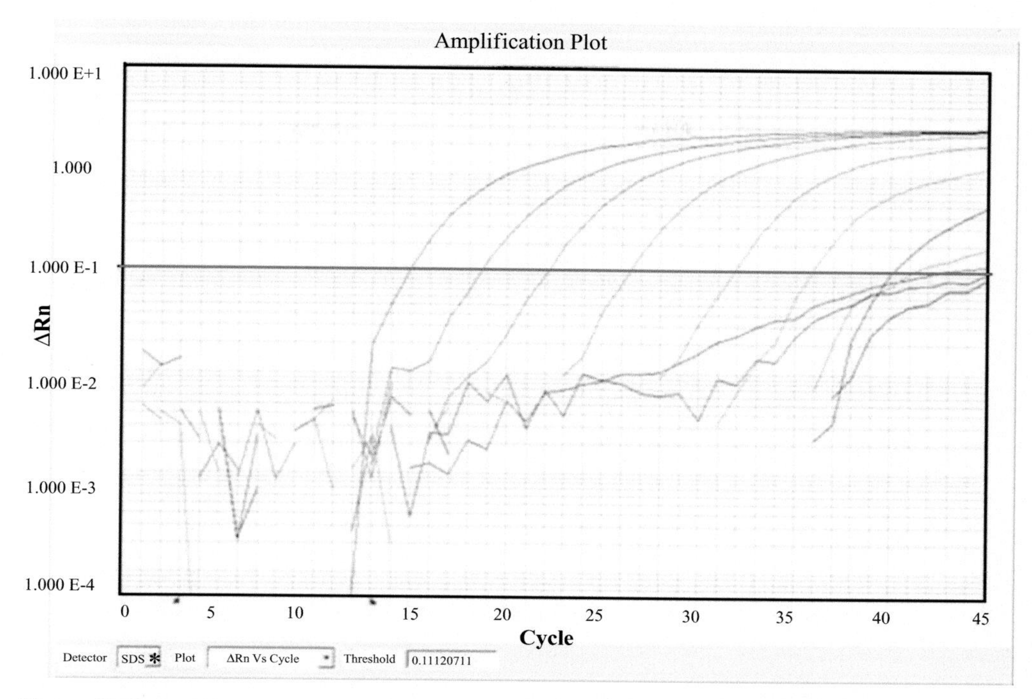

Figure 13–7. Detection of severe acute respiratory syndrome-coronavirus RNA by real-time quantitative reverse transcription–polymerase chain reaction (RT–PCR) for the nucleocapsid region of the viral genome. An amplification plot of △Rn, which is the fluorescence intensity over the background (*y*-axis) against the PCR cycle number (*x*-axis).[50] *See Color Plates.*

infection. Compared to the legacy single or nested PCR methods noted, the diagnostic application of the qPCR assays has certain advantages, such as: (1) faster and higher throughput; (2) no necessary handling of post-PCR products; (3) potentially higher sensitivity than traditional PCR (due to fluorescence signal amplification); (4) reduced possibility of contamination (because the amplified products are detected by measuring fluorescence in the reaction vessel without having to open the system); and (5) ability to quantify obtained result (not only a simple "positive" or "negative"). The variety and details of all the advantages of diagnostic qPCR assays have previously been reported in Mohamed et al.[47]

An RT-qPCR assay targeting SARS-CoV RNA has recently been shown to detect SARS accurately in different samples of pediatric patients during several different stages of infection.[48] Zhai et al. have developed a qPCR assay to detect SARS-CoV from feces, urine, and blood.[49] Chen et al. have also developed a qPCR assay to detect SARS-CoV from blood samples (Figure 13–7).[50]

RT-qPCR assay is far more sensitive than traditional RT-PCR. It can be also used to monitor the effect of antiviral agents. In addition to these current uses, it would also be valuable to explore the potentially damaging effect of giving steroids at a time when the viral load is still relatively high.

On the whole, despite the rapid discovery of the causative agent and the early development of diagnostic tests, further progress on the laboratory diagnosis of SARS has been somewhat slower than might have been expected. Although numerous PCR-based assays (some of which are technically superb) have been developed, there is still no test that can be used to rule out the diagnosis of SARS in a suspect case, due to the comparatively low virus excretion during the early course of SARS. In the current post-outbreak phase, thorough evaluation of suspect cases for other agents known to cause atypical pneumonia, such as influenza and parainfluenza viruses, *Legionella pneumophila, Mycoplasma pneumoniae,* and so forth, is even more important.

CONCLUSIONS

In summary, PCR-based methods will play a greater role in the earlier diagnosis and identification of SARS-CoV infection. Compared with conventional RT-PCR, qPCR is more sensitive, specific, and convenient to detect viruses that may be large-scale epidemics in the future.

REFERENCES

1. Lee N, Hui D, Wu A, Chan P, et al. (2003) A major outbreak of severe acute respiratory syndrome in Hong Kong. *New England Journal of Medicine* **348**: 1986–1994.
2. World Health Organization (WHO) (2003). Summary table of SARS cases by country, 1 November 2002–7 August 2003. Available at http://www.who.int/csr/sars/country/country2003_08_15.pdf.
3. SARS Expert Committee (2003) Report of the SARS Expert Committee. SARS in Hong Kong: from experience to action. October 2003. http://www.sars-expertcom.gov.hk/english/reports/reports.html.
4. Chan-Yeung M, Xu RH (2003) SARS: epidemiology. *Respirology* **8**: S9–S14.
5. World Health Organization (2003) Epidemic and pandemic alert and response (EPR). Summary of SARS and air travel. Available at www.who.int/entity/csr/sars/travel/airtravel/en/.
6. Meltzer MI (2004) Multiple contact dates and SARS incubation periods. *Emerging Infectious Diseases* **10**: 207–209.
7. World Health Organization (2003) Initiative for vaccine research (IVR). Available at http://www.who.int/vaccine_research/diseases/ari/en/index4.html.
8. Drosten C, Günther S, Preiser W, Van Der Werf S, Brodt HR, Becker S, et al. (2003) Identification of a novel coronavirus in patients with severe acute respiratory syndrome. *New England Journal of Medicine* **348**: 1967–1976.
9. Ksiazek TG, Erdman D, Goldsmith CS, Zaki S, Peret T, Emery S, et al. (2003) A novel coronavirus associated with severe acute respiratory syndrome. *New England Journal of Medicine* **348**: 1953–1966.
10. Peiris JS, Chu CM, Cheng VC, Chan KS, Hung IF, Poon LL, et al. (2003). Clinical progression and viral load in a community outbreak of coronavirus-associated SARS pneumonia: a prospective study. *The Lancet* **361**: 1767–1772.
11. World Health Organization (2003). Chapter 5. SARS: lessons from a new disease. Available at www.who.int/entity/whr/2003/en/Chapter5.pdf.
12. Marra MA, Jones SJ, Astell CR, Holt RA, Brooks-Wilson A, Butterfield YS, et al. (2003) The Genome sequence of the SARS associated coronavirus. *Science* **300**: 1399–1404.
13. Rota PA, Oberste MS, Monroe SS, Nix WA, Campagnoli R, Icenogle JP, et al. (2003) Characterization of a novel coronavirus associated with severe acute respiratory syndrome. *Science* **300**: 1394–1399.
14. Qin ED, Zhu QY, Yu M, Fan BC, Chang GH, Si BY, et al. (2003) A complete sequence and comparative analysis of a SARS-associated virus (Isolate BJ01). *Chinese Science Bulletin* **48**: 941–948.

15. Snijder EJ, Bredenbeek PJ, Dobbe JC, Thiel V, Ziebuhr J, Poon LL, et al. (2003) Unique and conserved features of genome and proteome of SARS-coronavirus, an early split-off from the coronavirus group 2 lineage. *Journal of Molecular Biology* **331**: 991–1004.
16. Holmes, K.V. (2003) Editorial perspective: SARS-associated coronavirus. *New England Journal of Medicine* **348**:1948–1951.
17. Stadler K, Masignani V, Eickmann M, Becker S, Abrignani S, Klenk HD, et al. (2003) SARS – beginning to understand a new virus. *Nature Reviews (Microbiology)* **1**: 209–218.
18. Fields BN, Knipe DM, Howley PM, Griffin DE (2001) Fields Virology. Fourth edition. Chapter 35. Philadelphia: Lippincott Williams & Wilkins.
19. Ziebuhr J (2004) Molecular biology of severe acute respiratory syndrome coronavirus. *Current Opinion in Microbiology* **7**: 412–419.
20. Groneberg DA, Hilgenfeld R, Zabel P (2005) Molecular mechanisms of severe acute respiratory syndrome (SARS). *Respiratory Research* **6**: 8.
21. Lai MM, Holmes KV (2005) Coronaviridae and their replication. In: *Fields-Virology* **1**: 1163–1185. Lippincott Williams and Wilkins, London.
22. Tan YJ, Lim SG, Hong W (2005) Characterization of viral proteins encoded by the SARS-coronavirus genome. *Antiviral Research* **65**: 69–78.
23. Zeng FY, Chan CW, Chan MN, Chen JD (2003) The complete genome sequence of severe acute respiratory syndrome coronavirus strain HKU-39849 (HK-39). *Experimental Biology and Medicine* **228**: 866–873.
24. Liang G, Chen Q, Xu J, Liu Y, Lim W, Peiris JS, et al. (2004) SARS Diagnosis Working Group. Laboratory diagnosis of four recent sporadic cases of community-acquired SARS, Guangdong Province China. *Emerging Infectious Diseases* **10**: 1774–1781.
25. Liang L, He C, Lei M, Li S, Hao Y, Zhu H, et al. (2005) Pathology of guinea pigs experimentally infected with a novel reovirus and coronavirus isolated from SARS patients. *DNA and Cell Biology* **24**: 485–490.
26. Guan Y, Zheng BJ, He YQ, Liu XL, Zhuang ZX, Cheung CL, et al. (2003) Isolation and characterization of viruses related to the SARS coronavirus from animals in southern China. *Science* **302**: 276–278.
27. Martina BE, Haagmans BL, Kuiken T, Fouchier RA, Rimmelzwaan GF (2003) SARS virus infection of cats and ferrets. *Nature* **425**: 915.
28. Anderson RM, Fraser C, Ghani AC, Donnelly CA, Riley S, Ferguson NM, et al. (2004) Epidemiology, transmission dynamics and control of SARS: the 2002–2003 epidemic. *Philosophical Transactions of the Royal Society of London. Series B, Biological Sciences* **359**: 1091–1105.
29. Shen Z, Ning F, Zhou W, He X, Lin C, Chin DP, et al. (2004) Superspreading SARS events, Beijing, 2003. *Emerging Infectious Diseases* **10**: 256–260.
30. Lu ZR, Qu LH (2004) Animal-to-human SARS-associated coronavirus transmission? *Emerging Infectious Diseases* **10**: 959.
31. Chen W, Yan M, Yang L, Ding B, He B, Wang Y, et al. (2005) SARS-associated coronavirus transmitted from human to pig. *Emerging Infectious Diseases* **11**: 446–448.
32. Hui DS, Chan MC, Wu AK, Ng PC (2004) Severe acute respiratory syndrome (SARS): epidemiology and clinical features. *Postgraduate Medical Journal* **80**: 373–381.
33. World Health Organization. Epidemic and Pandemic Alert and Response (EPR). Use of laboratory methods for SARS diagnosis. Available at http://www.who.int/csr/sars/labmethods/en/.
34. Chan PK, To WK, Ng KC, Lam RK, Ng TK, Chan RC, et al. (2004) Laboratory diagnosis of SARS. *Emerging Infectious Diseases* **10**: 825–831.
35. World Health Organization (2003), posting date. World Health Organization biosafety guidelines for handling of SARS specimens. Available at http://www.who.int/csr/sars/guidelines/en/.
36. Espy MJ, Uhl JR, Sloan LM, Rosenblatt JE, Cockerill FR 3rd, Smith TF (2002) Detection of vaccinia virus, herpes simplex virus, varicella-zoster virus, and Bacillus anthracis DNA by LightCycler polymerase chain reaction after autoclaving: implications for biosafety of bioterrorism agents. *Mayo Clinic Proceedings* **77**: 624–628.

37. Gao W, Tamin A, Soloff A, D'Aiuto L, Nwangebo E, Robbins PD, et al. (2003) Effects of a SARS-associated coronavirus vaccine in monkeys. *The Lancet* **362**(9399): 1895–1896.
38. Li G, Chen X, Xu A (2003) Profile of specific antibodies to the SARS-associated coronavirus. *New England Journal of Medicine* **349**: 508–509.
39. Centers for Disease Control and Prevention (2003) Prevalence of IgG antibody to SARS-associated coronavirus in animal traders – Guangdong Province, China, 2003. *MMWR Morbidity and Mortality Weekly Report* **52**(41): 986–987.
40. Poon LL, Chan KH, Wong OK, Yam WC, Yuen KY, Guan Y, et al. (2003) Early diagnosis of SARS coronavirus infection by real time RT-PCR. *Journal of Clinical Virology* **28**: 233–238.
41. Higuchi R, Dollinger G, Walsh PS, Griffith R (1992) Simultaneous amplification and detection of specific DNA sequences. *Biotechnology (NY)* **10**: 413–417.
42. Livak KJ, Flood SJ, Marmaro J, Giusti W, Deetz K (1995) Oligonucleotides with fluorescent dyes at opposite ends provide a quenched probe system useful for detecting PCR product and nucleic acid hybridization. *PCR Methods and Applications* **4**: 357–362.
43. Chen S, Yee A, Griffiths M, Larkin C, Yamashiro CT, Behari R, et al. (1997) The evaluation of a fluorogenic polymerase chain reaction assay for the detection of Salmonella species in food commodities. *International Journal of Food Microbiology* **35**: 239–250.
44. Peiris JS, Lai ST, Poon LL, Guan Y, Lam LY, Lim W, et al. (2003) Coronavirus as a possible cause of severe acute respiratory syndrome. *The Lancet* **361**: 1319–1325.
45. Poon LL, Wong OK, Chan KH, Luk W, Yuen KY, Peiris JS, et al. (2003) Rapid diagnosis of a coronavirus associated with severe acute respiratory syndrome (SARS). *Clinical Chemistry* **49**: 953–955.
46. Chen WJ, Xu ZY, Mu JS, Yang L, et al. (2004) Antibody response and viraemia during the course of severe acute respiratory syndrome (SARS)-associated coronavirus infection. *Journal of Medical Microbiology* **53**: 435–438.
47. Mohamed N, Belák S, Hedlund KO, Blomberg J (2006) Development of a rational diagnostic single tube qPCR for human caliciviruses, Norovirus genogroup I and II. *Journal of Virological Methods* **132**: 69–76.
48. Ng EK, Ng PC, Hon KL, Cheng WT, Hung EC, Chan KC, et al. (2003) Serial analysis of the plasma concentration of SARS coronavirus RNA in pediatric patients with severe acute respiratory syndrome. *Clinical Chemistry* **49**: 2085–2088.
49. Zhai J, Briese T, Dai E, Wang X, Pang X, Du Z, et al. (2004) Real-time polymerase chain reaction for detecting SARS coronavirus, Beijing, 2003. *Emerging Infectious Diseases* **10**(2): 300–303.
50. Chen WJ, Xu ZY, Mu JS, Yang L, et al. (2004) Real-time quantitative fluorescent reverse transcriptase-PCR for detection of severe acute respiratory syndrome-associated coronavirus RNA. *Molecular Diagnosis* **8**(4): 231–235.

14 The MMR vaccine, measles virus, and autism – A cautionary tale

Stephen A. Bustin

Although real-time reverse transcription–polymerase chain reaction (RT–qPCR) technology has become widely implemented in molecular diagnostics, it is worth pausing to consider the tremendous potential for real harm inherent in using such a sensitive and potentially easily contaminated assay in a clinical setting. Its central role in the controversy surrounding the triple measles, mumps, and rubella (MMR) virus vaccine, gut pathology, and autism serves as a textbook example of the enormous implications for the health of individuals that result from inappropriate use of this technology.

In 1996, a UK legal firm approached a gastroenterologist, Dr. Andrew Wakefield, who was then working at the Royal Free Hospital in London. He was asked to examine a group of children whose parents believed that their children's behavioral symptoms were directly caused by the MMR vaccine. In a 1998 *Lancet* publication, Wakefield reported on twelve autistic children with intestinal abnormalities, eight of whom had been supposedly affected after receiving the MMR vaccine.[1] Although stating that the findings did not prove an association between the MMR vaccine and the syndrome described, the article raised the possibility of a causal link between MMR, gut pathology, and autism. Wakefield later went on to speculate that the measles component of the vaccine had infected the children's intestines and in some way caused brain damage.

The reported detection of measles virus (MV) in the intestinal tissue of autistic children has been at the center of contentions of an association between MV and autism. It has been used to claim a link between the MMR vaccine and the development of a new, regressive form of autism in children. Hence the reliability

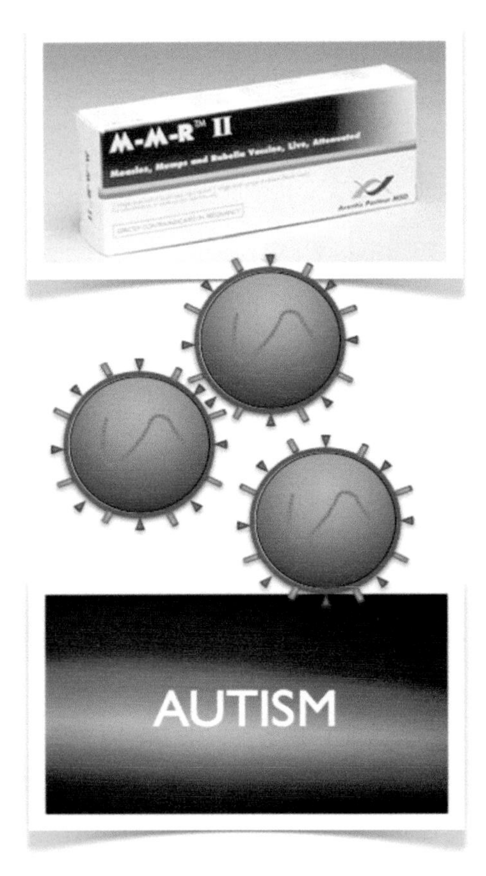

Figure 14–1. Measles, mumps, and rubella (MMR), measles virus (MV), and autism. Detection of MV (*middle*) is a prerequisite for a link between the triple MMR vaccine (l) and the development of a disorder on the autistic spectrum (lm). *See Color Plates.*

of the RT–qPCR data was a critical issue for two antivaccine campaigns, one in the UK, the other in the United States. In the UK, parents of more than 1,400 children participated in litigation against vaccine manufacturers. In the United States, the families of more than 4,800 children were claiming damages from the $2.5 billion government fund set aside to compensate people harmed by vaccination.

Probably as a direct effect of this controversy, MMR coverage in England fell from a peak of 92% in 1995 to 80% in 2003–2004, and the 2007–2008 statistics show an uptake rate of only 84% in England and Wales, well below the 95% required for herd immunity (http://www.ic.nhs.uk/statistics-and-data-collections/health-and-lifestyles/immunisation). In London, vaccine uptake was as low as 70% in 2003–2004 and even by 2006 stood at only 73% for the first dose at age two (2006 data from 22/31 primary care trusts). Coverage in some areas (e.g., Kensington and Chelsea) was as low as 52% with several other percentages in the high fifties to low sixties. In Greenwich, only 61% and 33% were vaccinated by their second and fifth birthdays, respectively. In 2006, a thirteen-year-old boy who had not received the MMR vaccine became the first person in the UK in fourteen years to die of measles.

A causal link requires a smoking gun and, unfortunately for Wakefield's conjectures, there were no concrete scientific data to support any link. The obvious requirement was for the demonstration of persistent MV infection of the intestine of autistic children (Figure 14–1); hence the excitement that greeted the publication in 2002 of an article purporting to have identified that smoking gun.[2] It described the use of an RT–qPCR assay to investigate the presence of persistent MV in the intestinal tissue of a cohort of children with a "new form of developmental disorder, ileocolonic lymphonodular hyperplasia." The article claimed that 75 of 91 patients with a histologically confirmed diagnosis of ileal lymphonodular hyperplasia and enterocolitis were positive for MV in their intestinal tissue compared with 5 of 70 control patients. It concluded that the data confirm

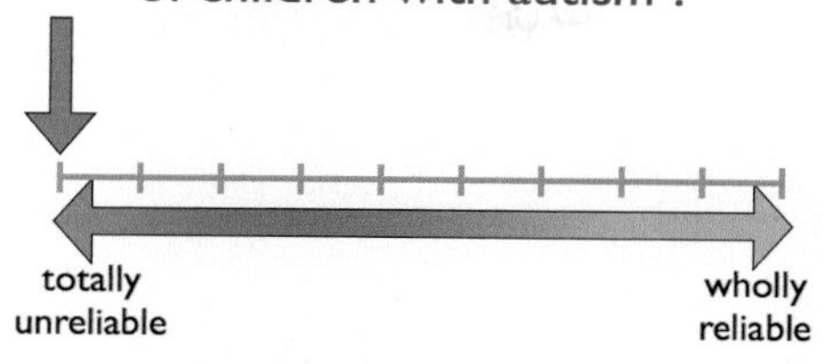

Figure 14–2. Reliability of quantitative polymerase chain reaction evidence. MV, measles virus; GI, gastrointestinal. *See Color Plates.*

an association between the presence of MV and gut pathology in children with a developmental disorder. These findings placed this publication at the center of the proceedings launched against the MMR vaccine manufacturers in the UK High Court and the Office of Special Masters of the U.S. Court of Federal Claims ("U.S. Vaccine Court").

The raw instrument-generated data underlying this publication were eventually made available to the author after an Irish court order was made forcing their release. This release permitted the detailed examination and reanalysis of all the individual RT–qPCR runs and revealed a catalogue of problems that included unclear data reporting, inappropriate data analysis, nonreproducibility of the assay, and evidence of widespread DNA contamination. Taken together, the reanalysis of the RT–qPCR data clearly shows no evidence for the presence of MV in the guts of children with developmental disorders (Figure 14–2). Instead, the assay at the time was detecting measles DNA, which, because MV does not naturally exist in DNA form, must have been due to laboratory contamination. Consequently, because the MMR/measles/autism conjecture requires persistent MV infection, this finding removes the scientific basis for any such association; indeed it proves the opposite. The following sections describe the details of these revelations.

MEASLES VIRUS

MV is an enveloped RNA virus; its genome consists of nonsegmented, negative sense, single-stranded RNA encased in the nucleocapsid (N) protein. Its envelope contains virus-encoded hemagglutinin (H) and fusion (F) glycoproteins embedded in the lipid bilayer, with the membrane or matrix (M) protein lying immediately below the membrane. MV attaches to the host cell through the interaction of the viral H and F glycoproteins with cellular receptors (Figure 14–3). After fusion of the virion with the cell membrane, the negative strand ribonucleoprotein (RNP) complex enters the cytoplasm where it acts as a template for both primary transcription of viral messenger RNAs (mRNAs) and replication of the negative-stranded genome RNA into positive antigenome RNA. This RNA in turn is reverse transcribed into negative-strand RNA, which is immediately encapsidated to generate negative-strand RNPs, which are transported to the cell membrane where they associate with viral M protein and the glycoproteins in lipid raft structures from which the RNPs bud to form new virus particles. The most important fact is that at no stage of its life cycles is there a DNA intermediate; hence the detection of MV by PCR must be preceded by an RT step.

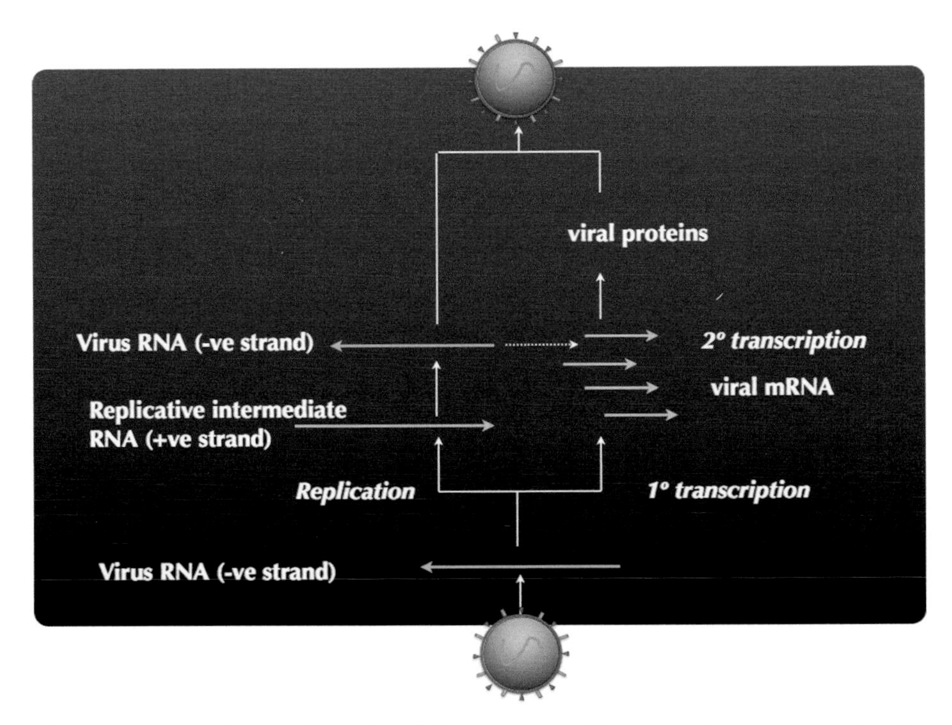

Figure 14–3. Measles virus life cycle. mRNA, messenger RNA, +ve, positive; −ve, negative. *See Color Plates.*

NOT JUST ONE PROBLEM, BUT ONE AFTER ANOTHER AFTER ANOTHER…

The first difficulty with the 2002 publication,[2] also known after its first author as the "Uhlmann paper," became apparent when the analysis of the raw data revealed that the results were obtained by targeting only the MV F gene. This revelation was in stark contrast to the article's abstract, which claimed detection of both H and F genes by RT–qPCR and provided a detailed description in the "Results" section of MV F-, N-, and H-gene primer optimization. The authors did indeed attempt to optimize the assays for both F and H genes, because a concordant test result, either positive or negative for two viral genes, would have given added confidence to the reliability of any results. However, they found that their H-gene assay was much more sensitive than their F-gene assay. Furthermore, there were instances when the F gene gave positive results, whereas the H-gene results were negative. However, rather than concluding that there were problems with their assays that required further investigation, they ignored the H-gene results and reported the F-gene results as positive for the presence of MV. Of course, all of this became apparent only when it was possible to examine the underlying raw data. So, problem number one is that the Uhlmann paper is misleading because it gives the impression that the results are based on data obtained from two viral genes, when in fact they are not.

Problem number two concerned the no template controls (NTCs). These are among the most important controls included with any assay, and are of particular

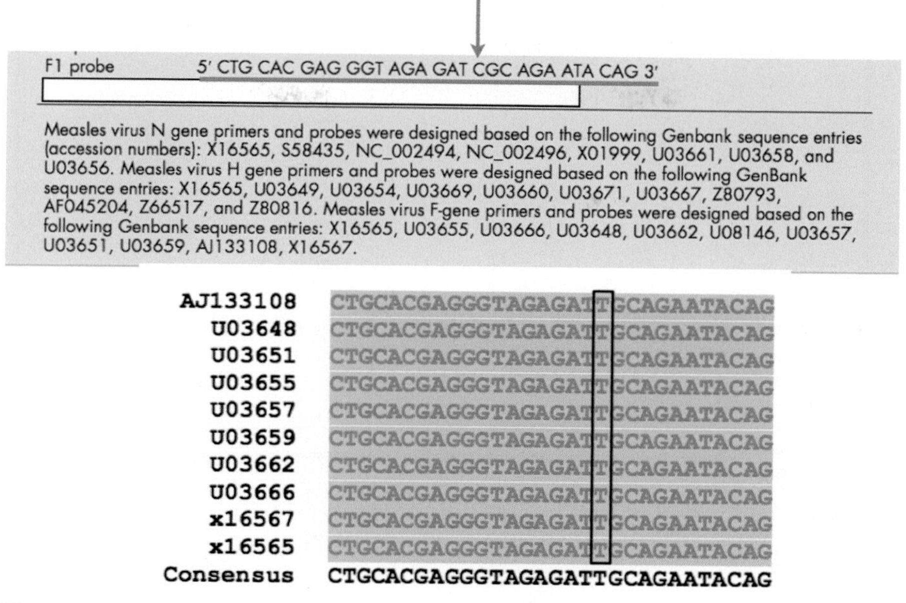

Figure 14–4. Identification of single base mismatch in the hydrolysis probe used for the detection of the measles virus (MV) F gene. Arrow points to the mistaken "C," which should clearly be a "T," if the authors had used the consensus sequences from the Genbank database, as claimed. *See Color Plates.*

importance when attempting to detect extremely low copy numbers of any target nucleic acid. The revelation that approximately 30% of NTCs gave positive results placed further serious doubt on the quality of data underlying this publication and for the first time raised the obvious prospect of contamination as an important culprit for the recording of the dubious data.

A third problem concerned the F-gene probe, which is reported as 5′-CTGCACGAGGGTAGAGATCGCAGAATACAG-3′ and was based on the following Genbank sequence entries: X16565, U03655, U03666, U03648, U03662, U08146, U03657, U03651, U03659, AJ133108, and X16567. An analysis of those sequence entries revealed a single mismatch of the F-gene probe with the consensus sequence of those Genbank entries (Figure 14–4). Although not necessarily terminal, this disclosure reveals a lack of care on the part of the authors and could have implications for the robustness of the assay. Certainly it does not inspire any confidence in the researchers or their tools, and it is totally unacceptable in an assay that is used in a diagnostic setting.

The next problem concerns the appropriate analysis of data. When recording a positive result, it is essential to ensure that the amplification plot that generates a positive quantification cycle (C_q) is the result of real amplification. For example, as shown in Figure 14–5A, a C_q can be recorded as positive when, in fact, there is an upward fluorescence drift that happens to cross the detection threshold and so generates that positive C_q. A simple adjustment of the threshold removes such spurious data and records a negative result (Figure 14–5A). In contrast, when the authors encountered such anomalies with samples from autistic children

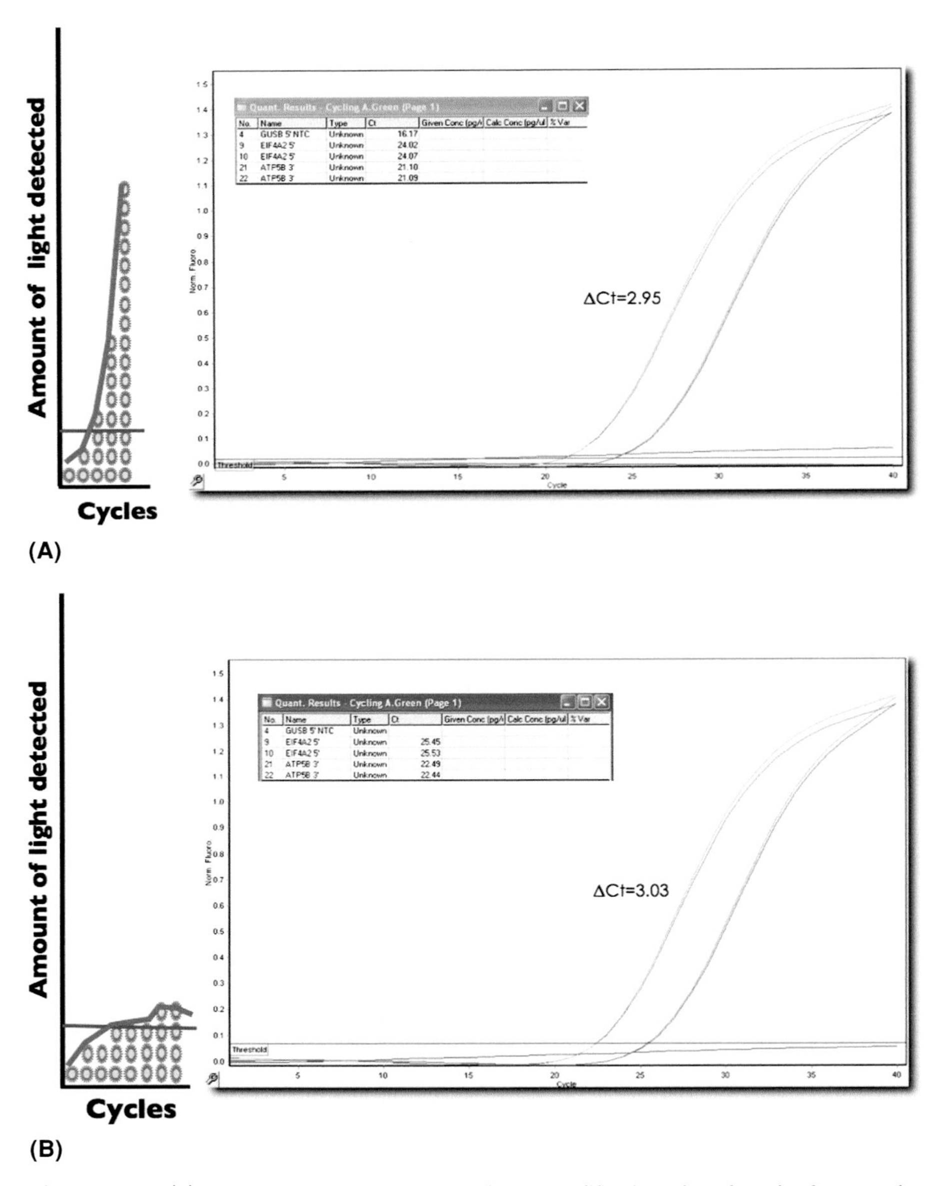

Figure 14–5. (A) Exponential amplification results in amplification plots that clearly cross the detection threshold of the real-time polymerase chain reaction (PCR) detection instrument. This positive amplification is evident from the two samples run in duplicate, which differ in their C_qs by 2.95. A third sample, shown in purple and barely crossing the threshold, does not show this pattern and is clearly spurious. **(B)** Adjusting the threshold more or less retains the ΔC_q value, but now shows a negative result for the spuriously amplifying sample. *See Color Plates.*

(Figure 14–6), they recorded them as positives. Consequently, the inappropriate analysis of the data generated false positives and resulted in the artificial inflation of the number of autistic children with supposed persistent MV infection.

Another problem came to light when analyzing the results of a study designed to test the reproducibility of the data obtained in Professor O'Leary's laboratory. Thirteen patient blood samples were shipped both to Professor O'Leary's

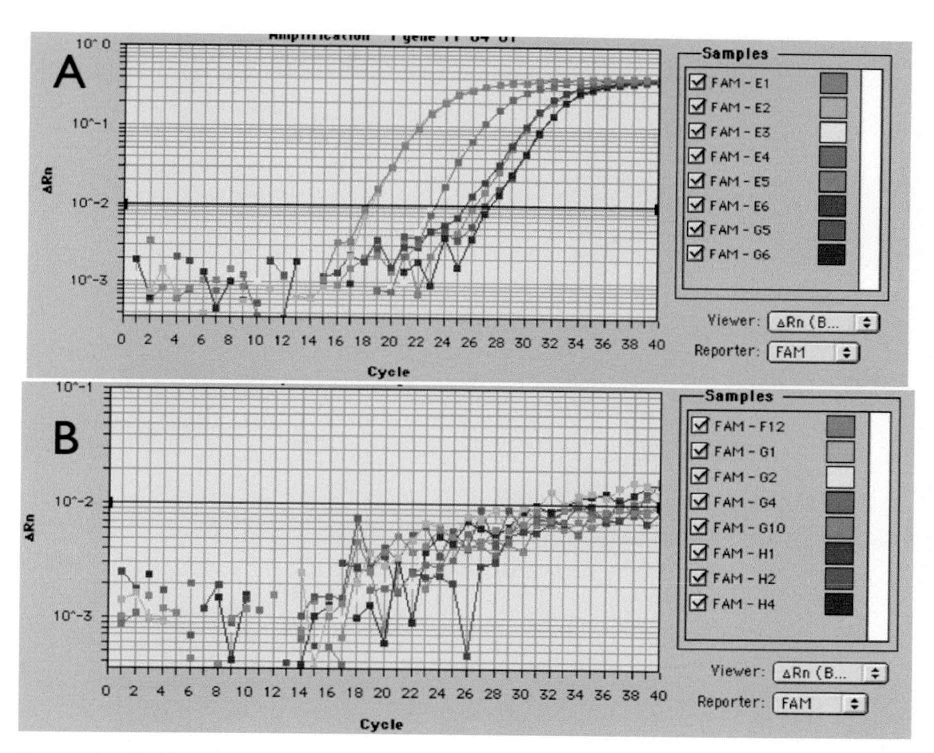

Figure 14–6. (A) Genuine amplification products. **(B)** Positive samples that should have been analyzed differently (i.e., by adjusting the threshold as suggested in Figure 14–5). *See Color Plates.*

laboratory and to Professor Cotter, who heads an independent laboratory carrying out routine RT–qPCR experiments. RNA was extracted at both locations, and RT–qPCR experiments were carried out (Figure 14–7A). All of Professor Cotter's samples returned negative results, whereas 3 of 13 samples were positive in Professor O'Leary's hands (Figure 14–8). The experiment was repeated, but with one crucial difference: Professor Cotter analyzed RNA extracted in Professor O'Leary's laboratory (Figure 14–7B). This time, Professor Cotter obtained positive results in 11 of 13 samples (Figure 14–8). Again, access to the raw data from both laboratories revealed that, in Professor Cotter's hands, the MV assay was more sensitive. This enhanced sensitivity explains the increased number of positives that he detected in O'Leary's RNA samples. Crucially, these experiments provide further strong evidence for contamination of the RNA samples extracted in Professor O'Leary's laboratory.

The next problem provides further evidence for contamination: Ironically the evidence is provided by experiments in Professor O'Leary's laboratory designed to ensure the generation of reliable data. Figure 14–9 shows how the laboratory's standard operating procedure (SOP) required every RNA sample to be analyzed for expression of a cellular reference gene, glyceraldehyde 3-phosphate dehydrogenase (GAPDH). The rationale behind this requirement is that only samples that contain GAPDH mRNA should be analyzed further, because those that are GAPDH negative are also not going to contain viral RNA. However, instead of

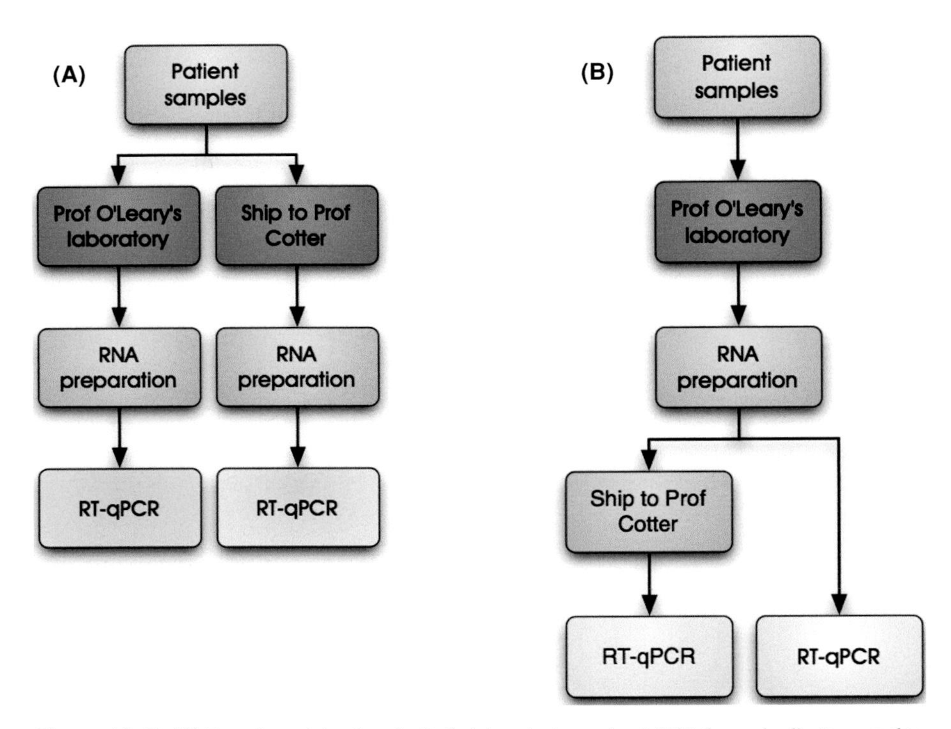

Figure 14–7. **(A)** Experimental setup 1: Both laboratories extract RNA from duplicate samples. **(B)** Experimental setup 2: O'Leary's laboratory extracts RNA from samples. Both laboratories analyze the same RNA samples. RT–qPCR, real-time reverse transcription–polymerase chain reaction. *See Color Plates.*

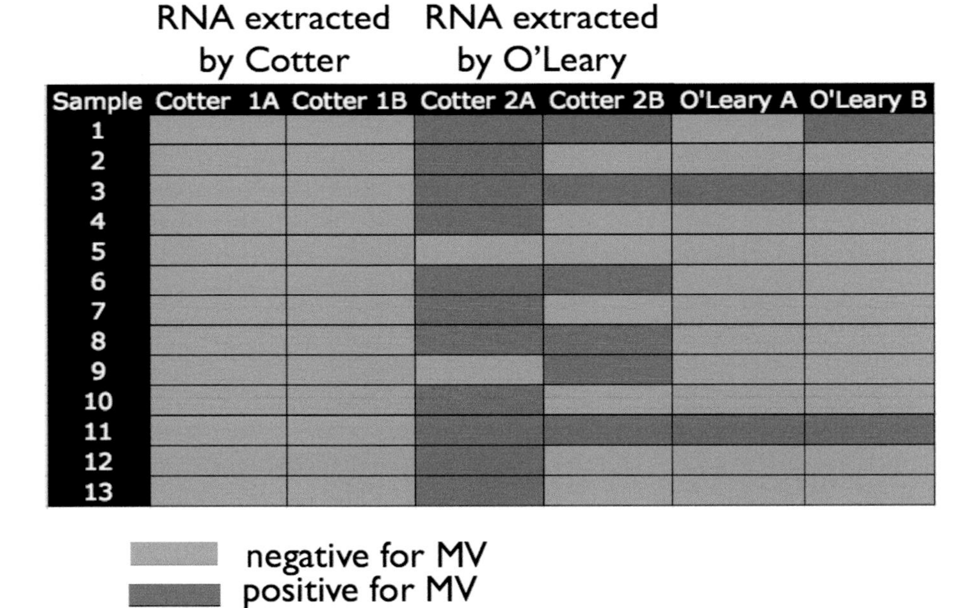

Figure 14–8. RNA extracted by the independent laboratory gives negative results for duplicate runs (Cotter 1A and 1B). RNA extracted in O'Leary's laboratory is positive, albeit discordantly, in the independent laboratory (Cotter 2A and 2B) as well as in O'Leary's laboratory (O'Leary A and B). MV, measles virus. *See Color Plates.*

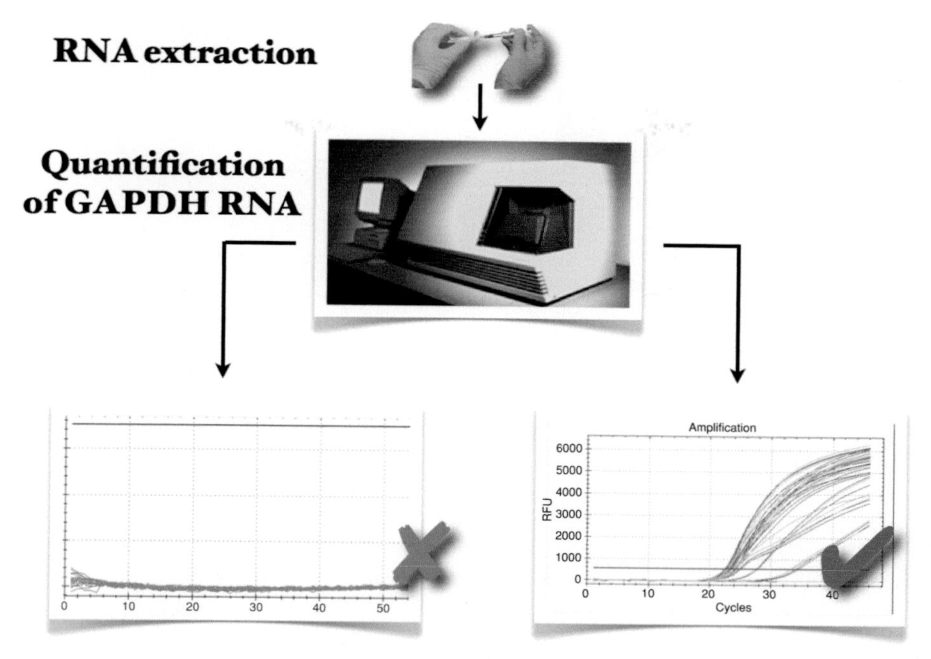

Figure 14–9. Schematic description of standard operating procedure requiring discarding of samples negative for glyceraldehyde 3-phosphate dehydrogenase (GAPDH; *red cross*) and further analysis for samples positive for GAPDH (*green check mark*). *See Color Plates.*

discarding GAPDH-negative samples, as required by their SOP, the authors did use those samples from autistic children and recorded positive results (Figure 14–10). Indeed, the average C_q is actually lower than that recorded for the samples positive for GAPDH. It really is remarkable that none of this started the alarm bells ringing.

A clear trend is beginning to emerge: The positive MV results recorded from autistic children are highly unlikely to be real. Instead, there is a variety of evidence that points strongly toward rampant contamination in the laboratory that, together with inappropriate data analysis and improper use of samples, is generating false-positive data. But, let us continue with the story.

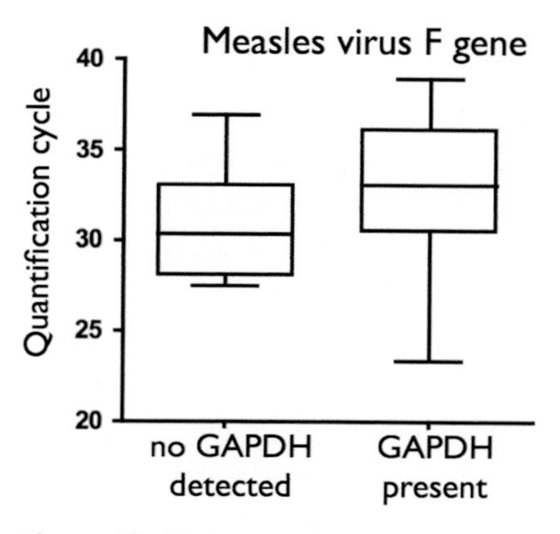

Figure 14–10. RNA analyzed from samples that do not contain RNA generate the same positive results as those from samples that do contain RNA.

The next problem concerns further inappropriate use of patient samples. The 2002 publication uses two kinds of biopsies: fresh/frozen and formalin-fixed. It is well known that formalin fixation alters nucleic acids, making them less amenable to RT and amplification.[3] Typically, samples amplified from formalin-fixed samples generate significantly lower quantification cycles and, hence, apparent copy

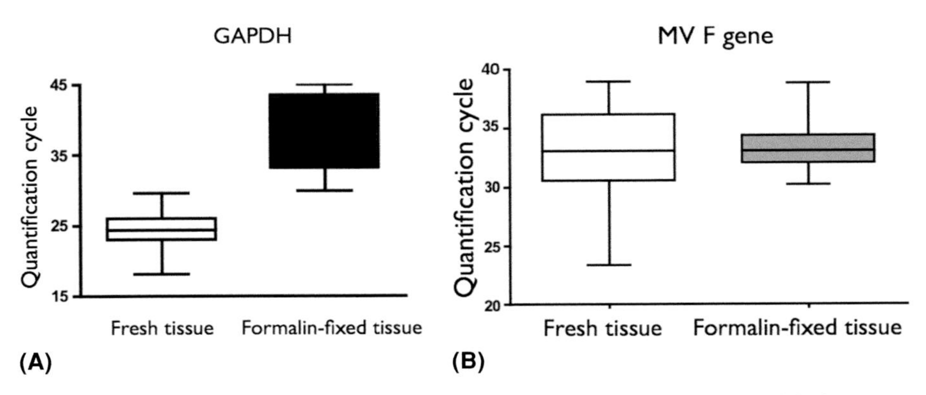

Figure 14–11. **(A)** Effects of formalin fixation on glyceraldehyde 3-phosphate dehydrogenase (GAPDH), with obvious difference in apparent copy numbers. **(B)** Effects of formalin fixation on measles virus (MV) F gene, with no effect on copy numbers.

numbers.[4] The significant qualitative differences between the RNAs makes it inappropriate to compare directly any RT–qPCR results.[5] This principle is beautifully illustrated by results from Professor O'Leary's own laboratory: When comparing the expression levels of the reference gene GAPDH, the average C_q recorded from fresh tissue samples was more than ten C_qs lower than the average recorded from formalin-fixed samples (Figure 14–11A). In complete contrast, the C_qs recorded for the MV F gene were virtually the same, with the MV F gene C_q actually slightly lower (i.e., apparent copy numbers slightly higher) than those from fresh/frozen samples (Figure 14–11B). He was hoisted with his own petard: Again, the data from his own laboratory provide the evidence for the assertion that Professor O'Leary's positive MV results are caused by contamination, in this case obviously introduced after the formalin-fixation process.

The coup de grâce is provided by the fortuitous discovery of unequivocal evidence that not only is there contamination, but that this contamination is DNA. Because MV does not exist as a DNA molecule in nature, any evidence that at least some of O'Leary's positives are the result of DNA amplification makes all his results unreliable and removes the smoking gun from the scene.

On two occasions O'Leary's laboratory accidentally omitted the RT step from two RT–qPCR runs. Standard curves behave as expected: In the absence of an RT step, the apparent C_qs for the F gene are nearly eight C_qs lower than for the same standards run with a preceding RT step (Figure 14–12A). This result is expected because Taq polymerase is not efficient at reverse transcribing RNA and must spend the first few cycles of the PCR assay inefficiently generating suitable DNA template. Consequently, the same amount of template will generate C_qs that are significantly higher than if an RT step had been included.

Fortunately, one of the RT-step–omitted runs contained four autistic patient samples. From O'Leary's own results, the expectation is that the C_qs of these four samples should be significantly higher than the C_qs from all other MV F-gene samples if the assay were detecting RNA. However, all four samples recorded positive C_qs that were in a similar range to most of the F-gene C_qs

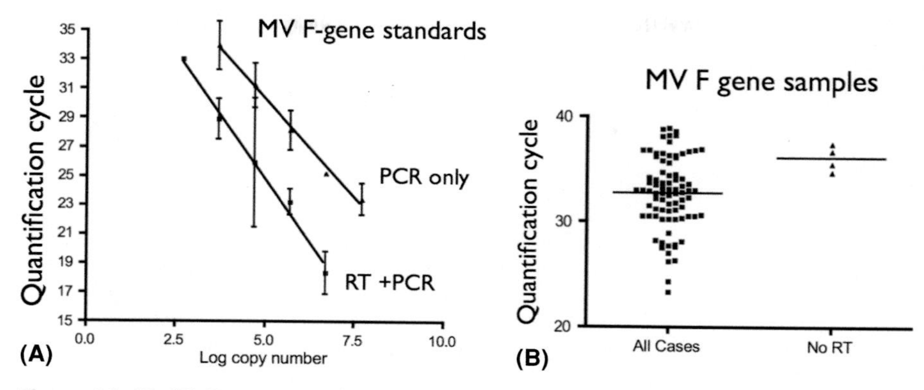

Figure 14–12. **(A)** F-gene standard curves with and without preceding reverse transcription (RT) step show clear difference in quantification cycles (C_qs). PCR, polymerase chain reaction. **(B)** Apparent copy numbers of samples from autistic children run without an RT step are broadly similar to those recorded from samples where the RT step was included.

recorded from runs that included the RT step (Figure 14–12B). These data provide incontrovertible evidence that the target detected by Professor O'Leary's laboratory is DNA.

THE FINAL PROOF

Several studies have attempted to reproduce the findings of the Uhlmann paper.[6–8] All failed to do so; instead they provide strong confirmation that contamination is the most likely cause of the positive findings. There were some technical differences, however, between the Uhlmann paper and the three more recent ones in the choice of tissue (gut vs. blood) or protocols (enzymes, real-time PCR chemistries). Therefore, although there was a strong suggestion that Professor O'Leary's laboratory was detecting contaminants, there was no proof. This situation has now changed with a recent publication, the authors of which include Professor O'Leary and Dr. Sheils, that has used the same methods, inter alia, as those originally published, to come to the conclusion that there is indeed no link between MV vaccine, autism, and enteropathy.[9] Astonishingly, there is no attempt to retract the original report, and the admission that these results are in direct contradiction of the previously reported ones are thoroughly disingenuous: "Our results differ with [their own] reports noting MV RNA in ileal biopsies of 75% of ASD vs. 6% of control children. Discrepancies are unlikely to represent differences in experimental technique because similar primer and probe sequences, cycling conditions and instruments were employed in this and earlier reports; furthermore, one of the three laboratories participating in this study performed the assays described in earlier reports. Other factors to consider include differences in patient age, sex, origin (Europe vs. North America), [gastrointestinal] disease, recency of MMR vaccine administration at time of biopsy, and methods for confirming neuropsychiatric status in cases and controls." Clearly, none of this applies: The obvious reasons for the different results obtained in the latest study are that (1) data analysis was carried out in an appropriate manner and (2) greater care was taken to avoid any possibility of contamination.

CONCLUSIONS

An exhaustive analysis of the experimental RT–qPCR data underlying the 2002 Uhlmann paper demonstrates numerous problems at every level of the RT–qPCR experimental setup, which results in the publication making misleading claims:

- 30% of negative controls are contaminated
- Assay design is incorrect
- Data are improperly analyzed
- Results cannot be independently reproduced
- Samples without RNA are positive for MV
- Fresh and archival samples gave the same result for MV
- Assay detects a DNA contaminant

As a result, there is no credible evidence for the presence of either MV genomic RNA or mRNA in the GI tracts (or blood samples) of any patient investigated by this laboratory. Consequently, this finding excludes any link between MV and, by extension, the MMR vaccine and autism. Unfortunately, this detailed analysis shows the RT–qPCR assay in a very bad light as it demonstrates how easy it is to get results that are not just wrong, but then take on a life of their own. Those results ultimately may have contributed to the unfortunate consequences of the low uptake rate for the MMR vaccine.

In February 2009, the U.S. Court of Federal Claims Office of Special Masters found that the MMR vaccine did not cause autism in any of the cases considered in three separate test cases representing more than 5,000 families. Special Master Hastings stated that "a key issue in this case concerns the reliability and validity of the laboratory testing that purported to find evidence of persisting measles virus in the intestinal tissue of . . . autistic children. After careful consideration, I conclude that the evidence indicates strongly that the testing in question was not reliable" (http://www.uscfc.uscourts.gov/node/5026). Special Master Vowell stated that there was "an overwhelming challenge to the reliability of Unigenetics' test results for measles virus" and concluded that "because of pervasive quality control problems at a now-defunct laboratory that tested a key piece of evidence, petitioners could not reliably demonstrate the presence of a persistent measles virus in . . . central nervous system. Petitioners failed to establish that measles virus can cause autism . . . ". Master Campbell-Smith found "that the laboratory practices at Unigenetics differed considerably not only from the standard practices for conducting PCR testing but also differed considerably from the operating procedures established within the laboratory" and that "the laboratory practices while conducting the PCR experiments in question were not scientifically sound, and the reported positive findings have not been replicated by researchers unaffiliated with the laboratories of either Dr. Wakefield or

Dr. O'Leary. Having carefully considered the record on this subject, the undersigned concludes that the published reports of findings of measles virus in the tissues of autistic children and the positive test results obtained from the Unigenetics laboratory were obtained through flawed laboratory practices and are therefore scientifically unreliable."

EPILOGUE

This sorry tale demonstrates with great clarity how the consistency and reliability of RT–qPCR assays depend on appropriate sample selection, template quality, assay design, and data analysis.[10] On the positive side, it has provided the impetus for an attempt to regulate the numerous, individual experimental protocols that can affect data reproducibility[11,12] and has led to the proposal of a series of guidelines, designated "minimum information for the publication of quantitative PCR experiments" (MIQE).[13] MIQE is a collection of procedures that describe the minimum information necessary for evaluation of qRT–PCR experiments. Included is a checklist to accompany the initial submission of a manuscript to the publisher. By providing all relevant experimental conditions and assay characteristics, reviewers can assess the validity of the protocols used. Full disclosure of all reagents, sequences, and analysis methods are necessary to enable other investigators to reproduce results. MIQE details should be published either in abbreviated form or as an online supplement. Following these guidelines will encourage better experimental practice, allowing reliable and unequivocal interpretation of qPCR results.

ACKNOWLEDGMENT

This work was carried out for the MMR vaccine litigation trial at the High Court of Justice in London. The author acted as an expert witness and was paid by the solicitors acting for the principal defendants SmithKline Beecham Plc and Smith Kline & French Laboratories Ltd, Merck & Co Inc, and Sanofi Pasteur MSD Ltd. The author also acted as an expert witness for the U.S. Department of Justice. The transcript of the detailed criticisms of the work underlying this article is available online (ftp://autism.uscfc.uscourts.gov/autism/transcripts/day08.pdf).

A witness has a simple, clear duty to assist the court. This duty overrides any obligation to the person from whom he has received instruction or payment. Analyses and conclusions are based on his qualifications and experience alone.

REFERENCES

1. Wakefield AJ, Murch SH, Anthony A, Linnell J, Casson DM, Malik M, et al. (1998) Ileal-lymphoid-nodular hyperplasia, non-specific colitis, and pervasive developmental disorder in children. *Lancet* **351**: 637–641.

2. Uhlmann V, Martin CM, Sheils O, Pilkington L, Silva I, Killalea A, et al. (2002) Potential viral pathogenic mechanism for new variant inflammatory bowel disease. *Molecular Pathology* **55**: 84–90.
3. Masuda N, Ohnishi T, Kawamoto S, Monden M, Okubo K (1999) Analysis of chemical modification of RNA from formalin-fixed samples and optimization of molecular biology applications for such samples. *Nucleic Acids Research* **27**: 4436–4443.
4. Koch I, Slotta-Huspenina J, Hollweck R, Anastasov N, Hofler H, Quintanilla-Martinez L, et al. (2006) Real-time quantitative RT-PCR shows variable, assay-dependent sensitivity to formalin fixation: implications for direct comparison of transcript levels in paraffin-embedded tissues. *Diagnostic Molecular Pathology* **15**: 149–156.
5. von Smolinski D, Leverkoehne I, von Samson-Himmelstjerna G, Gruber AD (2005) Impact of formalin-fixation and paraffin-embedding on the ratio between mRNA copy numbers of differently expressed genes. *Histochemistry and Cell Biology* **124**: 177–188.
6. Afzal MA, Ozoemena LC, O'Hare A, Kidger KA, Bentley ML, Minor PD (2006) Absence of detectable measles virus genome sequence in blood of autistic children who have had their MMR vaccination during the routine childhood immunization schedule of UK. *Journal of Medical Virology* **78**: 623–630.
7. D'Souza Y, Dionne S, Seidman EG, Bitton A, Ward BJ (2007) No evidence of persisting measles virus in the intestinal tissues of patients with inflammatory bowel disease. *Gut* **56**: 886–888.
8. D'Souza Y, Fombonne E, Ward BJ (2006) No evidence of persisting measles virus in peripheral blood mononuclear cells from children with autism spectrum disorder. *Pediatrics* **118**: 1664–1675.
9. Hornig M, Briese T, Buie T, Bauman ML, Lauwers G, Siemetzki U, et al. (2008) Lack of association between measles virus vaccine and autism with enteropathy: a case-control study. *PLoS ONE* **3**: e3140.
10. Nolan T, Hands RE, Bustin SA (2006) Quantification of mRNA using real-time RT-PCR. *Nature Protocols* **1**: 1559–1582.
11. Bustin SA (2005) Real-time, fluorescence-based quantitative PCR: a snapshot of current procedures and preferences. *Expert Review of Molecular Diagnostics* **5**: 493–498.
12. Bustin SA, Benes V, Nolan T, Pfaffl MW (2005) Quantitative real-time RT-PCR – a perspective. *Journal of Molecular Endocrinology* **34**: 597–601.
13. Bustin SA, Benes V, Garson J, Hellemans J, Hugett J, Kubista M, et al. (2009) The MIQE guidelines: minimum information for publication of quantitative real-time PCR experiments. *Clinical Chemistry* **55**: 611–622.

15 Noninvasive prenatal diagnosis using cell-free fetal nucleic acids in maternal plasma

Y. M. Dennis Lo

Prenatal diagnosis is now an established part of the modern obstetrics practice. For genetic and chromosomal analyses, however, conventional definitive methods for prenatal diagnosis would typically start with the invasive sampling of fetal materials, using procedures such as amniocentesis and chorionic villus sampling. These procedures are associated with a finite risk to the fetus.[1] Thus, over the last forty years, many researchers have attempted to develop methods for noninvasive prenatal diagnosis that do not carry such a risk. In particular, much effort has been spent on the development of noninvasive methods for screening certain chromosomal aneuploidies, especially trisomy 21. Approaches such as ultrasonography and serum biochemical screening have been developed for this purpose.[2] Although the recent developments in these approaches are remarkable, these methods essentially measure epiphenomena that are associated with chromosomal aneuploidies and do not analyze the core pathology of these disorders – namely, the actual chromosome abnormality.

To allow the direct analysis of this core pathology, a noninvasive source of fetal genetic material is needed. Investigators in this field have initially targeted fetal nucleated cells that may have entered into the maternal circulation, including trophoblasts,[3] lymphocytes,[4] and nucleated red blood cells.[5] However, the extreme rarity of such cells in the maternal circulation (of the order of a few cells per milliliter of blood) has been a major impediment to the development of the field.[6] Indeed, following a large clinical trial funded by the National Institute of Child Health and Human Development, the participating investigators concluded that "technological advances are needed before fetal cell analysis has clinical application as part of a multiple marker method for non-invasive prenatal screening."[7]

DISCOVERY OF CELL-FREE FETAL DNA IN MATERNAL PLASMA

In late 1996, two articles in the September 1996 issue of *Nature Medicine* caught my eye.[8,9] In these publications, the authors demonstrated that microsatellite alterations that occurred in certain tumors could be seen in the plasma or serum of a proportion of cancer patients. As the placenta of a fetus has certain similarities to a tumor, and is even referred to as "pseudomalignant" by some authors,[10] I wondered whether fetal DNA might also be present in the plasma and serum of pregnant women. This thinking represented a paradigm shift in the search of fetal genetic materials in maternal blood that had hitherto been focused on the cellular fraction of maternal blood. I had an additional simple thought: "As I have not yet seen a tumor which is as large as an 8 lb baby, I think that the chance of detecting cell-free fetal DNA, at least in the later stages of pregnancy, would be rather high!"

To demonstrate the presence of fetal DNA in maternal plasma and serum, my coworkers and I decided to use a target on the Y chromosome as a marker for male fetuses. We decided to use, as a detection system, a hot start polymerase chain reaction (PCR) system using Ampliwax,[11] instead of the contamination-prone nested PCR strategy. The most challenging step, however, was the choice of an extraction method for plasma and serum DNA. We decided to use a process that was as simple as possible, so that any laboratory would be able to reproduce the data. In the end, we used possibly the simplest method that we could find – namely, one that depended on the boiling of plasma and serum.[12]

The results of this combination of simple methods were beyond our imagination. We observed amplification products in the plasma and serum in a proportion of pregnant women after just a few optimization steps. When we compared the plasma/serum PCR results with the gender of the fetuses, it was obvious that Y chromosomal sequences were present in 70% to 80% of plasma/serum of pregnant women carrying male fetuses but was never observed in the samples of those carrying female fetuses. In essence, we have found fetal DNA in maternal plasma/serum![13] As we could observe a Y chromosomal signal in as few as 10 μL of boiled plasma/serum, we knew right from the beginning that cell-free fetal DNA had to be present in much higher concentrations than fetal nucleated cells in maternal blood. The latter would easily require milliliters of blood before robust detection could be observed.

REAL-TIME PCR: THE KEY TO ROBUST DETECTION AND QUANTITATION OF CELL-FREE FETAL DNA

Following the initial promising observations demonstrating the presence of fetal DNA in maternal plasma/serum, our next goals were to push the sensitivity of detection from 70% to 80% to close to 100% and to obtain quantitative data regarding the fractional and absolute concentrations of circulating fetal DNA. We were thus excited to hear about real-time quantitative PCR technology, which was just becoming available around that time.[14] We were fortunate to secure the necessary funding to install possibly the first of such machines (an Applied Biosystems 7700) in Southeast Asia in the summer of 1997.

In our first series of experiments on the real-time PCR machine, it was clear that the boiling method described in our first article on plasma DNA produced a DNA solution with PCR inhibitors that prevented accurate quantitation.[13] We thus promptly switched to alternative DNA extraction methods, finally using a column-based DNA extraction method (Qiagen) previously used for the detection of cancer DNA in plasma.[8] Following optimization using the new DNA extraction protocol and real-time PCR, the detection sensitivity of male fetal DNA in the plasma/serum of pregnant women carrying male fetuses was elevated to approach 100%, even during the first trimester, with no detection in women carrying female fetuses.[15]

The quantitative analysis of plasma/serum DNA demonstrated high absolute concentrations, with a mean value of 25 genome-equivalents/mL at 11 to 17 weeks of gestation, rising to a mean of 292 genome-equivalents/mL in the late third trimester.[15] The fractional concentrations of circulating fetal DNA were perhaps even more surprising, with a mean of 3% and 6% for the first/second trimester and the late third trimester, respectively. These surprisingly high values once again suggested that it should be relatively easy to develop robust assays for the detection of targets that were unique to the fetus.

The qualitative and quantitative results mentioned indicate that, for markers that are absolutely fetal specific (e.g., the Y chromosome of a male fetus), prenatal diagnosis using fetal DNA in maternal plasma would be accurate. These predictions have been shown to be correct by the many publications for gender prediction for sex-linked disorders and congenital adrenal hyperplasia using this approach.[16,17] In addition, fetal rhesus D (RhD) status has also been predicted with high accuracy by maternal plasma analysis using real-time PCR.[18] Indeed, this latter approach is so accurate that a number of centers are now offering this test diagnostically; at these centers, it has almost completely replaced conventional invasive tests for fetal RhD status.[19]

METHYLATION-SPECIFIC PCR AND RELATED TECHNOLOGIES: THE DEVELOPMENT OF FETAL EPIGENETIC MARKERS

The above-mentioned ability to measure circulating fetal DNA also has established the normative values for plasma fetal DNA concentrations in normal pregnancies, against which values from pathological pregnancies could be compared. Indeed, work in this direction has revealed that circulating fetal DNA concentrations are elevated in trisomy 21,[20] preeclampsia,[21,22] and preterm labor.[23] However, the translation of such observations into clinical tests for measuring circulating fetal DNA is complicated by the fact that these pilot studies have been based on the use of Y-chromosome markers. These markers are usable in only the 50% of pregnancies involving male fetuses. Thus, there is a genuine need for the development of "universal" fetal DNA markers, which can be used in all pregnancies.

However, using genetic markers, it is impossible to find a single marker that can differentiate fetal and maternal DNA 100% of the time. Realizing this limitation of genetic markers, we postulated in 2002 that it might be possible to exploit epigenetic differences between the fetal and maternal DNA to develop universal fetal DNA markers.[24] To detect such DNA methylation markers in maternal plasma, it was timely that a technique called methylation-specific PCR (MSP) was developed a few years prior to this work.[25] MSP is based on the prior treatment of DNA with bisulfite, which would convert unmethylated cytosine residues to uracil, while leaving methylated cytosine residues unchanged. Thus, following bisulfite treatment, DNA sequences of different methylation status will have different sequences, which can thus be amplified using different PCR primers. By using MSP on an imprinted locus that exhibited different methylation patterns (depending on whether it was inherited from the father or from the mother), we demonstrated that it was indeed possible to detect fetus-specific DNA methylation markers in maternal plasma.[24] Our report was essentially a proof-of-concept study, however, as the complexity of this system made it cumbersome to be used directly in routine diagnostics.

It took another three years before a nonimprinted locus was found. *SERPINB5* demonstrated a different pattern of DNA methylation between the placenta (hypomethylated) and maternal blood cells (hypermethylated).[26] The placenta was used because it was thought to be the major source of fetal DNA in maternal plasma.[27] In contrast, maternal hematopoietic cells were thought to be a major source of the nonfetal DNA in maternal plasma.[28] *SERPINB5* was then developed into the first universal fetal epigenetic marker for detection in maternal plasma. This developmental process was greatly facilitated by the development of a quantitative version of MSP, the so-called real-time MSP.[29] Real-time MSP allows quantitative analysis of DNA molecules possessing a particular DNA methylation pattern. Using real-time MSP, fetally derived hypomethylated *SERPINB5* sequences were detected in maternal plasma and were used as a marker for preeclampsia, irrespective of the gender of the fetus.[26]

MSP does have its own shortcomings, however, because the bisulfite conversion step destroys a great majority of the treated DNA.[30] This feature greatly reduces the number of fetal DNA molecules that are available for PCR detection. Realizing this disadvantage of MSP, Chan et al.[31] explored the use of methylation-sensitive restriction enzyme digestion followed by real-time PCR. As most methylation-sensitive restriction enzymes cut hypomethylated DNA, leaving hypermethylated

DNA intact, the *SERPINB5* gene, which is hypomethylated in the placenta, is not suitable for such analysis as the fetal sequence will be digested. Thus, one would need to find a target that is hypermethylated in the placenta but hypomethylated in maternal blood cells. Remembering that the field of fetal DNA in maternal plasma was started through the realization that the placenta has neoplastic features (the pseudomalignancy analogy),[10] we wondered whether this similarity might extend to the epigenetic level. As many tumor suppressor genes (TSGs) are hypermethylated in tumors, we tested if a number of such TSGs also might be hypermethylated in the placenta. Such a search led us to discover that a TSG, *Ras association domain family 1A* (*RASSF1A*), on chromosome 3 is hypermethylated in the placenta and unmethylated in maternal blood cells.[32] Through the use of methylation-sensitive restriction enzyme digestion followed by real-time PCR detection of the undigested hypermethylated DNA, this *RASSF1A* assay has been developed into a quantitative nonbisulfite-based technology for detecting fetal DNA in maternal plasma.[31] With further identification of other hypermethylated fetal DNA markers, this approach is expected to be applicable to an increasing number of markers.

MASS SPECTROMETRIC ANALYSIS OF PCR PRODUCTS FROM MATERNAL PLASMA FOR TRISOMY 21 DETECTION: IN SEARCH OF THE "HOLY GRAIL" OF PRENATAL DIAGNOSIS

As discussed earlier in this chapter, quantitative analysis of circulating fetal DNA has revealed that it represents a mean of 3% to 6% of all DNA in maternal plasma.[15] Although such fractional concentrations are sufficient for the robust detection of unique fetal targets such as Y chromosomal markers and the *RHD* gene, the fact that some 95% to 97% of the DNA in maternal plasma is maternally derived means that it would be a challenge to use maternal plasma DNA for the detection of fetal trisomy 21.

We reasoned that one way to do this was to target a subpopulation of fetal nucleic acids in maternal plasma that was virtually completely fetally derived. We could think of two such types of fetal nucleic acids: The first were DNA sequences bearing unique fetally specific DNA methylation patterns; the second were RNA molecules that were specifically expressed by fetal tissues. The first approach is theoretically possible, especially with the pilot studies described in the previous section that fetal epigenetic markers are detectable in maternal plasma.[26,31] In this regard, we have shown that trisomy 18 could be directly detected from maternal plasma by allelic ratio analysis of hypomethylated (i.e., fetus-specific) *SERPINB5* sequences, the so-called epigenetic allelic ratio approach.[33] For fetal trisomy 21 detection, however, the first step would be to identify sequences located on chromosome 21 that bear a fetally specific DNA methylation pattern. A number of such markers have indeed been found recently.[34]

The second approach, that based on plasma RNA detection, appears to be more accessible using current technologies. Nonetheless, when we first started our work on plasma RNA, there was a widespread belief that, because of the lability of RNA, it was not expected that plasma RNA would be a practical molecular diagnostic tool. Thus, in an early series of experiments, we showed that fetus-derived RNA molecules were detectable in the plasma of pregnant women.[35] Furthermore, plasma RNA was shown to be surprisingly stable,[36] an observation that might result from the fact that it was associated with subcellular particles in plasma.[37] In 2003, we showed that the placenta was an important source of fetal RNA in maternal plasma.[38] In this series of RNA-related work, real-time reverse transcriptase (RT)–PCR[39] and the use of the *Thermus thermophilus* polymerase,[40] which has both RT and DNA polymerase activities, have greatly facilitated the robustness of the assays.

The realization that placental RNA is detectable in maternal plasma is important because it has opened up a way for the rapid development of a large number of fetal RNA markers for noninvasive prenatal diagnosis. Thus, through the use of expression microarray technologies on placental tissues and mining for RNA species that are highly expressed in the placenta (but not in maternal blood cells), a number of new plasma RNA markers have become available in just a few months.[41]

For the plasma RNA approach to be usable for the noninvasive prenatal diagnosis of trisomy 21, the first step is to identify placental RNA markers that are expressed on chromosome 21. We have therefore mined our placental microarray data set for chromosome 21 transcripts that demonstrated a high absolute expression level in the placenta and that have a low relative expression in maternal blood cells.[42] At the top of our mined gene list was a gene called *placenta-specific 4* (*PLAC4*). Using one-step real-time RT–PCR, we showed that *PLAC4* messenger RNA (mRNA) was detectable in maternal plasma and was rapidly cleared following delivery.[42]

We then reasoned that to allow the prenatal detection of fetal trisomy 21 using plasma *PLAC4* mRNA, one would ideally need a "calibrator" that was individual specific. The best approach for "internal calibration" would be to use the *PLAC4* mRNA molecules transcribed from one chromosome 21 in comparison with those transcribed from the other chromosome(s) 21. Of course, this approach is usable only if the *PLAC4* mRNA molecules from one chromosome 21 are distinguishable from those transcribed from the other chromosome(s) 21; this situation would occur if a fetus is heterozygous for a single nucleotide polymorphism (SNP) in the *PLAC4* coding region. We called this the RNA–SNP allelic ratio approach.[42] Thus, if the fetus was normal, then the RNA–SNP allelic ratio would be 1:1. If the fetus has trisomy 21, then the allelic ratio would be 2:1 or 1:2.

We next needed an analytical method to accurately measure this RNA–SNP allelic ratio. We decided to use matrix-assisted laser desorption/ionization time-of-flight mass spectrometry (MALDI-TOF MS) for this purpose.[43] Through the use of MALDI-TOF MS on RT–PCR products of *PLAC4* transcripts from maternal plasma, trisomy 21 could be detected noninvasively with a sensitivity of 90% and a specificity of 96.5%.[42] These percentages imply that this plasma-based method is among the most accurate single-marker approach for the noninvasive detection of trisomy 21 yet developed. The main limitation at the moment is that through the use of just one SNP, only 45% of fetuses are heterozygous and thus amenable to this approach. With the use of more *PLAC4* SNPs and the discovery of more placenta-specific genes, it is hoped that increased population coverage would be forthcoming over the next few years. Furthermore, this method could also potentially be used for the detection of other trisomies important in prenatal diagnosis (e.g., trisomy 18 and trisomy 13).

DIGITAL PCR AND OTHER SINGLE-MOLECULE COUNTING METHODS

An alternative approach to the development of fetus-specific nucleic acid markers is the use of extremely precise detection methods that can detect a minor aberration in fetal chromosome dosage, even in DNA mixtures containing just a minor proportion of fetal DNA. One such method is digital PCR in which the nucleic acid sample to be analyzed is diluted to an extent in which half of a molecule, on average, would be amplified in each reaction.[44] When a large number of such digital PCRs are performed, each PCR would either contain one or zero template molecule. Thus, the concentration of the original, nonamplified material could be deduced by counting the number of positive amplifications. It has been proposed that digital PCR would allow the aneuploidy status of the fetus to be detected noninvasively, even without the prior separation of the fetal-derived and maternally derived DNA in maternal plasma.[45] It was further proposed that, in addition to digital PCR, other single-molecule detection methods, such as massively parallel genomic sequencing, could allow this diagnostic goal to be achieved.[45] Indeed, this prediction has recently been achieved by two groups, who used massively parallel genomic sequencing using the Illumina platform to sequence millions of plasma DNA molecules per case and obtained very robust detection of fetal chromosomal aneuploidies.[46,47] It was also recently demonstrated that such digital counting techniques also have application to the prenatal diagnosis of single gene disorders like β-thalassemia, through the precise measurement of the number of mutant alleles a fetus has inherited from its parents.[48]

CONCLUSIONS: PCR AS A KEY FACILITATOR IN THE PAST DECADE OF PROGRESS

Rapid progress in noninvasive prenatal diagnosis has been made since the discovery of cell-free fetal nucleic acids twelve years ago.[13,44] From this chapter, one can see that the PCR has been an essential tool in virtually every step of the development of this field. Thus, the first discovery of cell-free fetal DNA was facilitated through the sensitivity conferred by earlier PCR innovations such as hot start.[11] Real-time PCR played an essential role in demonstrating the high absolute and fractional concentrations of circulating fetal DNA[15] and allowed robust fetal DNA detection for gender prediction[16] and RhD determination.[18] Indeed, real-time PCR can be regarded as the key vehicle for bringing this technology from the bench into the clinic. The next phase of evolution in the field was dependent on PCR-based technologies for DNA methylation analysis – namely, MSP,[25] real-time MSP,[29] and methylation-sensitive restriction enzyme–mediated real-time PCR.[31] These developments have resulted in the development of the first series of universal fetal DNA markers that are not dependent on the gender or polymorphism status of the fetus.[26,31] In pursuit of the holy grail of noninvasive prenatal diagnosis (i.e., the detection of trisomy 21), PCR technologies for RNA analysis (namely, RT–PCR and its real-time variants)[39] and the analysis of PCR products by MALDI-TOF MS [43] have played a crucial role. These latter developments have culminated in a highly accurate new method for the noninvasive prenatal detection of trisomy 21, the RNA–SNP allelic ratio approach.[42] The very recent development of powerful single-molecule counting methods like digital PCR and massively parallel genomic sequencing has further enhanced the feasibility, population coverage, and robustness of noninvasive prenatal diagnosis.[49] It is hoped that the next decade will witness the increasing routine clinical application of these technologies, ultimately resulting in safer prenatal diagnosis of pregnant women worldwide. As before, I am confident that the PCR will continue to play a vital part in realizing this goal.

ACKNOWLEDGMENTS

The author wishes to thank his research team at The Chinese University of Hong Kong and collaborators for their contributions to the work described in this chapter. The author also wishes to thank the various funding bodies for supporting the work described here, including the University Grants Committee Areas of Excellence Scheme, Hong Kong Research Grants Council, the Innovation and Technology Fund, and the Li Ka Shing Foundation. The author holds patents and has filed patent applications on aspects of the diagnostic technology involving fetal nucleic acid analysis in maternal plasma. Part of this patent portfolio has been licensed to Sequenom, to which the author is a consultant and holds equities.

REFERENCES

1. Tabor A, Philip J, Madsen M, Bang J, Obel EB, Norgaard-Pedersen B (1986) Randomised controlled trial of genetic amniocentesis in 4606 low-risk women. *The Lancet* **1**: 1287–1293.
2. Malone FD, Canick JA, Ball RH, Nyberg DA, Comstock CH, Bukowski R, et al. (2005) First-trimester or second-trimester screening, or both, for Down's syndrome. *New England Journal of Medicine* **353**: 2001–2011.
3. Covone AE, Mutton D, Johnson PM, Adinolfi M (1984) Trophoblast cells in peripheral blood from pregnant women. *The Lancet* **2**: 841–843.
4. Herzenberg LA, Bianchi DW, Schroder J, Cann HM, Iverson GM (1979) Fetal cells in the blood of pregnant women: detection and enrichment by fluorescence-activated cell sorting. *Proceedings of the National Academy of Sciences of the United States of America* **76**: 1453–1455.

5. Bianchi DW, Flint AF, Pizzimenti MF, Knoll JH, Latt SA (1990) Isolation of fetal DNA from nucleated erythrocytes in maternal blood. *Proceedings of the National Academy of Sciences of the United States of America* **87**: 3279–3283.
6. Bianchi DW, Williams JM, Sullivan LM, Hanson FW, Klinger KW, Shuber AP (1997) PCR quantitation of fetal cells in maternal blood in normal and aneuploid pregnancies. *American Journal of Human Genetics* **61**: 822–829.
7. Bianchi DW, Simpson JL, Jackson LG, Elias S, Holzgreve W, Evans MI, et al. (2002) Fetal gender and aneuploidy detection using fetal cells in maternal blood: analysis of NIFTY I data. National Institute of Child Health and Development Fetal Cell Isolation Study. *Prenatal Diagnosis* **22**: 609–615.
8. Chen XQ, Stroun M, Magnenat JL, Nicod LP, Kurt AM, Lyautey J, et al. (1996) Microsatellite alterations in plasma DNA of small cell lung cancer patients. *Nature Medicine* **2**: 1033–1035.
9. Nawroz H, Koch W, Anker P, Stroun M, Sidransky D (1996) Microsatellite alterations in serum DNA of head and neck cancer patients. *Nature Medicine* **2**: 1035–1037.
10. Strickland S, Richards WG (1992) Invasion of the trophoblasts. *Cell* **71**: 355–357.
11. Chou Q, Russell M, Birch DE, Raymond J, Bloch W (1992) Prevention of pre-PCR mis-priming and primer dimerization improves low-copy-number amplifications. *Nucleic Acids Research* **20**: 1717–1723.
12. Emanuel SL, Pestka S (1993) Amplification of specific gene products from human serum. *Genetic Analysis, Techniques and Applications* **10**: 144–146.
13. Lo YM, Corbetta N, Chamberlain PF, Rai V, Sargent IL, Redman CW, et al. (1997). Presence of fetal DNA in maternal plasma and serum. *The Lancet* **350**: 485–487.
14. Heid CA, Stevens J, Livak KJ, Williams PM (1996) Real time quantitative PCR. *Genome Research* **6**: 986–994.
15. Lo YM, Tein MS, Lau TK, Haines CJ, Leung TN, Poon PM, et al. (1998) Quantitative analysis of fetal DNA in maternal plasma and serum: implications for noninvasive prenatal diagnosis. *American Journal of Human Genetics* **62**: 768–775.
16. Rijnders RJ, Van Der Schoot CE, Bossers B, de Vroede MA, Christiaens GC (2001) Fetal sex determination from maternal plasma in pregnancies at risk for congenital adrenal hyperplasia. *Obstetrics and Gynecology* **98**: 374–378.
17. Costa JM, Benachi A, Gautier E (2002) New strategy for prenatal diagnosis of X-linked disorders. *New England Journal of Medicine* **346**: 1502.
18. Lo YM, Hjelm NM, Fidler C, Sargent IL, Murphy MF, Chamberlain PF, et al. (1998) Prenatal diagnosis of fetal RhD status by molecular analysis of maternal plasma. *New England Journal of Medicine* **339**: 1734–1738.
19. Finning KM, Martin P, Summers J, Massey E, Poole G, Daniels G. (2008) Effect of high throughput RHD typing of fetal DNA in maternal plasma on use of anti-RhD immunolglobulin in RhD negative pregnant women: prospective feasibility study. *British Medical Journal* **336**: 816–818.
20. Lo YM, Lau TK, Zhang J, Leung TN, Chang AM, Hjelm NM, et al. (1999) Increased fetal DNA concentrations in the plasma of pregnant women carrying fetuses with trisomy 21. *Clinical Chemistry* **45**: 1747–1751.
21. Lo YM, Leung TN, Tein MS, Sargent IL, Zhang J, Lau TK, et al. (1999) Quantitative abnormalities of fetal DNA in maternal serum in preeclampsia. *Clinical Chemistry* **45**: 184–188.
22. Levine RJ, Qian C, Leshane ES, Yu KF, England LJ, Schisterman EF, et al. (2004) Two-stage elevation of cell-free fetal DNA in maternal sera before onset of preeclampsia. *American Journal of Obstetrics and Gynecology* **190**: 707–713.
23. Leung TN, Zhang J, Lau TK, Hjelm NM, Lo YM (1998) Maternal plasma fetal DNA as a marker for preterm labour. *The Lancet* **352**: 1904–1905.
24. Poon LL, Leung TN, Lau TK, Chow KC, Lo YM (2002) Differential DNA methylation between fetus and mother as a strategy for detecting fetal DNA in maternal plasma. *Clinical Chemistry* **48**: 35–41.

25. Herman JG, Graff JR, Myohanen S, Nelkin BD, Baylin SB (1996) Methylation-specific PCR: a novel PCR assay for methylation status of CpG islands. *Proceedings of the National Academy of Sciences of the United States of America* **93**: 9821–9826.
26. Chim SS, Tong YK, Chiu RW, Lau TK, Leung TN, Chan LY, et al. (2005) Detection of the placental epigenetic signature of the maspin gene in maternal plasma. *Proceedings of the National Academy of Sciences of the United States of America* **102**: 14753–14758.
27. Flori E, Doray B, Gautier E, Kohler M, Ernault P, Flori J, et al. (2004) Circulating cell-free fetal DNA in maternal serum appears to originate from cyto- and syncytio-trophoblastic cells. Case report. *Human Reproduction* **19**: 723–724.
28. Lui YY, Chik KW, Chiu RW, Ho CY, Lam CW, Lo YM (2002) Predominant hematopoietic origin of cell-free DNA in plasma and serum after sex-mismatched bone marrow transplantation. *Clinical Chemistry* **48**: 421–427.
29. Lo YM, Wong IH, Zhang J, Tein MS, Ng MH, Hjelm NM (1999) Quantitative analysis of aberrant p16 methylation using real-time quantitative methylation-specific polymerase chain reaction. *Cancer Research* **59**: 3899–3903.
30. Grunau C, Clark SJ, Rosenthal A (2001) Bisulfite genomic sequencing: systematic investigation of critical experimental parameters. *Nucleic Acids Research* **29**: E65–5.
31. Chan KC, Ding C, Gerovassili A, Yeung SW, Chiu RW, Leung TN, et al. (2006) Hypermethylated RASSF1A in maternal plasma: a universal fetal DNA marker that improves the reliability of noninvasive prenatal diagnosis. *Clinical Chemistry* **52**: 2211–2218.
32. Chiu RW, Chim SS, Wong IH, Wong CS, Lee WS, To KF, et al. (2007) Hypermethylation of RASSF1A in human and rhesus placentas. *American Journal of Pathology* **170**: 941–950.
33. Tong YK, Ding C, Chiu RW, Gerovassili A, Chim SS, Leung TY, et al. (2006) Noninvasive prenatal detection of fetal trisomy 18 by epigenetic allelic ratio analysis in maternal plasma: theoretical and empirical considerations. *Clinical Chemistry* **52**: 2194–2202.
34. Chim SS, Jin S, Lee TY, Lun FM, Lee WS, Chan LY, et al. (2008) Systematic search for placental DNA-methylation markers on chromosome 21: toward a maternal plasma-based epigenetic test for fetal trisomy 21. *Clinical Chemistry* **54**: 500–511.
35. Poon LL, Leung TN, Lau TK, Lo YM (2000) Presence of fetal RNA in maternal plasma. *Clinical Chemistry* **46**: 1832–1834.
36. Tsui NB, Ng EK, Lo YM (2002) Stability of endogenous and added RNA in blood specimens, serum, and plasma. *Clinical Chemistry* **48**: 1647–1653.
37. Ng EK, Tsui NB, Lam NY, Chiu RW, Yu SC, Wong SC, et al. (2002) Presence of filterable and nonfilterable mRNA in the plasma of cancer patients and healthy individuals. *Clinical Chemistry* **48**: 1212–1217.
38. Ng EK, Tsui NB, Lau TK, Leung TN, Chiu RW, Panesar NS, et al. (2003) mRNA of placental origin is readily detectable in maternal plasma. *Proceedings of the National Academy of Sciences of the United States of America* **100**: 4748–4753.
39. Bustin SA (2000) Absolute quantification of mRNA using real-time reverse transcription polymerase chain reaction assays. *Journal of Molecular Endocrinology* **25**: 169–193.
40. Myers TW, Gelfand DH (1991) Reverse transcription and DNA amplification by a *Thermus thermophilus* DNA polymerase. *Biochemistry* **30**: 7661–7666.
41. Tsui NB, Chim SS, Chiu RW, Lau TK, Ng EK, Leung TN, et al. (2004) Systematic microarray-based identification of placental mRNA in maternal plasma: towards noninvasive prenatal gene expression profiling. *Journal of Medical Genetics* **41**: 461–467.
42. Lo YM, Tsui NB, Chiu RW, Lau TK, Leung TN, Heung MM, et al. (2007) Plasma placental RNA allelic ratio permits noninvasive prenatal chromosomal aneuploidy detection. *Nature Medicine* **13**: 218–223.
43. Tang K, Fu DJ, Julien D, Braun A, Cantor CR, Koster H (1999) Chip-based genotyping by mass spectrometry. *Proceedings of the National Academy of Sciences of the United States of America* **96**: 10016–10020.
44. Vogelstein B, Kinzler KW (1999) Digital PCR. *Proceedings of the National Academy of Sciences of the United States of America* **96**: 9236–9241.

45. Lo YM, Lun FM, Chan KC, Tsui NB, Chong KC, Lau TK, et al. (2007) Digital PCR for the molecular detection of fetal chromosomal aneuploidy. *Proceedings of the National Academy of Sciences of the United States of America* **104**: 13116–13121.
46. Chiu RW, Chan KC, Gao Y, Lau VY, Zheng W, Leung TY, et al. (2008) Noninvasive prenatal diagnosis of fetal chromosomal aneuploidy by massively parallel genomic sequencing of DNA in maternal plasma. *Proceedings of the National Academy of Sciences of the United States of America* **105**: 20458–20463.
47. Fan HC, Blumenfeld YJ, Chitkara U, Hudgins L, Quake SR (2008) Noninvasive diagnosis of fetal aneuploidy by shotgun sequencing DNA from maternal blood. *Proceedings of the National Academy of Sciences of the United States of America* **105**: 16266–16271.
48. Lun FM, Tsui NB, Chan KC, Leung TY, Lau TK, Charoenkwan P, et al. (2008) Noninvasive prenatal diagnosis of monogenic diseases by digital size selection and relative mutation dosage on DNA in maternal plasma. *Proceedings of the National Academy of Sciences of the United States of America* **105**: 19920–19925.
49. Chiu RW, Cantor CR, Lo YM (2009) Noninvasive prenatal diagnosis by single molecule counting technologies. *Trends in Genetics* **25**: (in press).

16 Polymerase chain reaction–based analyses of nucleic acids from archival material

Ulrich Lehmann

The qualitative and quantitative analysis of genetic material obtained from so-called "archival" tissue specimens (see next section for a definition of "archival") is nowadays a routine operation in modern molecular laboratories thanks to improvements in the extraction methods of nucleic acids and subsequent amplification technology.

Today it is not only possible to analyze localized alterations like point mutations or the expression level of single genes in archival samples, but also to analyze large stretches of genomic material (up to several million base pairs of deoxyribonucleic acid [DNA] sequence) and to assess the expression level of thousands of genes in parallel. For nearly all applications, a polymerase chain reaction (PCR)-based amplification step of the extracted nucleic acid is necessary, and many investigations of this kind are virtually impossible without a powerful amplification technology like PCR. Therefore, "analysis of genetic material from archival specimens" is nearly synonymous with the title of this chapter.

First the word "archival" is explained a bit more in detail and the advantages of archival specimens in general are summarized. Then, several areas of basic and clinical research as well as routine diagnostics are described for which the ability to analyze genetic material extracted from archival specimens represents a major technological advancement, and that is aptly titled a "revolution."

WHAT DOES "ARCHIVAL" MEAN?

The term "archival" in the title of this chapter has a specific meaning in the context of processing and storing tissue samples from human beings or animals for the purpose of histopathological examination, but it is also used in both a somewhat broader sense and a much broader sense.

Its specific meaning refers to tissue samples fixed in a buffered solution of formaldehyde (also known as formalin) and subsequently embedded in paraffin, a wax that melts at approximately 55°C and is therefore solid at room temperature. These samples are called "formalin-fixed and paraffin-embedded," abbreviated as FFPE. The processing of tissues in this way enables the generation of very thin tissue sections that can be stained and subsequently analyzed under the microscope. The formalin fixation leads to a good preservation of morphological details that allows for the detection of the slightest aberrations in, for example, nuclear morphology or tissue architecture by an experienced pathologist. This visual inspection is still the cornerstone of surgical pathology. In addition, the presence or absence of specific proteins can be assessed by incubating these sections with suitable antibodies and analyzing the staining pattern under the microscope (a technique called "immunohistochemistry"). One example with important clinical implications is the detection of the receptor for the female sex hormone estrogen on breast cancer cells. This receptor cannot be seen by conventional microscopy but can be detected by immunohistochemistry. If the cells of a breast tumor specimen turn out to be positive for the estrogen receptor, an anti-estrogen therapy can be initiated.

In the somewhat broader sense, "archival" refers to all tissue biopsies or body fluids that have been fixed in one way or the other and stored subsequently. "Fixation" refers to any process that irreversibly stops all biochemical reactions inside the cells and the surroundings tissue; fixation prevents the "rotting" of a given sample. Air drying is, for example, one of the oldest and most simple fixation procedures, widely used since ancient times for preserving food but also still used in the laboratory to preserve, for example, blood smears on glass slides.

In a broader, somewhat looser sense it refers also to all "old" tissue samples stored in the deep freezer in the laboratory, air-dried by nature and stored at a dry place (e.g., a cave), or tissue frozen by nature and stored below 0°C by nature (e.g., in Alpine glaciers or the permafrost of Siberia or Alaska).

ADVANTAGES OF ARCHIVAL SPECIMENS

Fixation of human tissue samples in formalin and subsequent embedding in paraffin offers several advantages in the daily practice of medical diagnostics: The procedure is well established, and all parties involved (surgeons, general practitioners, health care staff, laboratory technicians, pathologists, etc.) are familiar with it. The paraffin blocks can be stored for decades without the necessity for special equipment. A darkened storage room with normal ambient temperature and humidity and good ventilation is sufficient. Also, the collection of samples from different institutions for large-scale scientific studies or the exchange of specimens for consultation of an expert is convenient.

Therefore, the demonstration that the study of DNA and ribonucleic acid (RNA) from FFPE tissue is possible (the famous "unlocking of the archives") represents a major breakthrough in medical research. It allows the amalgamation of the accumulated morphological knowledge of decades with new molecular and cell biological findings and uses the "dormant knowledge" collected over the last decades in the archives of the pathology departments all over the world. In addition to making this treasure accessible, the change of a decades-old infrastructure for sampling and processing human tissue specimens is not necessary for future studies, thanks to this technological advancement.

Furthermore, all tissue samples containing bone or cartilage (e.g., bone marrow trephines) have to be fixed to enable the preparation of thin sections displaying all relevant morphological details, a prerequisite for an accurate diagnosis. That means that for certain areas of medical diagnosis only "archival" specimens are at hand.

RECONSTRUCTION OF A PANDEMIC (SPANISH FLU)

The influenza pandemic of 1918 is estimated to have killed 20 to 50 million human beings worldwide, making it the worst infectious pandemic in history. It cost much more lives than the First World War, with approximately 10 million casualties from 1914 through 1918. A repetition of this pandemic in today's globalized world is one of the most frightening (but not so unlikely) scenarios. Therefore, it is of utmost importance (and of great scientific interest) to figure out why this particular influenza virus was so infectious and efficient in killing human beings. Because influenza is evolving quite rapidly and changes infectivity and deadliness principally every year, today's strains deliver only the blueprint, with the crucial details unknown. What are required are samples from infected patients from 1918 that are suitable for molecular analysis. Fortunately, the FFPE blocks from two American soldiers, which were positive for influenza RNA, survived in the archive of the Armed Forces Institute of Pathology (AFIP). In addition, a team of AFIP pathologists was able to retrieve frozen tissue samples from an Inuit woman, who died from influenza and was buried in 1918 in a mass grave in permafrost in Alaska. These extremely precious tissue samples represent the two types of "archival material" currently stored in tissue banks systematically: FFPE and snap-frozen cryo specimens. Therefore, the methodology for the detailed molecular analysis of the genetic information of the influenza virus contained within these specimens was well developed. RNA was extracted from these samples, and eventually it was possible to decipher the whole influenza genome. In close collaboration with several other research groups in the United States, it was possible to reconstruct in 2005 viable influenza virus particles with the genetic information of the Spanish flu virus from 1918. These experiments not only provided valuable insights into the biology and infectivity of the deadliest influenza strain known so far, which might help in the development of new vaccines and the prevention of a repetition of something similar to the pandemic of 1918, but also sparked a lively and controversial discussion of whether the gain in medical knowledge outweighs the risk of recreating an extinct deadly virus. This interesting and important discussion is well beyond the scope of this chapter, but undoubtedly, this example shows the power of analyzing and manipulating nucleic acids extracted from archival material, which is impossible without the technique of PCR.

MOLECULAR ARCHAEOLOGY

Another area of great interest for the application of PCR-based analysis of nucleic acids from archival material is the molecular genetic examination of mummies, deceased human beings air-dried by nature and stored in a dry place (e.g., a cave) or preserved by accident in a glacier, or even older human remains like the Neanderthal bones. The exact determination of the origin and ethnicity of a hunter found in the Austrian Alps gives, for example, new and exciting insights into the range of hunting and trading activities in this area of Europe approximately 5,000 years ago.

The ability to analyze thousands of base pairs of DNA extracted from Neanderthal bones and the comparison of these sequences with present-day human beings as well as with the primates most closely related to us for which extensive sequence data are available (chimpanzees) contribute to a refinement of the genealogical tree of modern humans. A long-standing question was whether the Neanderthals contribute to modern variation or whether they represent a distinct population driven to extinction by modern humans invading Europe from the Southeast. The sequencing projects pursued independently by two groups from Germany and the United States provide strong support for the latter model. Even if the molecular data will not solve all questions concerning, for example, the mating behavior of the (nevertheless closely related) Neanderthals and *Homo sapiens* unequivocally, they clearly contribute to a much more detailed picture of the evolution of modern human beings. These advances rely heavily on the experiences with the analysis of somehow damaged genomic material extracted from archival biopsies and the methodology developed for this purpose.

For research projects like these described, extreme care has to be taken that the genetic material isolated from these specimens and amplified by PCR is not a contamination from more recent times or even from laboratory staff involved in recovering, processing, or analyzing the samples. Initial reports that reached newspaper headlines about the analysis of insect DNA isolated from flies enclosed in amber approximately 20 million years ago turned out to be seriously flawed and represented the analysis of much more recent contaminations of these specimens, confirming scientists who raised severe doubts from the beginning that DNA can survive 20 million years even when enclosed in amber.

THE IMPORTANCE OF TISSUE ARCHIVES IN MODERN MEDICINE

Today many malignant diseases can be diagnosed much earlier than they could have been twenty or even fifty years ago, thanks to the progress in medical knowledge and the improvement of diagnostic procedures (e.g., high-resolution imaging systems). As a consequence, many patients are treated at a much earlier stage of their disease. This early detection and treatment is a major factor in reducing the overall mortality for many types of disease. The disadvantage of this progress from which we all benefit is that tissue samples from, for example, advanced and untreated tumors are virtually not available any more. These samples from an advanced stage of the disease from a patient who has not yet received any treatment, however, can provide invaluable insights into the natural course of this particular disease. Moreover, the comparison of tissue samples from treated and untreated patients gives important clues about the reaction of a given tumor to a certain therapeutic agent.

Fortunately, the "unlocking" of the archives of the pathology departments all over the world provides a nearly inexhaustible source of specimens from all stages of disease and also many samples from patients who did not receive treatment because it was not yet available.

These archives also represent an invaluable source of material for retrospective studies, which enable the quick evaluation of new molecular findings in large patient cohorts. If, for example, a new mutation triggering breast cancer development is identified in a mouse model, the prevalence (frequency of occurrence) of this mutation in human breast cancer patients can be assessed easily and quickly by analyzing existing collections of breast cancer specimens with all the clinical data and histopathological diagnoses at hand. Without the accessibility of the FFPE archives for these kind of studies, it would be necessary to collect fresh tissue specimens over decades. In this context it should be stressed that for many purposes the prospective collection under tightly controlled conditions (selection of patients, collection and processing of specimens, etc.) is an absolute prerequisite. Nevertheless, a fast retrospective study using archival material might help researchers decide whether to make the effort to perform a prospective study.

"SMALL" MEANS "ARCHIVAL"

The described medical progress leading to much earlier diagnosis of many types of disease also results in a reduction of the amount of tissue collected for diagnostic (and research) purposes, simply because the infected tissue area or the tumor is still quite small. Also the attempts to minimize the "injury" of the patients for diagnostic (and curative) procedures (called "minimal invasive diagnostics") leads to the sampling of smaller and smaller tissue specimens using, for example, ultrasound-guided fine-needle core biopsy technology. All these small tissue specimens are primarily or exclusively fixed in formalin and embedded in paraffin because only this way of tissue processing provides, for reasons mentioned earlier in this chapter, high-quality morphological preservation of the specimens, which is required for the identification of the subtle morphological changes.

Another meaning of the subheading of these paragraphs refers to the fact that most precancerous alterations or early stages of cancer development are by definition small and can be identified unequivocally only in formalin-fixed and paraffin-embedded material.

Again, only the adaptation of PCR-based molecular techniques to the analysis of archival human tissue samples made these two types of "small tissue specimens" accessible for DNA- and RNA-focused research.

PATIENTS MAY BENEFIT FROM FUTURE PROGRESS

Unlocking the archives enables also the application of methods or knowledge not yet available at the time of collection. This knowledge may be a newly discovered tumor marker for the more precise classification of a malignancy (guiding therapeutic decisions); the identification of a tumor cell–specific genetic alteration, which might serve as a specific target for a newly developed therapeutic agent; or a newly discovered virus as a cause for a disease with a hitherto unknown etiology. Unlocking the archives means that many patients, from whom biopsies where taken in the past and stored as FFPE specimens, may benefit from the progress in medical knowledge and diagnostic procedures.

ANALYSIS OF ARCHIVAL MATERIAL IN DAILY PRACTICE

As already mentioned, the technical advances concerning the nucleic acids extracted from archival material enable molecular studies (in qualitative and quantitative terms) on the very same material that is used routinely for morphological evaluation. Nowadays, the well-established morphological diagnostic procedures are complemented and improved by these methods on a routine basis in all advanced institutions, like university departments or large communal hospitals. Important examples ("important" in terms of impact for the individual patient as well as meaning "frequently applied") follow.

One example is the identification of specific structural changes in the genes coding for antibodies and other molecules important for the regulation of the immune response (in technical terms: immunoglobulin heavy chain gene and T-cell receptor gene rearrangements), which is a specific and sensitive tool for the identification and proper classification of hematological malignancies originating from B or T lymphocytes.

In addition, many human malignancies are characterized by single base pair exchanges ("point mutations") in genes coding for proteins with essential regulatory functions. The detection of such a point mutation can confirm the morphological diagnosis and might guide therapeutic decisions. Nowadays, several promising compounds specifically targeting these mutated regulatory proteins are evaluated in clinical trials.

CONCLUSIONS

As outlined in this chapter, the possibility of analyzing the genetic material of stored tissue samples provides unexpected opportunities for basic research and diagnostic procedures alike.

Thanks to these technological improvements, we have learned a great deal about the Spanish flu of 1918, which might help us in preventing the next pandemic and develop effective therapeutics. We also know much more about the relationship of human beings deceased thousands of years ago to human beings still alive. In addition to answering questions from many areas of basic research, the technological improvements described in this chapter have had a direct impact on the diagnosis and therapy of patients. These technological advancements establish a link between stored patient samples and the progress of science and technology for the benefit of these patients as well as for future generations.

ACKNOWLEDGMENTS

The author would like to thank his colleagues Verena Bröcker and Danny Jonigk, who read an earlier version of this chapter and contributed valuable comments and suggestions.

SUGGESTED READING

The list of suggested reading provides examples from the primary scientific literature intended to guide the reader interested in delving into this subject in much more detail.

The following two articles deal with legal and ethical questions concerning the collection and storage of human tissue for future use in research or diagnostics, a topic beyond the scope of this chapter but nevertheless important if dealing with archival material:

Bauer K, Taub S, Parsi K (2004) Ethical issues in tissue banking for research: a brief review of existing organizational policies. *Theoretical Medicine and Bioethics* **25**: 113–142.

Caulfield T (2004) Tissue banking, patient rights, and confidentiality: tensions in law and policy. *Medicine and Law* **23**: 39–49.

The following reviews give a more technical overview about PCR-based analyses of archival biopsies and provide many references to original research articles:

Lehmann U, Kreipe H (2001) Real-time PCR analysis of DNA and RNA extracted from formalin-fixed and paraffin-embedded biopsies. *Methods* **25**: 409–418.

Lewis F, Maughan NJ, Smith V, Hillan K, Quirke P (2001) Unlocking the archive–gene expression in paraffin-embedded tissue. *Journal of Pathology* **195**: 66–71.
Srinivasan M, Sedmak D, Jewell S (2002) Effect of fixatives and tissue processing on the content and integrity of nucleic acids. *American Journal of Pathology* **161**: 1961–1971.

An example is given here of a primary research article demonstrating the feasibility of PCR-based quantitative gene expression analysis using FFPE specimens (contains also many references to previous work from other groups):

Antonov J, Goldstein DR, Oberli A, Baltzer A, Pirotta M, Fleischmann A, et al. (2005) Reliable gene expression measurements from degraded RNA by quantitative real-time PCR depend on short amplicons and a proper normalization. *Laboratory Investigation* **85**: 1040–1050.

This article describes a recent methodological breakthrough, the extraction of intact proteins from archival biopsies:

Becker KF, Schott C, Hipp S, Metzger V, Porschewski P, Beck R, et al. (2007) Quantitative protein analysis from formalin-fixed tissues: implications for translational clinical research and nanoscale molecular diagnosis. *Journal of Pathology* **211**: 370–378.

17 Microarrays and quantitative real-time reverse transcriptase–polymerase chain reaction

Elisa Wurmbach

Since the introduction of gene expression microarray technology, the number of applications and publications based on it has grown enormously.[1,2] Nowadays, there is almost no institution or university in the field of molecular biology that has no genomic facility helping to apply this technique. Microarrays, an ordered assembly of thousands of probes, have the ability to allow the simultaneous determination of the expression levels of thousands of genes.[3] This technique was used to describe gene programs that underlie various cellular processes, such as immunity and hormone responses,[4,5] as well as to refine classifications of neoplasias,[6,7] and to define diagnostic molecular markers for diseases.[8,9] However, one drawback of the microarray technique is that, the more genes are tested, the higher the risk of identifying false positives as a result of random effects.[10] Furthermore, biological and technical variations, including the microarray design, can affect the precision of microarray results.[11] More difficult situations are found when working with complex multicellular tissue samples as compared to cell line experiments. The outcome of these microarray experiments can result in low fold changes and low signal intensities for differentially expressed genes, which makes it difficult to detect regulated genes reliably. Therefore, the identification of differentially expressed genes requires independent confirmation. Quantitative real-time reverse transcriptase–polymerase chain reaction (qPCR) is the method of choice because of its broad range of linearity, high sensitivity, and reproducibility and because it can be easily adapted to test several hundreds of transcripts.[12,13] Microarray techniques identify candidate genes that are regulated or unregulated under the experimental conditions tested. Genes that are not differentially expressed in microarray experiments can serve as normalization genes for qPCR. This chapter gives an overview of the commonly used microarray platforms, their correlations, and the corroboration of microarray results by qPCR. It also demonstrates how microarrays and qPCR complement each other.

MICROARRAYS

A number of different types of microarrays are available, and new technologies are continuously being introduced in an attempt to improve throughput and sensitivity. Most commonly used are the commercially available short oligonucleotide Affymetrix GeneChips, in-house-manufactured complementary deoxyribonucleic acid (cDNA) arrays, and long oligomer arrays.[14–16]

cDNA microarrays are manufactured in specialized laboratories and microarray facilities based on the protocols originally developed in the Brown laboratory.[2] In principal, the generation of cDNA microarrays includes amplification of selected cDNAs via PCR, the deposition and cross-linking of these products onto glass slides in well-ordered defined grids. An uncomplicated experimental design is the examination of the effects of a treatment in a cell line, for example, comparing treatment versus nontreatment. Ribonucleic acid (RNA) is extracted from treated and untreated cell lines, reverse transcribed and fluorescently labeled, most commonly by incorporating Cy3- and Cy5-conjugated nucleotides. Indirect labeling, whereby 5-[3-aminoallyl]-2′-deoxyuridine-5′-triphosphate (aminoallyl-dUTP) is incorporated during the cDNA synthesis, followed by Cy-dye coupling in a second step, is preferred over direct labeling using Cy-coupled nucleotides, because the fluorescent Cy-dyes cause steric problems for the reverse transcriptase, resulting in low incorporation and sequence bias due to the different sizes of Cy3 and Cy5. Dye swap experiments using the indirect labeling approach yielded similar results.[17] In contrast, swapping dyes using direct labeling led to different results for some genes.[18] The labeled cDNAs are pooled and used for the hybridization onto the cDNA array. After washing to remove excess dye and scanning the arrays at appropriate wavelengths (532 nm for Cy3 and 635 nm for Cy5), overlaying the scans (colored, Cy3 commonly in green and Cy5 in red) reveals which transcript is differentially expressed due to the treatment based on the resulting color, displayed as shades of yellow, red, and green. Yellow spots represent equal amounts of both dyes, meaning no regulation. Depending on the fluorophore, which was used for labeling the treated samples, upregulated genes will appear in the corresponding color. For example, if the cDNA of the treated cell line was labeled with Cy5, red spots will identify upregulation.[17] After scanning, the cDNA microarray data need to be corrected for overall differences in the signal intensities of the two wavelengths measured on each slide.[19]

cDNA microarrays are relatively easy to design, but the fabrication requires specialized equipment and expertise to produce chips of highest quality. Criteria for quality include sequence verification of the PCR products, avoidance of contaminations, and controlling for potential concentration variations of cDNAs deposited onto the slide, all of which affect the reliability of results.[5,20] Therefore, cDNA arrays manufactured in different laboratories can vary in their outcomes.

Even if manufactured to the highest standard, cDNA arrays have inherent disadvantages. Because the deposited probe generated by PCR is double-stranded, the complementary strand of the spotted probe can compete with the labeled sample during the process of hybridization, and the attached DNA strand may not be easily accessible. Both effects can result in lower signal intensities and may reduce the overall sensitivity of the array.[21] Furthermore, the density of the attached cDNAs is critical for hybridizations. If the cDNA density is too high, steric hindrance can slow the hybridization,[22] whereas low density may lead to early saturation, limiting the detection of high fold changes.[23]

The use of long oligonucleotides (50- to 80mers) spotted onto glass slides may improve the performance of gene expression microarrays. They are handled for hybridizations similarly to PCR products, but they do not have the complementary strand and can be spotted easily at the same concentration. Furthermore, they can be chosen carefully for specificity to the gene of interest and thereby avoiding cross-hybridizations. Long oligomers and oligo-arrays are available from several companies (Clontech, Operon, Agilent, Applied Biosystems, NimbleGen). To lower the price per hybridization, a reuse kit was introduced (NimbleGen). This kit removes previously hybridized samples allowing for up to three hybridizations on each chip. However, the company does not guarantee the reuse of the arrays.

The most established commercially available microarray platform is produced by Affymetrix. Each transcript is represented by multiple single-stranded DNA short oligomers (25mers). The 25mers are individually synthesized by photolithography directly onto the chip,[24] and each oligonucleotide has a sister spot containing a mismatch in the middle region. The hybridization conditions are chosen to distinguish between perfect- and mismatch. The resulting perfect- to mismatch ratio is used to determine whether a gene is expressed, which leads to a present or absent call.[25] Biotin labeled, fragmented cRNA is used for hybridizations, the product of reverse transcription of the extracted RNA from samples, followed by in vitro transcription. In contrast to the cDNA arrays, Affymetrix uses a one-color scheme (i.e., hybridizations are performed under noncompetitive conditions). Therefore, comparison of treatments requires at least two hybridizations. The hybridization procedure is highly standardized, which leads, in principle, to reproducible and comparable results.

CodeLink™ (GE Healthcare) developed a three-dimensional gel matrix, which holds the spotted probe away from the surface of the slide to reduce steric hindrances during the process of hybridization.[26] Similar to Affymetrix, the CodeLink™ platform uses small oligonucleotide (30mer) arrays. However, in contrast to Affymetrix, each transcript is represented by only one oligomer, the synthesis of which occurs before deposition, allowing for purification and validation of the probe.

CORRELATION OF MICROARRAY DATA FROM DIFFERENT PLATFORMS

The efficiency of high-throughput microarray experiments depends on the reliability of the microarray technology used to screen thousands of genes in parallel, as well as on the experimental design, including RNA extraction and quality control, labeling, hybridization, and data analysis, which can vary greatly from laboratory to laboratory. To test the concordance of different microarray platforms, Tan et al.[27] used identical RNA preparations, avoiding variations from the extraction, on three commercially available high-density microarray platforms (Affymetrix [25mer oligos], Agilent [cDNA], and CodeLink™ [30mer oligos]). The RNA was labeled and hybridized according to the manufacturer's protocol. The intra-slide correlation was very good (>0.9). However, the results for the inter-slide comparisons were disappointing. Correlations in gene expression levels showed considerable divergence across the different platforms. The average correlation was 0.53, and the range for all three comparisons was between 0.48 and 0.59. The authors assumed that the cross-platform differences arose from intrinsic properties of the microarrays themselves and/or from the processing and analytical steps of these arrays, which may include probe sequence differences, variations in labeling and hybridization conditions, and other factors that derived from an overall lack of industrial standards across multiple technologies. The best way to build confidence in microarray results as a consequence of these outcomes is to validate microarray results by using qPCR.

One year later, Järvinen et al.[28] published the outcome of a similar microarray platform comparison. They compared short oligonucleotide slides (Affymetrix) with commercial cDNA arrays (Agilent) and with custom-made cDNA arrays. In agreement with Tan et al., they found the intra-slide correlation as very good (>0.94). The inter-slide correlation was better than reported by Tan et al. (0.67),[27] whereby the commercial available arrays were better (0.78–0.86) than the custom-made arrays (0.62–0.76). Similar correlations were reported for the inter-platform comparisons of Affymetrix, long oligonucleotide-arrays (70mers from Operon), and cDNA arrays (0.71–0.79).[29]

The microarray quality control (MAQC) project was initiated to address concerns about the reliability of different microarray platforms. It is led by scientists at the U.S. Food and Drug Administration (Rockville, MD), involving 137 participants from 51 organizations. Recently, they compared six commercially available microarray platforms (Applied Biosystems [60mer oligo, one-color microarray], Affymetrix [25mer oligonucleotides, one-color microarray], Agilent [60mer oligo, two- and one-color microarrays], Eppendorf [200–400mers nucleotides, one-color microarray], GE Healthcare [CodeLink™ platform: 30mer oligonucleotides imbedded in a three-dimensional gel matrix, one-color microarray], and Illumina [79mers, which includes a 50mer probe and a 29mer address allowing unambiguous identification, attached to silicon beads, one-color microarray]), as well as a custom-made platform generated at the National Cancer Institute using long oligonucleotides (Operon).[30] They considered the ranks of log ratios as highly correlated (0.69–0.87) and found that all platforms were detecting similar

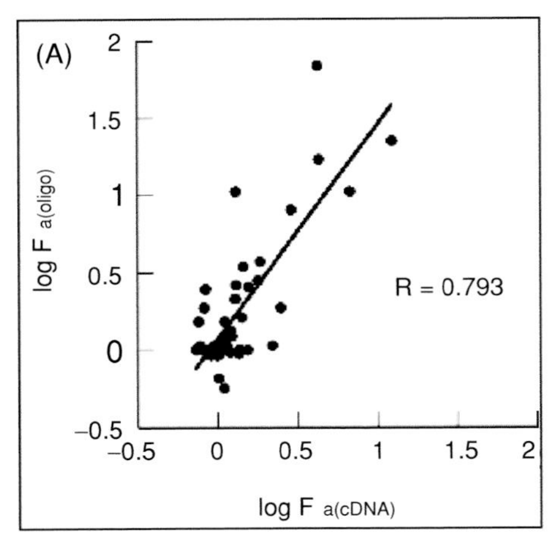

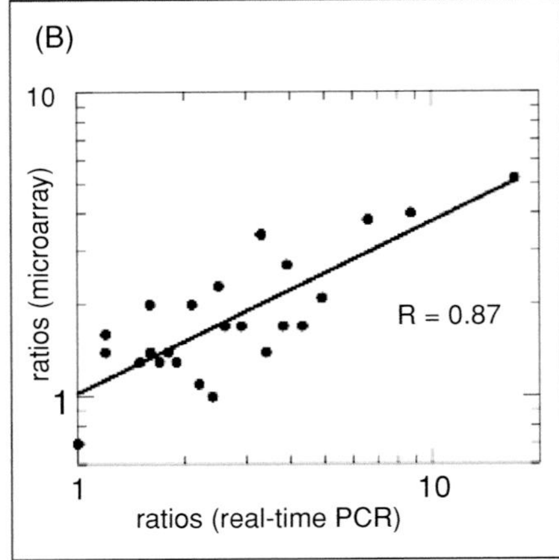

Figure 17–1. Correlation of fold-change measurements. **(A)** Comparison of the log-transformed fold-change values for forty-seven genes determined by the short oligo-array (Affymetrix) and those determined by complementary DNA (cDNA) array.[23] **(B)** The fold changes obtained with both techniques, cDNA array and quantitative real-time reverse transcriptase–polymerase chain reaction (qPCR), for genes showing fewer than 20 fold changes by qPCR were plotted.[5]

changes in gene abundance. They concluded that the results were generally reproducible and comparable across platforms, even when the platforms used probes with sequence differences as well as unique protocols for labeling and expression detection.

MICROARRAY AND qPCR: FOLD CHANGES

Following from these comparisons between different microarray platforms, one may assume that they lead to different results for the same experiment. This assumption will be considered in more detail. Most commonly, the results from microarray experiments are given as fold changes of up- or downregulation due to treatment in comparison to control (nontreatment). Yuen et al.[23] also compared different microarray platforms: an in-house–generated cDNA array with the Affymetrix platform. For the experimental design, a pituitary gland–derived mouse cell line (LβT2) expressing the gonadotropin releasing hormone (GnRH) receptor was used, either mock treated or treated with GnRH. The resulting fold changes of forty-seven transcripts that were present on both arrays were compared (Figure 17–1A). The correlation coefficient was 0.79, which equals roughly the average of all the comparisons above. However, most fold changes are obviously different, but both platforms identified the same genes as being highly or moderately regulated. The rank order of the regulated genes was similar.

As already mentioned, qPCR can be used to validate microarray data. Yuen et al.[23] used qPCR for generating reference measurements, because of its reproducibility, higher sensitivity and large measurement range. The comparison of the microarray data to the qPCR data revealed that the performance of both

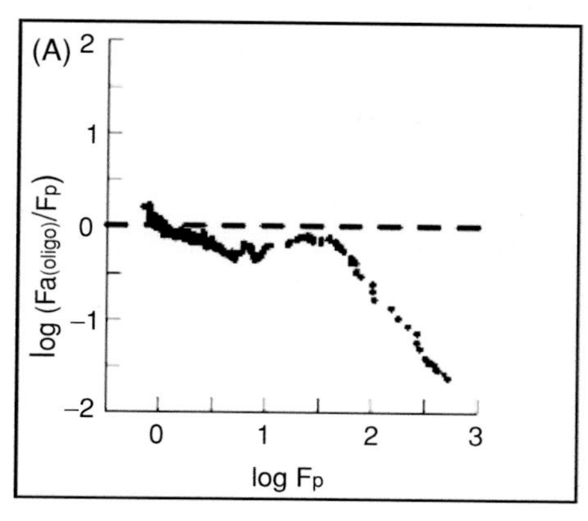

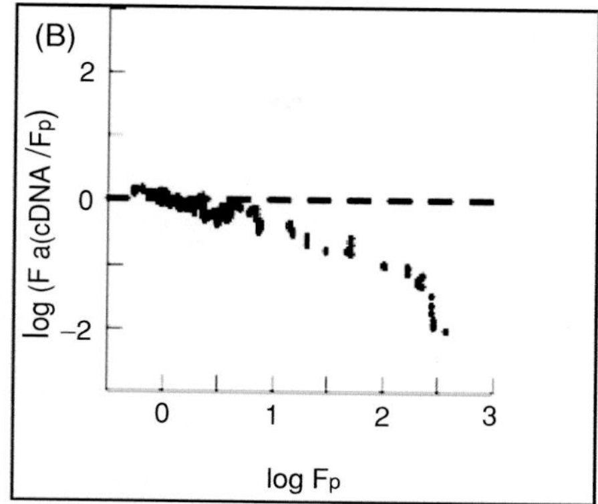

Figure 17–2. Bias of microarray data. **(A)** Ratios obtained by short oligo-array (Affymetrix) and by quantitative real-time reverse transcriptase–polymerase chain reaction (qPCR) [$\log(F_{a(oligo)}/F_p)$] are plotted against the ratio determined by qPCR ($\log F_p$). Data are plotted as a moving average (the mean of the first ten [$\log(F_{a(oligo)}/F_p)$] values is plotted as the first point, the mean of the second through eleventh values is plotted as the second point, etc.). **(B)** Ratios obtained by complementary DNA (cDNA) array and by qPCR [$\log(F_{a(cDNA)}/F_p)$] are plotted as a moving average against the ratio determined by qPCR ($\log F_p$). **(A)** and **(B)** Fp: fold change obtained by qPCR; $F_{a(oligo)}$: fold change, Affymetrix platform; $F_{a(cDNA)}$: fold change, cDNA platform. Figure is from Yuen et al. (2002), *Nucleic Acids Research* **30**: 248.

platforms in identifying regulated and nonregulated genes was identical. Of the forty-seven genes analyzed, sixteen genes were identified as being upregulated on both platforms. Upregulation was defined as a fold change greater than or equal to 1.3. Upregulation of these sixteen genes was verified by qPCR. However, performing qPCR on all forty-seven genes revealed an additional gene as being upregulated that was not detected by either platform (Affymetrix and the cDNA array). Importantly, no unregulated gene was falsely identified as being regulated by microarrays. This result demonstrates that both platforms were equally sensitive, when sensitivity is defined as correctly identifying upregulated genes.

In addition, accurately identifying upregulated genes, both platforms yielded measures of relative expression that correlate well with qPCR data. Again, the fold changes were not identical, and both the cDNA microarray and the Affymetrix platform showed a marked tendency to underestimate the differential expression (Figure 17–1B). Figure 17–2 compares the fold changes of the microarray platforms with qPCR. This presentation makes data interpretation obvious: Similar fold changes of microarray and qPCR would scatter the points around the dashed line. Lower fold changes for the microarray data place the points below the dashed line, and higher fold changes would position them above the dashed line. Obviously, both platforms underestimate the fold changes compared to qPCR. Figure 17–2, A and B, presents the data as a moving average. (The mean of the first ten [$\log(F_{a(oligo)}/F_p)$] values is plotted as the first point, the mean of the second through the eleventh values is plotted as the second point, etc.) This presentation makes the overall trends more apparent. No simple pattern can be observed in the bias of individual transcripts for the Affymetrix data (Figure 17–2A). The cDNA array, in contrast, showed a power scale increase with increasing fold

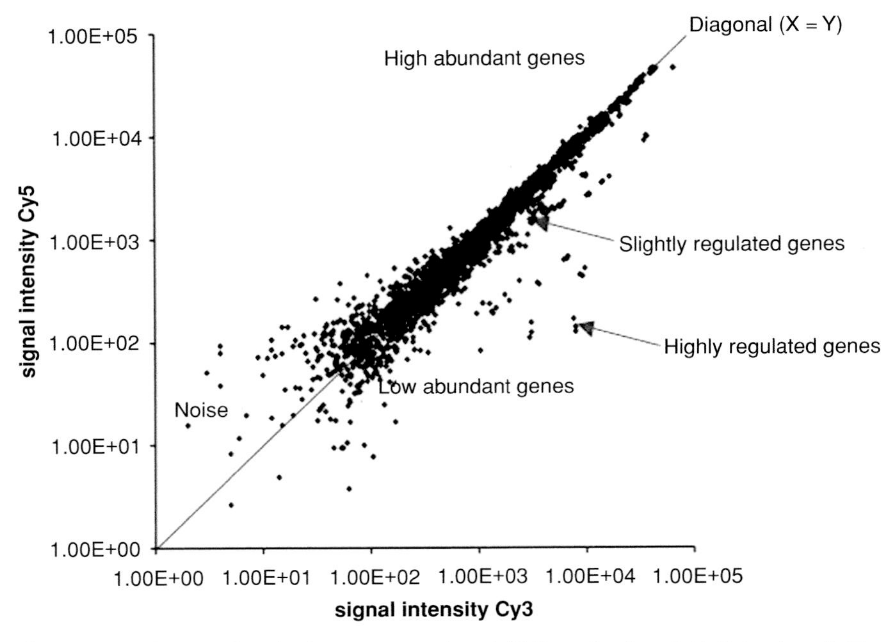

Figure 17–3. Scatter plot of microarray experiment (Cy3: no treatment, Cy5: treatment). Signal intensity correlates with the abundance of transcripts. Unregulated genes appear on the diagonal, and regulated genes are off the diagonal. The stronger the regulation, the farther away the points are located off the diagonal.

changes, causing a linear deviation of the log-transformed data (Figure 17–2B). Obviously, there appears to be a level of fold change (F_P in Figure 17–2B) above which no further increase of the cDNA microarray values occur, which may reflect saturation of the cDNA microarray.

In summary, both microarray platforms were equally sensitive in identifying regulated genes and were in concordance with the general direction of regulation (up- or downregulation). However, the level of regulation varied, but rarely affected the rank order of gene regulation.

UNREGULATED GENES

Normalization in qPCR corrects for errors in sample quantification that can arise by experimental differences, such as differences in RNA quality and quantity as well as the cDNA synthesis. Internal endogenous reference genes with constant expression in cells or tissues under investigation and that do not respond to the treatment can be used for normalization.[13,31] In general, microarrays are used to find differentially expressed genes, which are a small percentage of the total (i.e., most genes in microarray experiments show no regulation as shown in scatter plots). Figure 17–3 shows a scatter plot of a microarray experiment, during which the signal intensities of each gene from both channels (Cy3 and Cy5) are plotted against each other. Abundance of transcripts correlates with signal intensity (i.e., highly abundant transcripts are represented by high signal intensities, and unexpressed genes appear within the noise at low signal intensities). Regulated genes are off the $x = y$ line. The stronger the regulation, the more distant they appear from the diagonal. Nonregulated genes appear on the diagonal ($x = y$ line).

Good candidate genes for normalization are the nonregulated genes ($x = y$ line) with similar abundance to the genes of interest. Hence, microarray analysis can be used to find both regulated and unregulated genes. The treatment is reflected in the differential expression of genes, and the unaffected genes provide candidates for reference genes.

In a simple microarray experiment, during which one condition is tested (e.g., treatment vs. nontreatment), it is relatively easy to find good candidate genes for normalization. The situation becomes more complex in the case of multiple comparisons. Multiple comparisons involve more than two stages and may even include several repeats for each stage. Common examples are time course experiments and the analysis of several stages of a disease, including many cancer studies. Data analysis of these microarray experiments is more complicated. Nevertheless, regulated genes identified by microarray analysis need to be corroborated by qPCR. Finding reference genes that do not respond to the treatment and show constant expression throughout all stages becomes more complicated than for a one-comparison approach. For example, it is not possible to generate scatter plots for four or more comparisons. One way to find appropriate genes that can be used for normalization is to exclude differentially expressed genes.[32] Another approach uses the standard deviation for the fold changes across all samples.[32] The use of more than one gene for normalization is robust and advantageous, because the genes can compensate for slight differences in their expression as they are not coregulated.[33] The candidate genes for normalization must then be tested for their usefulness for normalization in qPCR validation, because their ranking from the most stably to the least stably expressed gene may be different for microarray and qPCR data.

Effects of inappropriate genes used for normalization are shown in Figure 17–4. The data are based on a multiple comparison microarray experiment, including eight stages of hepatocarcinogenesis with mostly ten biological repeats.[9] The four genes – neuregulin 1 (NRG1), hyaluronan-mediated motility receptor (HMMR), primase polypeptide 1 (PRIM1), and interleukin-1 receptor-associated kinase 1 (IRAK1) – were identified by microarray analysis as being significantly differentially expressed between precancer and cancer stages.[9] The qPCR data were normalized to the combination of the two genes' ribosomal protein L41 (RPL41) and splicing factor, arginine/serine-rich 4 (SFRS4), which were found to be most stably expressed throughout all stages of hepatocellular carcinogenesis, as well as to four single genes – glyceraldehyde 3-phosphate dehydrogenase (GAPDH), actin, beta (ACTB), TATA binding protein (TBP), and ribosomal protein S20 (RPS20).[32] Figure 17–4 shows that, for NRG1, the different reference genes caused different resulting fold changes, but the overall pattern was similar. However, the

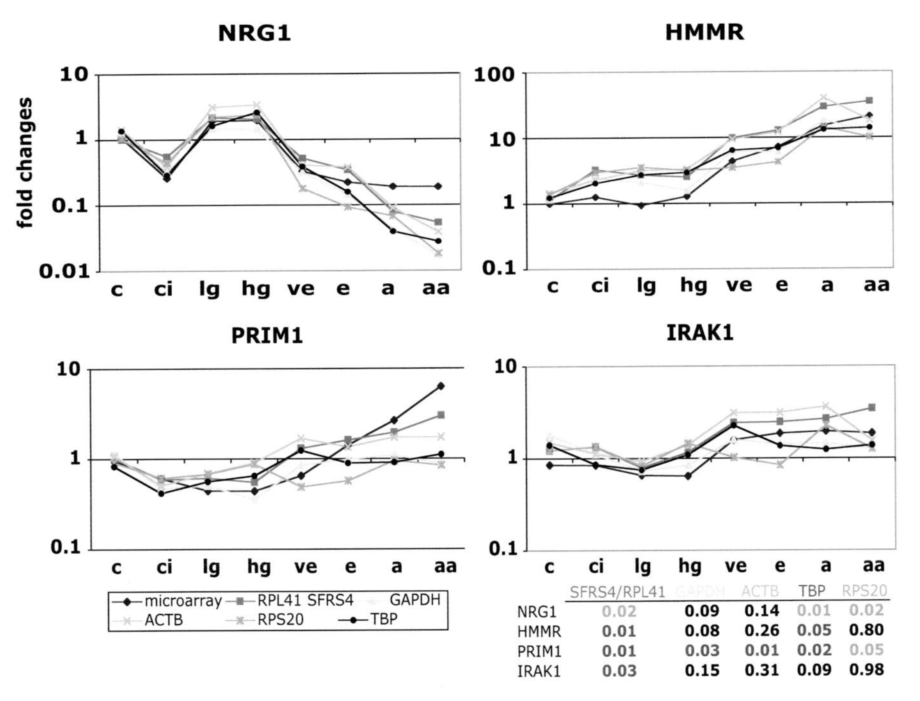

	SFRS4/RPL41	GAPDH	ACTB	TBP	RPS20
NRG1	0.02	0.09	0.14	0.01	0.02
HMMR	0.01	0.08	0.26	0.05	0.80
PRIM1	0.01	0.03	0.01	0.02	0.05
IRAK1	0.03	0.15	0.31	0.09	0.98

Figure 17–4. Effects of reference genes used for normalization: Relative expression of neuregulin 1 (NRG1), hyaluronan-mediated motility receptor (HMMR), primase polypeptide 1 (PRIM1), and interleukin-1 receptor-associated kinase 1 (IRAK1) for all stages of hepatitis C virus (HCV)-induced hepatocellular carcinoma (HCC). Quantitative real-time reverse transcriptase–polymerase chain reaction (qPCR) data were normalized to ribosomal protein L41 (RPL41) and splicing factor, arginine/serine-rich 4 (SFRS4; *pink*), to glyceraldehyde 3-phosphate dehydrogenase (GAPDH; *yellow*), to actin, beta (ACTB *light blue*), to ribosomal protein S20 (RPS20; *green*), and to TATA binding protein (TBP; *brown*). Microarray data are shown in *dark blue*. Fold changes are indicated on the *y* axis. Stages of hepatocarcinogenesis: c = control; ci = cirrhosis; lg = low-grade dysplasia; hg = high-grade dysplasia; ve = very early HCC; e = early HCC; a = advanced HCC; aa = very advanced HCC. Table shows *p* values for the change in gene expression from high-grade dysplasia to very early HCC for NRG1, HMMR, PRIM1, and IRAK1 (*rows*) when normalized to the genes indicated (*columns*). Significant ($p < 0.05$) upregulation between these stages is indicated in *red*; downregulation in *green*. Figure and table are from Waxman S, Wurmbach E (2007), *BMC Genomics* **8**: 243. *See Color Plates.*

significant downregulation between precancer and cancer could not be verified when GAPDH or ACTB was used for normalization. Similarly, the fold changes for HMMR were over- or underestimated and again, the significant upregulation between precancer and cancer could only be confirmed when normalized with RPL41/SFRS4 or TBP. In contrast to NRG1 and HMMR, more dramatic effects were found for the differentially expression of PRIM1 and IRAK1. Normalization using inappropriate genes could lead to misinterpretation of the data (e.g., using RPS20 for normalization resulted in a downregulation of PRIM1 and IRAK1 instead of an upregulation between precancer and cancer). These results show the importance of using appropriate genes for normalization in qPCR and that using at least two genes from different pathways has the advantage that the genes can compensate for slight differences in their expression.

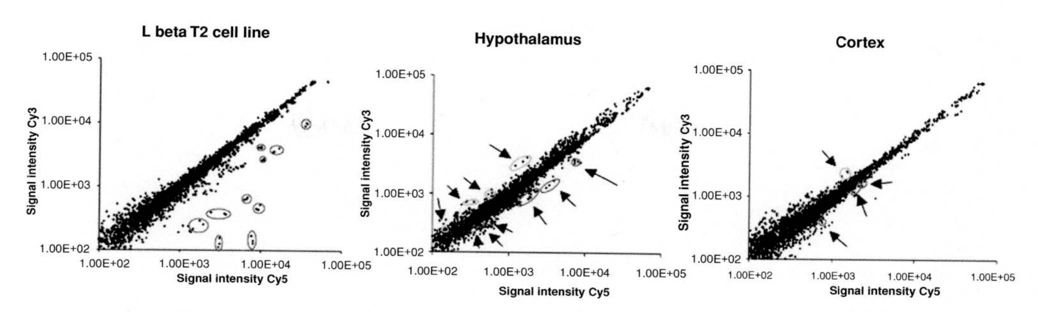

Figure 17–5. Scatter plots of microarray assays from tissues with varying complexity: Plotted are the signal intensities of the LβT2 cell line (Cy3: vehicle, Cy5: gonadotropin releasing hormone [GnRH]), the hypothalamus (Cy3: euglycemic, Cy5: hypoglycemic), and the cortex (Cy3: saline, Cy5: 2,5-dimethoxy 4-iodoamphetamine [DOI]) experiments. The triplicates confirmed by quantitative real-time reverse transcriptase–polymerase chain reaction (qPCR) are marked with *red circles* indicate upregulation, and those marked with *green circles* indicate downregulation. Figures adapted from Wurmbach E, et al. (2002), *Neurochemical Research* **27**: 1027–1033. *See Color Plates.*

COMPLEX TISSUES – DILUTION EFFECTS

Treatment of a cell line ideally affects each cell the same way and accordingly should result in the same response. Using tissue samples for microarray experiments increases the biological variation of the sample and therefore the complexity. It is likely that not all cell types of a given tissue sample react in the same way. If a particular experimental treatment affects only a subset of cells within the tissue analyzed, the resulting fold change may appear lower in microarray experiments than that observed in homogenous samples. This effect is explained by the expression of the same genes in unaffected cells. In addition, when analyzing tissue samples, differentially expressed genes may be expressed in only a subset of cells, resulting in low signal intensities on the microarray. This effect can make it difficult to reliably measure differentially expressed genes. These limitations, reduced fold changes and lower signal intensities, are referred to as "dilution effects."[17,34] To demonstrate these effects in microarray experiments, the impact of treatment was examined in two types of tissues (hypothalamus and cortex) and compared to the cell line experiment (LβT2 cells were treated with GnRH). In the hypothalamus, hypoglycemia-associated autonomic failure was studied. Mice were fasted, then received an insulin injection. RNA was extracted from dissected hypothalami.[35] In the more complex somatosensory cortex, serotonergic hallucinogens were studied. Mice received injections of 2,5-dimethoxy 4-iodoamphetamine (DOI). RNA was extracted from dissected somatosensory cortex samples.[36] The cDNA microarrays mentioned above for the cell line experiment were also used for both of these experiments.

Figure 17–5 shows representative examples of the effect on samples with increasing complexity in the outcome of microarray data. The strongest regulation can be seen in the cell line experiment (LβT2). Obviously, the degree of regulation is much lower in the hypothalamus experiment, and differentially expressed genes in the cortex experiment show only slight regulation and even

Table 17–1. Summary of regulated gene confirmation

Source	Arrays used	Gene candidates tested by qPCR	PCR confirmed genes: all candidates	PCR confirmed genes: Fc > 1.6 on array	PCR confirmed genes: Fc < 1.6 on array
LβT2 cells	3	26	23 (88.5%)	17/17 (100%)	6/9 (66.7%)
Hypothalamus	5	16	12 (75%)	9/12 (75%)	3/4 (75%)
Cortex	7	14	4 (28.6%)	3/7 (42.9%)	1/7 (14.3%)

lower signal intensities. To identify regulated genes, more experimental replicate microarray data were needed with the increasing complexity of the tissues tested, and fewer candidate genes were found (Table 17–1). qPCR corroboration confirmed almost 90% of the candidate genes in the cell line experiment, 75% in the hypothalamus experiment, but only 28% (four genes) in the cortex experiment.

These experiments reveal the limitations of microarray approaches. Due to the dilution effects, the resulting lower fold changes and the decrease in signal intensities restricted the reliable detection of differentially expressed genes. These effects can be so strong that the resulting fold changes and signal intensities for most potentially regulated genes are less than the sensitivity of microarrays.

The dilution effects can be overcome at least in part by dissecting the cells of interest from the surrounding material (e.g., finer physical dissection). Laser capture microdissection, used in combination with a microscope, can be useful to isolate specific cells. Despite improvements in the technology over the past years, the possibility of RNA degradation is higher with additional procedures. RNA degradation increases the signal-to-noise ratio of microarray data and can lead to lower sensitivity and less reliable results. In addition, microdissection might not produce enough starting material for microarray experiments. If so, cells from more tissue samples can be pooled or the RNA can be amplified.[37] However, RNA amplification may introduce bias in the RNA distribution.[38] The approach that is best for analyzing tissues depends on the complexity of the tissue sample and the sensitivity of the microarray platform.

Using complex tissue samples stresses the need for corroboration of the identified candidate genes with other methods. qPCR may be appropriate because of its broader dynamic range and the need for only small amounts of starting material. In contrast to microarrays, qPCR is not so strongly affected by partially degraded RNA, because qPCR relies only on a small intact part of the mRNA.

CONCLUSIONS

The substantial advantage of using microarrays is the possibility of studying the regulation of transcripts in a massively parallel way. Very often, microarrays are used to screen the transcriptome for differential expression, which can detect affected genes that might be novel within the experimental context. Mostly, the design of microarray experiments is given by a comparison (e.g., treatment vs.

nontreatment [control]), resulting in relative quantifications (fold changes). Rarely are absolute values the outcome of microarray experiments.

The results from microarray experiments can be ranked, however, from high to low fold changes. Usually, this ranking is similar for different platforms, despite different fold changes as outcomes. In principal, the latter applies also for qPCR data.

Complexity is introduced by analyzing tissue samples instead of cell lines that do not behave uniformly upon treatment. The effects, referred to as dilution effects, contribute to smaller fold changes and/or to lower signal intensities.

Performing microarray experiments leads to genes affected and not affected by treatment. Both types of genes can and should be identified for the corroboration by qPCR. The unregulated genes build a pool of normalization genes for the following qPCR, essential for the confirmation of regulated gene detected by the microarray approach. Thus, the combination of microarrays and qPCR allows reliable and accurate detection of differentially expressed genes.

ACKNOWLEDGMENTS

Many thanks to Andreas Jenny, Joseph Delaney, Alexander Grishin, Zoran Budimlija, and Wayne Shreffler for discussions and careful reading of the manuscript.

REFERENCES

1. Jaluria P, Konstantopoulos K, Betenbaugh M, Shiloach J (2007) A perspective on microarrays: current applications, pitfalls, and potential uses. *Microbial Cell Factories* **6**: 4.
2. Schena M, Shalon D, Davis RW, Brown PO (1995) Quantitative monitoring of gene expression patterns with a complementary DNA microarray. *Science* **270**: 467–470.
3. van Bakel H, Holstege FC (2004) In control: systematic assessment of microarray performance. *EMBO Reports* **5**: 964–969.
4. Boutros M, Agaisse H, Perrimon N (2002) Sequential activation of signaling pathways during innate immune responses in Drosophila. *Developmental Cell* **3**: 711–722.
5. Wurmbach E, Yuen T, Ebersole BJ, Sealfon SC (2001) Gonadotropin releasing hormone receptor-coupled gene network organization. *Journal of Biological Chemistry* **276**: 47195–47201.
6. Alizadeh AA, Eisen MB, Davis RE, Ma C, Lossos IS, Rosenwald A, et al. (2000) Distinct types of diffuse large B-cell lymphoma identified by gene expression profiling. *Nature* **403**: 503–511.
7. Golub TR, Slonim DK, Tamayo P, Huard C, Gaasenbeek M, Mesirov JP, et al. (1999) Molecular classification of cancer: class discovery and class prediction by gene expression monitoring. *Science* **286**: 531–537.
8. van't Veer LJ, Dai H, van de Vijver MJ, He YD, Hart AA, Mao M, et al. (2002) Gene expression profiling predicts clinical outcome of breast cancer. *Nature* **415**: 530–536.
9. Wurmbach E, Chen YB, Khitrov G, Zhang W, Roayaie S, Schwartz M, et al. (2007) Genome-wide molecular profiles of HCV-induced dysplasia and hepatocellular carcinoma. *Hepatology* **45**: 938–947.
10. Dudoit S, Popper Shaffer J, Boldrick JC (2002) Multiple hypothesis testing in microarray experiments. UC Berkeley Division of Biostatistics Working Paper Series.

11. Churchill GA (2002) Fundamentals of experimental design for cDNA microarrays. *Nature Genetics* **32** Suppl: 490–495.
12. Bustin SA, Benes V, Nolan T, Pfaffl MW (2005) Quantitative real-time RTPCR – a perspective. *Journal of Molecular Endocrinology* **34**: 597–601.
13. Wong ML, Medrano JF (2005) Real-time PCR for mRNA quantitation. *BioTechniques* **39**: 75–85.
14. Cowell JK, Hawthorn L (2007) The application of microarray technology to the analysis of the cancer genome. *Current Molecular Medicine* **7**: 103–120.
15. Lee N, Saeed A (2007) Microarrays: An Overview. Volume 353, Second edition. Totowa, NJ: Humana Press.
16. Rosok O, Sioud M (2007) Discovery of differentially expressed genes: technical considerations. *Methods in Molecular Biology (Clifton, N.J.)* **360**: 115–129.
17. Wurmbach E, Yuen T, Sealfon SC (2003) Focused microarray analysis. *Methods* **31**: 306–316.
18. Taniguchi M, Miura K, Iwao H, Yamanaka S (2001) Quantitative assessment of DNA microarrays – comparison with Northern blot analyses. *Genomics* **71**: 34–39.
19. Quackenbush J (2002) Microarray data normalization and transformation. *Nature Genetics* **32** Suppl: 496–501.
20. Auburn RP, Kreil DP, Meadows LA, Fischer B, Matilla SS, Russell S (2005) Robotic spotting of cDNA and oligonucleotide microarrays. *Trends in Biotechnology* **23**: 374–379.
21. Wang Y, Li Y, Liu S, Shen W, Jiang B, Xu X, et al. (2005) Study on the dynamic behavior of a DNA microarray. *Journal of Nanoscience and Nanotechnology* **5**: 1249–1255.
22. Peterson AW, Heaton RJ, Georgiadis RM (2001) The effect of surface probe density on DNA hybridization. *Nucleic Acids Research* **29**: 5163–5168.
23. Yuen T, Wurmbach E, Pfeffer RL, Ebersole BJ, Sealfon SC (2002) Accuracy and calibration of commercial oligonucleotide and custom cDNA microarrays. *Nucleic Acids Research* **30**: e48.
24. Chee M, Yang R, Hubbell E, Berno A, Huang XC, Stern D, et al. (1996) Accessing genetic information with high-density DNA arrays. *Science* **274**: 610–614.
25. Lockhart DJ, Dong H, Byrne MC, Follettie MT, Gallo MV, Chee MS, et al. (1996) Expression monitoring by hybridization to high-density oligonucleotide arrays. *Nature Biotechnology* **14**: 1675–1680.
26. Ramakrishnan R, Dorris D, Lublinsky A, Nguyen A, Domanus M, Prokhorova A, et al. (2002) An assessment of Motorola CodeLink microarray performance for gene expression profiling applications. *Nucleic Acids Research* **30**: e30.
27. Tan PK, Downey TJ, Spitznagel EL Jr, Xu P, Fu D, Dimitrov DS, et al. (2003) Evaluation of gene expression measurements from commercial microarray platforms. *Nucleic Acids Research* **31**: 5676–5684.
28. Järvinen AK, Hautaniemi S, Edgren H, Auvinen P, Saarela J, Kallioniemi OP, et al. (2004) Are data from different gene expression microarray platforms comparable? *Genomics* **83**: 1164–1168.
29. Petersen D, Chandramouli GV, Geoghegan J, Hilburn J, Paarlberg J, Kim CH, et al. (2005) Three microarray platforms: an analysis of their concordance in profiling gene expression. *BMC Genomics* **6**: 63.
30. Shi L, Reid LH, Jones WD, Shippy R, Warrington JA, Baker SC, et al. (2006) The MicroArray Quality Control (MAQC) project shows inter- and intraplatform reproducibility of gene expression measurements. *Nature Biotechnology* **24**: 1151–1161.
31. Schmittgen TD, Zakrajsek BA (2000) Effect of experimental treatment on housekeeping gene expression: validation by real-time, quantitative RT-PCR. *Journal of Biochemical and Biophysical Methods* **46**: 69–81.
32. Waxman S, Wurmbach E (2007) De-regulation of common housekeeping genes in hepatocellular carcinoma. *BMC Genomics* **8**: 243.
33. Vandesompele J, De Preter K, Pattyn F, Poppe B, Van Roy N, De Paepe A, et al. (2002) Accurate normalization of real-time quantitative RT-PCR data by geometric averaging of multiple internal control genes. *Genome Biology* **3**: 34.

34. Wurmbach E, Gonzalez-Maeso J, Yuen T, Ebersole BJ, Mastaitis JW, Mobbs CV, et al. (2002) Validated genomic approach to study differentially expressed genes in complex tissues. *Neurochemical Research* **27**: 1027–1033.
35. Mastaitis JW, Wurmbach E, Cheng H, Sealfon SC, Mobbs CV (2005) Acute induction of gene expression in brain and liver by insulin-induced hypoglycemia. *Diabetes* **54**: 952–958.
36. Gonzalez-Maeso J, Yuen T, Ebersole BJ, Wurmbach E, Lira A, Zhou M, et al. (2003) Transcriptome fingerprints distinguish hallucinogenic and nonhallucinogenic 5-hydroxytryptamine 2A receptor agonist effects in mouse somatosensory cortex. *Journal of Neuroscience* **23**: 8836–8843.
37. Ginsberg SD (2005) RNA amplification strategies for small sample populations. *Methods* **37**: 229–237.
38. Marko NF, Frank B, Quackenbush J, Lee NH (2005) A robust method for the amplification of RNA in the sense orientation. *BMC Genomics* **6**: 27.

18 Polymerase chain reaction in the detection of genetic variation

Pui-Yan Kwok

IN THE BEGINNING

When the polymerase chain reaction (PCR) burst onto the scene in the mid-1980s, its usefulness for genetic analysis was immediately recognized. Indeed, the first publication of the PCR method was on its use in the prenatal diagnosis of sickle cell anemia.[1] When the use of thermostable deoxyribonucleic acid (DNA) polymerases[2] and programmable thermocyclers made PCR a commonly used method in the laboratory, the detection of genetic variation became a much easier enterprise. Instead of relying on laborious approaches such as restriction fragment length polymorphism (RFLP) analysis[3] or DNA sequencing of complementary DNA clones[4] to detect genetic variation, PCR allowed the "extraction" of a specific locus of the genome and produced sufficient quantities of it for further analysis. The main contributions of PCR to the detection of genetic variation are in three areas: amplification of small, unique regions of the genome harboring DNA sequence variants; discrimination of allelic differences between genomes; and amplification of products of other allelic discrimination reactions for detection by conventional means.

In the early days of the PCR revolution, the main obstacles to the deployment of PCR for genetic variation were the paucity of genomic sequence information for PCR primer design, the relatively high cost of oligonucleotide synthesis, and the laborious procedures used in DNA sequencing. Fortunately, automated DNA synthesis and DNA sequencing instruments became available in the early 1990s and the initial genomic mapping phase of the Human Genome Project provided the impetus to produce genetic markers based on PCR. As the speed of DNA sequencing and oligonucleotide synthesis increased while their cost went down, PCR became the principal approach to genetic analysis.

STRPs

The first set of genetic markers based on PCR amplification was the short tandem repeat polymorphism (STRP) marker[5,6] used in genome-wide linkage analysis. The STRP markers were developed by probing small insert clones of human DNA with short tandem repeat sequences, sequencing the ends of the inserts for unique DNA sequences for PCR assay design, and amplifying the loci across a panel of individuals to identify those PCR products that harbor the STRPs. Genotyping the STRPs is straightforward. Following the PCR assay, the PCR products are run on a sequencing gel to size them against a standard. With automated DNA sequencers and software tools for sizing, the STRPs were used successfully in many linkage studies in the 1990s.

SNPs

Despite the success of the STRPs, it became quite clear early on that the STRPs were not the most abundant sequence variation in the genome and that they were usually not disease-causing mutations. The most common sequence variation in the human genome is the single nucleotide polymorphism (SNP). The SNP is the basis for the RFLP markers, and many genetic disorders result from single-base-pair (bp) mutations. Systematic study of the human genome shows that, when two human chromosomes are compared to each other, an SNP is found in every 1,000 bp scanned.[7] A large number of genotyping methods have been developed for detecting SNPs, and almost all of them involve the use of PCR.[8]

It is not easy to distinguish between two alleles at a specific location of the 3-billion-bp human genome. With PCR, however, the complexity of the problem is reduced to one that involves DNA fragments that are just a few hundred base pairs in length. Given the small DNA size, multiple allelic discrimination methods can be used to produce genotype information. In addition, PCR amplification generates a large number of DNA molecules, making it possible to use less sensitive detection modalities. Finally, PCR incorporating an allelic discrimination reaction involves two levels of specificity, that of genomic location and allelic identity, yielding a robust assay for genetic analysis.

DGGE, SSCP, AND DHPLC

There are a number of allelic discrimination methods that use PCR-amplified DNA as substrate. Some of them detect changes in the physical properties of the DNA fragments based on allelic differences. Three examples of these methods are denaturing gradient gel electrophoresis (DGGE),[9] single-strand conformation polymorphism (SSCP) analysis,[10] and denaturing high performance liquid chromatography (DHPLC).[11] In each of these methods, the 1-bp difference in a PCR product confers a change in a physical property – such as the denaturant concentration at which DNA denatures or the three-dimensional conformation of single-stranded DNA – that causes allelic forms of the same locus to behave differently during gel electrophoresis or liquid chromatography. These particular methods are especially useful in the search for previously unknown variations because the allelic discrimination reaction does not require probes that are at or near the site of variation.

CCM, RFLP, AND TaqMan

Other allelic discrimination methods that use PCR-amplified DNA as templates are based on the formation of distinguishable products from the allelic sequences. Some examples are chemical cleavage of mismatch (CCM),[12] RFLP,[13] minisequencing (single-base extension),[14] oligonucleotide ligation assay (OLA),[15] and the TaqMan® assay.[16] The CCM method uses chemicals that selectively cleave heteroduplex DNA and is therefore another method useful in identifying unknown sequence variations. PCR-based RFLP relies on restriction enzymes that cut the PCR product at the restriction site(s) (if present), thereby generating DNA fragments as predicted by the genotype of the DNA sample assayed. The CCM and RFLP methods are analyzed by gel electrophoresis. The single-base extension, OLA, and TaqMan® assays yield products that can be analyzed by automated means without gel electrophoresis. For example, single-base extension products are analyzed by mass spectrometry[17,18] or the change in fluorescence polarization[19]; OLA can be analyzed by an enzyme-linked immunosorbent (ELISA) assay[20]; and the TaqMan® assay can be analyzed by fluorescence intensity change.[21] In the single-base extension reaction, a dye terminator is incorporated in an allele-specific manner by DNA polymerase with a primer designed to anneal immediately upstream for the polymorphic site. Detecting the identity of the incorporated base allows one to infer the allele present in the PCR product. Similarly, in the OLA, allele-specific probes and reporter probes straddling the polymorphic site are ligated by DNA ligase only if there is a perfect match between the probes and the template. The TaqMan® assay uses the 5′-nuclease activity of Taq DNA polymerase to cleave a fluorogenic probe that anneals only to a perfectly matching template. Cleavage of the probe releases the fluorescent label from a quencher and makes it possible to observe a fluorescent signal when the reaction mixture is excited by an appropriate light source.

SEQUENCING

PCR amplification is of course the first step in the ultimate genotyping reaction: DNA sequencing.[22] PCR primers, used one at a time, can be used as sequencing primers to obtain sequencing data from the test samples. If the PCR product is too large to be sequenced with just the PCR primers, internal primers can be used to complete the coverage. By aligning the sequences from a number of individuals, one can easily identify sequence variations of any kind, be it SNPs, insertions, or deletions. Both Sanger sequencing[23] and pyrosequencing[24] can be done with PCR products. With the advent of "next generation sequencing," millions of bases can be obtained from a DNA sample in one experiment for sequence variation detection. "Digital PCR,"[25,26] in which DNA fragments are diluted to a point where only one DNA molecule is in a reaction compartment such as tiny aqueous droplets in an emulsion,[27] is key to the success of the massively parallel sequencing approaches employed in the "next generation sequencers." Indeed, millions of sequence variations were found in the first complete human genome sequenced using a "next generation sequencer."[28]

MULTIPLEX PCR

In some genotyping methods, multiplex PCR is employed to amplify tens and even up to hundreds of loci simultaneously followed by allelic discrimination steps and detection.[29] For assays consisting of twenty to thirty loci, the simplest detection method is by matrix-assisted laser desorption/ionization time-of-flight mass spectrometry (MALDI-TOF MS), in which mass labels are tagged onto the markers such that the allelic products of all the markers can be differentiated by mass. Other assays are done on arrays with locus-specific capture probes, allele-specific probes, or primers useful for single-base extension reactions.

ALLELIC-SPECIFIC PCR

In addition to amplifying a genetic locus for further analysis, PCR contributes to genetic variation detection by acting as the allelic discrimination reaction itself. The allelic discrimination reaction is done simply by designing PCR primers that are "allele specific," usually with the 3′ end of one of the primers ending with one of the allelic bases. This way, only one of the alleles will amplify whereas the other allele, having a mismatch at the 3′ end of a PCR primer, will not. The presence or absence of a PCR product (of the expected size) becomes the readout for the genotype. If one uses gel electrophoresis to determine whether a PCR product is formed, two PCR assays (one for each allele) have to be set up to obtain the genotype of a sample. If one uses melting point analysis as the detection method, the assay can be done in one reaction vessel by adding a "guanine–cytosine (GC) clamp" to one of the allele-specific primers so that the two products formed will have different melting curves.[30]

Allele-specific PCR is most useful in obtaining haplotypes for two markers that are in proximity to each other. By designing two sets of allele-specific primers (one set for each marker) and using them in different combinations, one can determine the haplotypes of any DNA sample based on which combinations of primers produce the expected PCR products.[31]

The final contribution of PCR to genetic variation analysis is its trivial use to amplify DNA fragments fitted with adaptors using a set of universal primers. Developed initially as a means to amplify the "signal" (meaning the product) of the allelic discrimination reaction, it has become a key step in massively parallel genotyping and DNA sequencing. Indeed, this application of PCR decreases the cost of large-scale genotyping and sequencing to the point that is affordable to the average laboratory.

GOLDENGATE AND MIP

The power of PCR as a signal amplification method was largely responsible for the success of the International HapMap Project. The two highly multiplexed SNP-genotyping methods used in the project developed by Illumina and ParAllele – namely, the GoldenGate assay[32] and the Molecular Inversion Probe (MIP) assay,[33] – depended heavily on the use of universal PCR amplification of the initial allele discrimination reaction that produced a small number of products when SNP-specific probes were incubated directly with genomic DNA. In the GoldenGate assay, allele-specific probes were employed in a primer extension step followed by ligation of the extended probe to a reporter probe. Both probes bear universal priming sequences so that the successfully extended *and* ligated probes become substrates of the universal PCR. Sufficient amounts of the ligated products result from the universal amplification step to allow for their capture onto beads coated with locus-specific probes and detection by fluorescence imaging. In the MIP assay, a long oligonucleotide is designed with its two ends' bearing sequences complementary to those flanking a polymorphic site, leaving a one-base gap. In four parallel reactions, each incubated with one of the four possible bases, the gap is filled by DNA polymerase and the probe is circularized by DNA ligase. The circular probe is "inverted" by cleavage at a linker site such that the universal PCR priming sequences are now pointing toward the polymorphic site, thereby allowing universal PCR to occur. In both the GoldenGate and MIP assays, thousands of SNPs can be typed simultaneously because, with universal PCR to amplify the signal, only a small number of successful events are needed for each SNP for the experiment to work.

In the realm of massively parallel genomic analysis, PCR amplification of DNA fragments fitted with adaptors bearing universal priming sequences finds use in two settings. The first setting is in genome-wide SNP analysis. In this application, genomic DNA is completely digested by a restriction enzyme followed by adaptor ligation. Universal PCR serves two purposes in this application: reducing the complexity of the genome by amplifying the smaller fragments (<1,000 bp) and producing enough copies of the fraction of the genome where the restriction sites are within 1,000 bp of each other. Probes for the detection of SNPs predicted to be contained in this amplified fraction of the genome are synthesized on a microarray and hybridization of the PCR amplified, small restriction fragments allows one to determine which alleles of the hundreds of thousands of SNPs are present in the DNA sample being tested.[34] In the latest version of the microarray, copy number probes are also placed on the microarray so that copy number variations can be detected.

In the second setting, genomic DNA is sheared and the resulting fragments are ligated to adaptors. The "library" of these genomic fragments is spread across either a flow cell or into small emulsion droplets with the goal of separating them into single fragments. The individual DNA fragments are then amplified by universal PCR by priming off the sequences found on the adaptors. The amplified DNA fragments form clones of the individual DNA molecules and can be sequenced by pyrosequencing,[35] stepwise primer extension,[36] or stepwise sequencing by ligation. Currently, up to 2 billion bases of sequence data can be generated in one experiment on these commercialized platforms.

OUTLOOK

In the two decades since the development of PCR, detection of genetic variation grew in all directions. The influence of and dependence on PCR are universal. In conjunction with developments in other areas of the genetic and genomic fields, such as DNA sequencing and the completion of the Human Genome Project and the HapMap Project, detection of genetic variation can be done at one specific locus at a time, as in clinical diagnosis of particular genetic disorders, or comprehensively using a genome-wide SNP array. In a few years, complete sequencing of the human genome may prove to be an affordable and efficient way to detect genetic variation in the entire genome. PCR will no doubt be a key step in the experimental strategy that makes whole-genome sequencing possible.

REFERENCES

1. Saiki RK, Scharf S, Faloona F, Mullis KB, Horn GT, Erlich HA, et al. (1985) Enzymatic amplification of beta-globin genomic sequences and restriction site analysis for diagnosis of sickle cell anemia. *Science* **230**: 1350–1354.
2. Saiki RK, Gelfand DH, Stoffel S, Scharf SJ, Higuchi R, Horn GT, et al. (1988) Primer-directed enzymatic amplification of DNA with a thermostable DNA polymerase. *Science* **239**: 487–491.
3. Botstein D, White RL, Skolnick M, Davis RW (1980) Construction of a genetic linkage map in man using restriction fragment length polymorphisms. *American Journal of Human Genetics* **32**: 314–331.
4. Chang JC, Kan YW (1979) Beta 0 thalassemia, a nonsense mutation in man. *Proceedings of the National Academy of Sciences of the United States of America* **76**: 2886–2889.

5. Weber JL, May PE (1989) Abundant class of human DNA polymorphisms which can be typed using the polymerase chain reaction. *American Journal of Human Genetics* **44**: 388–396.
6. Litt M, Luty JA (1989) A hypervariable microsatellite revealed by in vitro amplification of a dinucleotide repeat within the cardiac muscle actin gene. *American Journal of Human Genetics* **44**: 397–401.
7. Sachidanandam R, Weissman D, Schmidt SC, Kakol JM, Stein LD, Marth G, et al.; International SNP Map Working Group (2001) A map of human genome sequence variation containing 1.42 million single nucleotide polymorphisms. *Nature* **409**: 928–933.
8. Landegren U, Nilsson M, Kwok PY (1998) Reading bits of genetic information: methods for single-nucleotide polymorphism analysis. *Genome Research* **8**: 769–776.
9. Fischer SG, Lerman LS (1983) DNA fragments differing by single base-pair substitutions are separated in denaturing gradient gels: correspondence with melting theory. *Proceedings of the National Academy of Sciences of the United States of America* **80**: 1579–1583.
10. Orita M, Iwahana H, Kanazawa H, Hayashi K, Sekiya T (1989) Detection of polymorphisms of human DNA by gel electrophoresis as single-strand conformation polymorphisms. *Proceedings of the National Academy of Sciences of the United States of America* **86**: 2766–2770.
11. Underhill PA, Jin L, Lin AA, Mehdi SQ, Jenkins T, Vollrath D, et al. (1997) Detection of numerous Y chromosome biallelic polymorphisms by denaturing high-performance liquid chromatography. *Genome Research* **7**: 996–1005.
12. Cotton RG, Rodrigues NR, Campbell RD (1988) Reactivity of cytosine and thymine in single-base-pair mismatches with hydroxylamine and osmium tetroxide and its application to the study of mutations. *Proceedings of the National Academy of Sciences of the United States of America* **85**: 4397–4401.
13. Higuchi R, von Beroldingen CH, Sensabaugh GF, Erlich HA (1988) DNA typing from single hairs. *Nature* **332**: 543–546.
14. Syvänen AC, Aalto-Setälä K, Harju L, Kontula K, Söderlund H (1990) A primer-guided nucleotide incorporation assay in the genotyping of apolipoprotein E. *Genomics* **8**: 684–692.
15. Landegren U, Kaiser R, Sanders J, Hood L (1988) A ligase-mediated gene detection technique. *Science* **241**: 1077–1080.
16. Livak KJ, Flood SJ, Marmaro J, Giusti W, Deetz K (1995) Oligonucleotides with fluorescent dyes at opposite ends provide a quenched probe system useful for detecting PCR product and nucleic acid hybridization. *PCR Methods and Applications* **4**: 357–362.
17. Ross P, Hall L, Smirnov I, Haff L (1998) High level multiplex genotyping by MALDI-TOF mass spectrometry. *Nature Biotechnology* **16**: 1347–1351.
18. Tang K, Fu DJ, Julien D, Braun A, Cantor CR, Köster H (1999) Chip-based genotyping by mass spectrometry. *Proceedings of the National Academy of Sciences of the United States of America* **96**: 10016–10020.
19. Chen X, Levine L, Kwok PY (1999) Fluorescence polarization in homogeneous nucleic acid analysis. *Genome Research* **9**: 492–498.
20. Nickerson DA, Kaiser R, Lappin S, Stewart J, Hood L, Landegren U (1990) Automated DNA diagnostics using an ELISA-based oligonucleotide ligation assay. *Proceedings of the National Academy of Sciences of the United States of America* **87**: 8923–8927.
21. Livak KJ (2003) SNP genotyping by the 5′-nuclease reaction. *Methods in Molecular Biology (Clifton, N.J.)* **212**: 129–147.
22. Kwok PY, Duan S (2003) SNP discovery by direct DNA sequencing. *Methods in Molecular Biology (Clifton, N.J.)* **212**: 71–84.
23. Sanger F, Donelson JE, Coulson AR, Kössel H, Fischer D (1973) Use of DNA polymerase I primed by a synthetic oligonucleotide to determine a nucleotide sequence in phage

fl DNA. *Proceedings of the National Academy of Sciences of the United States of America* **70**: 1209–1213.

24. Ronaghi M, Karamohamed S, Pettersson B, Uhlén M, Nyrén P (1996) Real-time DNA sequencing using detection of pyrophosphate release. *Analytical Biochemistry* **242**: 84–89.
25. Sykes PJ, Neoh SH, Brisco MJ, Hughes E, Condon J, Morley AA (1992) Quantitation of targets for PCR by use of limiting dilution. *BioTechniques* **13**: 444–449.
26. Vogelstein B, Kinzler KW (1999) Digital PCR. *Proceedings of the National Academy of Sciences of the United States of America* **96**: 9236–9241.
27. Dressman D, Yan H, Traverso G, Kinzler KW, Vogelstein B (2003) Transforming single DNA molecules into fluorescent magnetic particles for detection and enumeration of genetic variations. *Proceedings of the National Academy of Sciences of the United States of America.* **100**: 8817–8822.
28. Wheeler DA, Srinivasan M, Egholm M, Shen Y, Chen L, McGuire A, et al. (2008) The complete genome of an individual by massively parallel DNA sequencing. *Nature* **452**: 872–876.
29. Ohnishi Y, Tanaka T, Ozaki K, Yamada R, Suzuki H, Nakamura Y (2001) A high-throughput SNP typing system for genome-wide association studies. *Journal of Human Genetics* **46**: 471–477.
30. Germer S, Holland MJ, Higuchi R (2000) High-throughput SNP allele-frequency determination in pooled DNA samples by kinetic PCR. *Genome Research* **10**: 258–266.
31. Sarkar G, Sommer SS (1991) Haplotyping by double PCR amplification of specific alleles. *BioTechniques* **10**: 436–440.
32. Oliphant A, Barker DL, Stuelpnagel JR, Chee MS (2002) BeadArray technology: enabling an accurate, cost-effective approach to high-throughput genotyping. *BioTechniques* Suppl: 56–58, 60–61.
33. Hardenbol P, Banér J, Jain M, Nilsson M, Namsaraev EA, Karlin-Neumann GA, et al. (2003) Multiplexed genotyping with sequence-tagged molecular inversion probes. *Nature Biotechnology* **21**: 673–678.
34. Kennedy GC, Matsuzaki H, Dong S, Liu WM, Huang J, Liu G, et al. (2003) Large-scale genotyping of complex DNA. *Nature Biotechnology* **21**: 1233–1237.
35. Margulies M, Egholm M, Altman WE, Attiya S, Bader JS, Bemben LA, et al. (2005) Genome sequencing in microfabricated high-density picolitre reactors. *Nature* **437**: 376–380.
36. Hillier LW, Marth GT, Quinlan AR, Dooling D, Fewell G, Barnett D, et al. (2008) Whole-genome sequencing and variant discovery in *C. elegans*. *Nature Methods* **5**: 183–188.

19 Polymerase chain reaction: A blessing and a curse for ancient deoxyribonucleic acid research

Michael Hofreiter and Holger Römpler

The analysis and use of ancient deoxyribonucleic acid (DNA) is intimately linked with the polymerase chain reaction (PCR). Although the first ancient DNA sequences were uncovered before the invention of PCR, ancient DNA research, like many other fields in molecular biology, only began to develop after this technique became established. From initial short fragments that could be amplified via PCR, ancient DNA research has evolved into a field in which complete mitochondrial genomes and genomic shotgun sequences of several megabases can be amplified and analyzed using modern variations of PCR, such as multiplex or emulsion PCR. These achievements became possible by PCR's extraordinary sensitivity, which allows amplification from as little as a single target molecule. However, this sensitivity has a dark side, because PCR also frequently amplifies contaminating DNA. Consequently, spectacular errors, such as the presumed amplification of several-million-year-old dinosaur DNA from bone or insect DNA from amber fossils, have plagued the field almost from its beginnings. In this chapter, we explore how PCR has been a blessing for the advancement of ancient DNA research, while also addressing its limitations, which seem much like a curse.

HISTORY OF ANCIENT DNA AND PCR

Ancient DNA research started more than twenty years ago, with the sequencing of two short mitochondrial DNA (mtDNA) fragments from the extinct quagga.[1] This sequencing was achieved even in the absence of PCR; the first article to describe

PCR was published almost exactly one year later.[2] However, the field of ancient DNA gathered momentum only after PCR had become an established technique. As early as 1988, studies were reported that used PCR to amplify ancient DNA from such diverse sources as the 140-year-old quagga previously sequenced in 1984,[3] 7,000-year-old human brain tissue,[4] and 1,000-year-old maize remains.[5] Since then, PCR has been the key technique in ancient DNA research because of its extraordinary sensitivity: It can amplify DNA fragments from as little as a single molecule.[6,7] Such sensitivity is crucial in ancient DNA research, because ancient DNA is invariably degraded and usually present only in small amounts.[8,9] However, PCR's high sensitivity has its drawbacks. Already one of the first publications on ancient DNA, which involved the direct cloning of DNA from an Egyptian mummy,[10] is likely to have been adversely affected by contamination with modern DNA.[11] This issue has become more vexed with the use of the highly sensitive PCR.

After the initial successful amplification of DNA obtained from ancient plants[5] and animals,[3,12] researchers started to investigate specimens of ever increasing age. Thus, DNA sequences from several-million-year-old plant remains were reported as early as 1990,[13] followed by reports of 125-million-year-old insect DNA[14] and 65-million-year-old dinosaur DNA.[15] However, it was soon realized that some of these results could not be correct, as it is unlikely that DNA could survive for such a long time.[16] Still, as recently as 2002, sequences from 415-million-year-old bacteria were published, even in prestigious journals.[17] Nonetheless, a recent study has rejected all claims of ancient DNA exceeding 1 million years of age.[18]

As a consequence of these problems and to avoid misleading results, many researchers became rather conservative in their estimates of the maximum age of retrievable ancient DNA. For a while it was assumed that DNA could not survive for longer than 100,000 years anywhere in the geological record.[16] Some researchers imposed an even more rigorous limit of approximately 10,000 years,[19] based on theoretical considerations. Moreover, several researchers have tried to enforce rigorous criteria that must be fulfilled before seriously contemplating any ancient DNA study.[20] Such an approach is neither appropriate for all studies,[8,21] nor does it prevent spurious results in every situation.[21,22] Fortunately, this debate has cooled in recent years, and experimental data have shown that DNA preservation depends heavily on environmental conditions,[23] with permafrost environment yielding amplifiable DNA as old as half a million years.[24,25]

The increased availability of genuine, very old, and amplifiable DNA is mirrored by significant advances in PCR technology. For example, multiplex PCR[26] was used to amplify the first complete mitochondrial genome from a Pleistocene species, the mammoth,[27] and has allowed the first complete extinct nuclear gene to be analyzed.[28] In addition, a new sequencing technique involving emulsion PCR[29] resulted in the sequencing of 13 million base pairs (bp) of nuclear mammoth DNA sequences[30] and more than 1 million bp of Neanderthal DNA.[31] Thus, PCR has become a key technology in ancient DNA research, and, on balance, has turned out to be more a blessing than a curse.

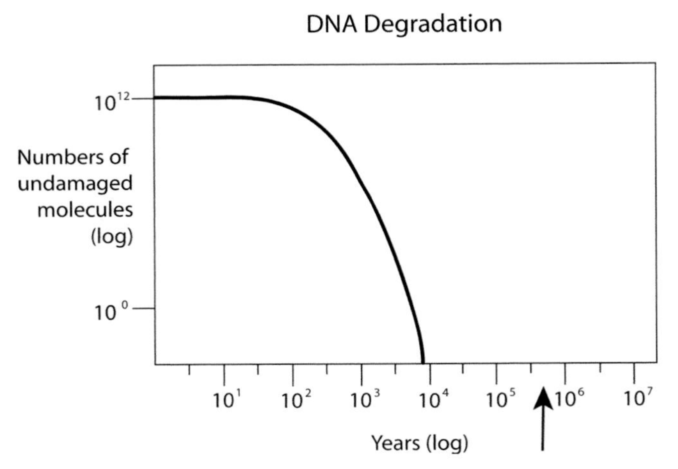

Figure 19–1. Decay of deoxyribonucleic acid (DNA) over time. At 15°C and neutral pH, it will take only 5,000 years before no undamaged 800-bp molecule would remain from 10^{12} starting molecules.[19] However, as chemical reactions proceed more slowly at lower temperatures, ancient DNA of greater age has been recovered from colder environments. Arrow indicates the age of the oldest authenticated ancient DNA recovered to date.

SENSITIVITY AND THE ISSUE OF CONTAMINATION

One of the most astonishing features of PCR is its extraordinary potency: Doubling the number of PCR cycles (e.g., from 20 to 40) generates a 1-million-fold increase in product molecules. Thus, even if only traces of target DNA remain, these can be amplified. In fact, analyses of damage patterns show that PCR amplification of ancient DNA often originates from just a single molecule of DNA.[6] This sensitivity is clearly necessary because in most fossils only traces of DNA (if any) remain because DNA decays exponentially (Figure 19–1). However, this sensitivity also causes the risk of contamination, because PCR does not discriminate between endogenous ancient DNA and contaminating modern DNA. Rather, modern DNA is more likely to be amplified, as it is usually undamaged and likely to be present in higher concentrations. Although PCR primers specific to the desired species can sometimes be designed, this is not always possible or necessary. For example, if mammoth mtDNA is targeted, one can easily design specific primers using the mitochondrial genome sequences from the two elephant species alive today. Moreover, mammoth and elephant mtDNA sequences are easily distinguishable. Furthermore, elephant DNA is unlikely to occur as contamination in the average laboratory, although it is not impossible (see box in Figure 19–2). This issue becomes more complicated when fossil DNA from species related to human life, such as domestic animals and humans themselves, is analyzed. In fact, human DNA can be found on almost every fossil bone.[32,33] Strict precautions and substantial efforts must be undertaken to recover uncontaminated bones.[31,34] This problem is not limited to human contamination. Contamination of bones, and probably also reagents and disposables, with DNA from certain domestic animals such as pigs, cattle, or chickens is also encountered fairly regularly.[35,36]

We encountered a particularly interesting example of contamination when testing a new, nondestructive extraction technique[85] on chimpanzee teeth obtained from Ivory Coast. In the first round of extraction (by simply soaking the chimpanzee teeth in a buffer and subsequently processing the buffer), we found, using general vertebrate primers, the sequences of a monkey species and a tortoise species in addition to chimpanzee and human sequences (the usual contamination caused by handling of the specimens by humans). Initially, we considered the possibility that the monkey sequence could derive from the chimps' diet, because this population is known to eat monkeys,[86] and the DNA could remain on their teeth. It seemed rather unlikely, however, that chimps would have eaten a tortoise. After consulting with the field researchers who collected the dead chimps, the explanation turned out to be much simpler. In the hut, where the chimp skeletons had been stored prior to their shipment to our laboratory, monkey skeletons and tortoise carapaces also had been stored. This mixed storage was enough to transfer DNA from one species to another (Figure 19–2). Therefore, physical contact or handling of one species immediately after another can be sufficient to transfer amounts of DNA that PCR can detect later on. Similar mechanisms may explain the detection of – apart from the target species – tiger, vole, deer, and goat DNA on a quagga tooth from the Berlin Natural History Museum (Figure 19–2).

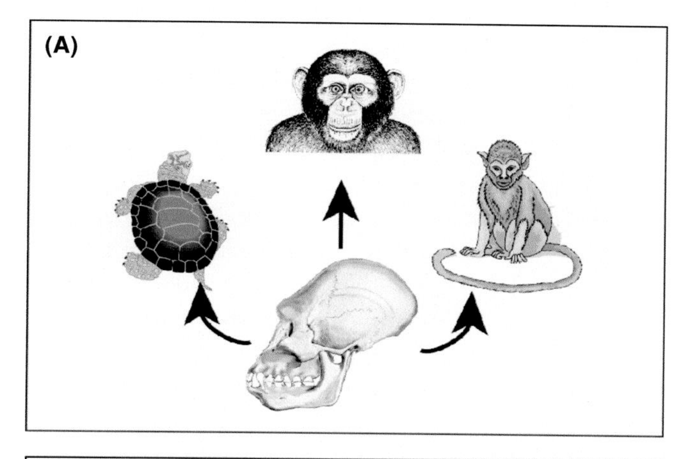

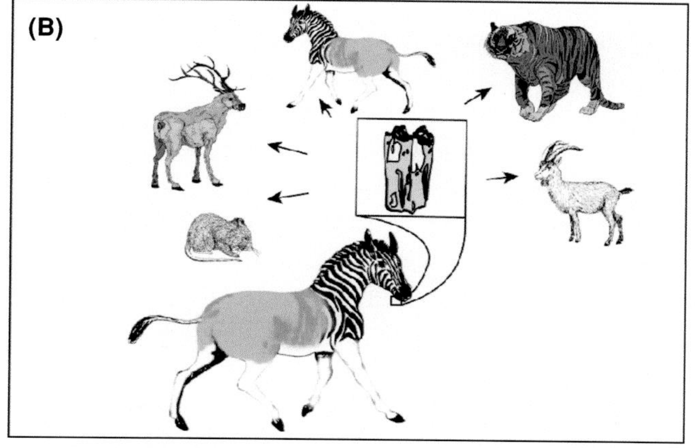

Figure 19–2. Two examples of exotic contamination detected on museum samples. **(A)** Apart from chimpanzee deoxyribonucleic acid (DNA), tortoise and monkey DNA were recovered from a chimpanzee tooth. **(B)** Tiger, deer, vole, and goat sequences in addition to quagga sequences could be recovered from a single quagga tooth from the Berlin Natural History Museum. For details, see box.

Unfortunately, contamination is not restricted to these species. During several years of research in this field, we and others have found a whole menagerie of contaminating species, ranging from rather common animals, such as mice, to exotic species like tigers and jaguars. A number of contaminant species together with the species on whose bones they were detected and possible mechanisms of contamination are discussed separately (see box in Figure 19–2).

Similar problems were probably responsible for the spectacular cases of contamination in early ancient DNA studies that led to false reports of ancient DNA sequences claimed to be millions of years old. The first of these studies reporting DNA exceeding 1 million years of age was published in 1990, reporting alleged Miocene plant sequences,[13] and was closely followed by a similar study approximately two years later.[37] However, these studies were performed without considering the dangers of contamination; therefore, they lacked appropriate controls. Moreover, several of the early studies were conducted in laboratories where modern DNA from related species had been handled previously, without taking precautions against contamination, such as working in dedicated separate laboratories to which no modern DNA is transferred.[18] Not surprisingly, it was shown later that the fossils analyzed contained only bacterial DNA, but no higher plant DNA.[38] Other studies reporting DNA sequences of similar geological age suffered the same fate. For example, sequences supposedly derived from a 125-million-year-old insect preserved in amber[14] were actually a mixture of modern yeast and fly sequences,[39] and the reported dinosaur sequences from Woodward et al.[15] were nothing more than contaminating modern human DNA sequences.[40] The most spectacular reports on supposed millions-of-years-old ancient DNA in recent years were no doubt claims that 250-million-year-old[41] and 415-million-year-old[17] bacterial DNA was amplified. Vreeland et al.[41] even claimed that the 250-million-year-old bacteria had been cultivated. Notwithstanding the fact that such results are unlikely, given everything we know about DNA stability, these "ancient" sequences look like perfectly modern bacterial DNA sequences.[42,43] Given their alleged age, one would expect that they had accumulated significantly fewer substitutions than modern bacteria and also that these ancient sequences would occupy a basal position in a phylogenetic tree. Unfortunately, the fact that neither is the case strongly argues that the sequences (and viable bacteria) were the product of contamination with modern bacteria or their DNA. In fact, none of the geologically ancient DNA sequences published to date pass a relative-rate test that assesses the number of substitutions on the branches leading to these sequences.[18] Taken together, we must conclude that all reports of ancient DNA sequences exceeding an age of 1 million years are most likely artifacts created by contamination.

These data also highlight a critical problem with ancient DNA research in general – namely, detecting sample contamination.[21] Some researchers have propagated strict criteria for avoiding contamination.[20,44,45] These measures fail, however, when the contamination occurs on the samples themselves.[21,22] Sample contamination can neither be detected during extraction, nor with PCR controls, nor by reproducing PCR experiments in a second laboratory. The failure of these measures to detect sample contamination does not mean that it is impossible to detect. Cloning PCR products and sequencing multiple clones is a powerful tool for detecting heterogeneous DNA sources. With sufficient care, even analysis of DNA from modern human fossils is possible.[34,46] In a recent study, Haak et al.[34] analyzed DNA from a number of Neolithic humans. Two lines of evidence indicate that their data are probably correct. First, they took several samples from different parts of each skeleton and accepted a sequence only if they could consistently detect it in different parts of the skeleton. Second, one of the major haplogroups (sets of related and therefore similar sequences) that they detected exists in modern populations only in low frequency. This study demonstrates that ancient DNA analyses can work around contamination problems, even in modern human samples, if the right precautions are taken.

THE ADVANTAGES OF PCR

When researchers avoid contamination pitfalls, the sensitivity and specificity of PCR provides a myriad of possibilities for studying ancient DNA. This fact immediately became obvious with the first ancient DNA studies using PCR. To obtain two clones containing quagga DNA sequences, Higuchi and colleagues[1] had to hybridize as many as 25,000 phage plaques with mtDNA from modern zebra – and even then the resulting sequence contained two errors. To correct these errors, Pääbo and Wilson[3] performed just a single PCR. Only two years later, Thomas et al.[47] published the first population genetics study using museum specimens up to approximately eighty years old. The authors sequenced DNA from as many as forty-three museum specimens, a task that would have taken a lifetime without PCR. However, not only has the number of specimens investigated increased (culminating so far in more than 400 fossilized bison samples sequenced in a single study[48]), but so have the ages of the samples investigated. Whereas the first pre-PCR ancient DNA sequences were as young as 140 years, this age increased to approximately 1,000 years by 1988 with the report of Pre-Columbian maize sequences[5] and to 3,300 years with the Moa sequences reported in 1992,[49] culminating in an age of at least 500,000 years for recently published sequences from a Greenland ice core[24] (Table 19–1). Because ancient human DNA sequences may be the result of contamination, we excluded them from Table 19–1. Although sequences older than those in Table 19–1 have been claimed, as noted earlier in this chapter, none of the DNA sequences exceeding 1 million years in age is likely to be accurate. Interestingly, the oldest replicated DNA sequences published to date, those from the Greenland ice core, even pass the relative-rate test, strongly suggesting that they are indeed ancient.

Finally, the amount of sequence data that can be obtained has increased in a similar way. The initial study using PCR amplified as few as 79 bp, but by the mid-1990s, several studies reported ancient DNA sequences exceeding 1 kb[50–52] – albeit pieced together from several fragments. In 2001, two groups[53,54] reported the sequencing of several complete mitochondrial genomes, each with a length

Table 19–1. Increasing age of authenticated ancient DNA analyzed

Year	Species	Age	Reference
	Pre-PCR		
1984	Quagga	140	(1)
	Post-PCR		
1988	Maize	1,000	(5)
1992	Moa	3,000	(49)
1993	Horse	25,000	(87)
1994	Cave bear/mammoth	40,000	(59, 88, 60)
2001	Cave bear	130,000	(89)
2003	Plants (permafrost sediment)	300,000	(25)
2007	Plants, insects (ice core)	500,000	(24)

of approximately 16,000 bp, from the extinct Moa, flightless birds living on New Zealand before they were exterminated by humans approximately 1,000 years ago.[55] These studies all used standard PCR, but new variants of PCR have been introduced to ancient DNA research in recent years. First came multiplex PCR, a technique long known[56] but rarely used with ancient DNA. When applied as a two-step process to ancient DNA, it not only successfully amplified the first mitochondrial genome of a Pleistocene animal, the mammoth,[27] but it also made possible the first sequencing of a complete nuclear gene of an extinct animal.[28] The principle of this variant of PCR is simple. In the first step, multiple primer pairs are thrown together and all fragments are amplified simultaneously (Figure 19–3). However, the PCR is stopped after approximately 30 cycles and then diluted to serve as a template in multiple second-step reactions that amplify each primer pair individually using another 30 cycles. At least sixty fragments can simultaneously be amplified using the same amount of ancient DNA extract as normally used in a single amplification. Apart from conserving extract from often precious samples, this method also decreases the likelihood of contamination as it requires fewer PCRs directly using ancient DNA template, which are most susceptible to contamination.[26] It also increases the data output, as many more fragments (and therefore also samples) can be processed in parallel.

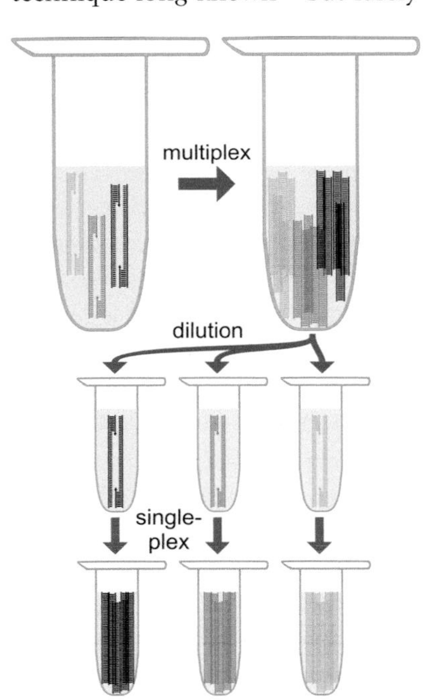

Figure 19–3. Principle of multiplex polymerase chain reaction. In the first step, multiple primer pairs are simultaneously used for amplification. This reaction is stopped after approximately 30 cycles, then the product is diluted and used as a template in single-plex amplifications where each fragment is amplified individually for another 30 cycles.

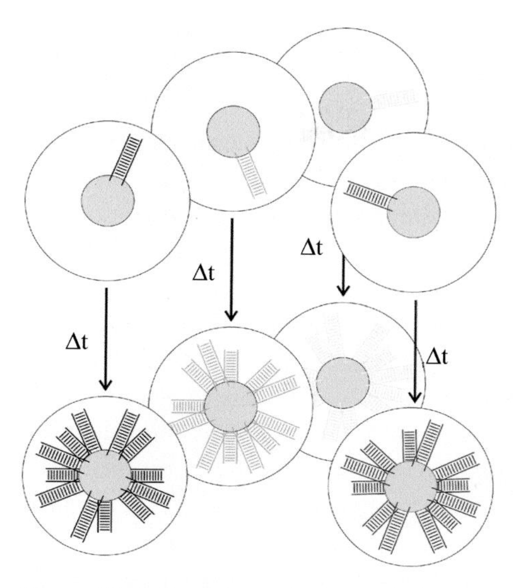

Figure 19–4. Emulsion polymerase chain reaction used to obtain deoxyribonucleic acid (DNA) for parallel sequencing. DNA fragments are bound individually to beads, each of which is captured in a water droplet in a water-in-oil emulsion. Thus, each DNA fragment has its own reaction container, in which it is amplified without interference from other fragments.

Another even more efficient technology in terms of data output was introduced with direct shotgun sequencing preceded by emulsion PCR.[29] In this technique, individual ancient DNA fragments are attached to beads, which are then captured in the water droplets of a water-in-oil emulsion. PCR is performed in these emulsified water droplets, where each individual molecule is amplified in its own private pico-reactor (Figure 19–4). In this way, hundreds of thousands of fragments can be amplified. Because the PCR products are bound to the beads, they can later be sequenced in parallel on picotiter plates. Using this technique, as many as 13 million bp of mammoth sequence have been recovered to date.[30] This technique's efficiency – including the emulsion PCR step – compared to other techniques not employing PCR can be demonstrated by comparing two studies on genomic sequences from Neanderthal remains. The first study cloned ancient DNA extract directly into bacteria, and then constructed a genomic library.[57] The authors recovered approximately 60,000 bp of nuclear Neanderthal DNA. In contrast, in the second study,[31] using a different aliquot of the same extract and the above-mentioned parallel sequencing technique, as many as 1 million bp of nuclear Neanderthal DNA were sequenced. In fact, this technique is so efficient that genome-sequencing projects are underway for both the mammoth and the Neanderthal.

PRACTICAL APPLICATIONS OF PCR IN ANCIENT DNA

The applications of PCR in ancient DNA research are almost as widespread as in work with modern DNA – except that ancient DNA is highly degraded, requiring amplification to be performed in short, overlapping fragments. Even though PCR has allowed amplification of 900 bp[30] and even 1,600-bp fragments[58] from exceptionally well-preserved specimens, most samples contain only extremely short DNA fragments.[8,32] However, as the above-described advances in PCR techniques mostly ameliorate this limitation, this concern is not as great as it was years ago.

The first and still quite commonly used application for PCR in ancient DNA research is to decipher the phylogenetic relationships between extinct species and their living relatives. For numerous extinct species, their phylogenetic position among living relatives has by now been determined.[8,22] One of the best examples for such studies is the mammoth. Since the first sequences of this species were published in 1994,[59,60] more than a dozen publications have tried to resolve its relationship to living elephants. This issue remained contentious until recently, when two groups independently reported complete mammoth mitochondrial genome sequences and concluded that it is more closely related to the Asian than to the African elephant.[27,61]

Another species group whose phylogenetic position has been firmly established only through ancient DNA sequence analysis are the Moa.[49,53,54] Moa were also the first group of extinct species whose within-group phylogenetic relationships were resolved. This endeavor began in 2003, when two groups reported that molecular sexing of Moa birds using sequences on the X and the Y chromosomes had revealed that three supposed species, differentiated by size differences, in fact represented males and females of a single species with a pronounced size dimorphism.[62,63] This study was followed by a large-scale analysis of mtDNA sequences that, according to the authors, established the existence of fourteen different Moa species.[64]

The European cave bear, *Ursus spelaeus*, represents another extensively studied extinct species. As for the Moa, DNA analyses have revealed some unexpected results. Sequencing mtDNA from two Austrian caves revealed that, although the caves are less than 10 km apart, and no physical barriers separate them, no gene flow had occurred for more than 15,000 years between the bear populations occupying the two caves.[65] As gene flow across much greater geographical distances has been found for cave bears, the best explanation for this result is that some type of reproductive isolation existed between the two populations that subsequently became elevated to species rank.[66] Interestingly, in another cave in Germany, evidence of both sequence groups was found in the same cave, but this time they were temporally separated.[67] Thus, although almost a hundred cave bear sequences have been published to date,[68] much more data will be required to reveal the complex population – and likely speciation – history of this extinct species.

Such a large-scale analysis has already been completed for the steppe bison, *Bison priscus*.[48] Using 600-bp-long mitochondrial sequences from more than 200 specimens that were radiocarbon dated via accelerator mass spectrometry (the complete study included genetic data from more than 400 specimens), the authors could detect a steady population growth beginning at the penultimate glacial maximum and continuing until about 30,000 years ago, when bison populations started to plummet across Eurasia and North America. In Eurasia, the species eventually became extinct, whereas in North America it evolved into the America bison, *Bison americanus*, a tiny genetic subsample of their pre-extinction diversity. Apart from its population history, Shapiro et al.[48] also inferred the bison's migration pattern across the Bering Strait. To study migration patterns it is, however, not always necessary to use such large amounts of data. For example, using DNA sequences from only twenty-three modern spotted hyenas and twenty-six extinct cave hyenas, Rohland et al.[69] could show that cave hyenas in Eurasia were the products of at least three independent migrations from Africa.

The power of PCR also has been demonstrated in DNA analysis of ancient vegetation taken from dung and sediment. Its potential for amplifying DNA sequences from Pleistocene dung was first shown in 1998,[70] using samples from a 20,000-year-old ground sloth coprolite. By amplifying not only DNA from this extinct species (the Shasta ground sloth, *Nothrotheriops shastensis*), but also from the plant remains in its dung, Poinar et al. could reconstruct the animal's diet and also identify the vegetation around the cave. Two years later, the authors expanded this study and analyzed DNA from several coprolites of varying ages (spanning almost 20,000 years) to show how vegetation changes can be traced.[71] Eventually, in 2003, two groups[25,72] showed that DNA can be amplified directly from sediment. One of these studies extended the upper age limit for amplifiable ancient DNA to approximately 300,000 to 400,000 years.[25] Recently, this age has been further increased to approximately half a million years.[24] Willerslev et al. not only extended the age range for ancient DNA analyses, with their sequences originating from the Greenland ice core Dye 3, they also showed that southern central Greenland supported full-grown forests some 500,000 years ago. These results raise the exciting possibility of reconstructing past ecosystems in great detail, even in the absence of macrofossils.

Another quite recent development within ancient DNA research is genome sequence analysis. The first publication on this topic involved cloning ancient DNA directly into bacterial vectors, without using PCR at all.[73] This process enabled the study's authors to recover some 27,000 bp of nuclear DNA sequences from the extinct cave bear. Although this was a major achievement, the pace of development is evident from the fact that only six months later Poinar et al.[30] published a data set two orders of magnitude larger, some 13 million bp of random shotgun DNA sequences of the extinct mammoth. Poinar et al. used a combination of emulsion PCR and direct sequencing, a technique which has subsequently also been used to sequence more than 1 million bp of Neanderthal DNA.[31]

Shotgun sequencing ancient DNA, however, poses some inherent problems of its own, requiring careful analysis of any resulting data. First, shotgun sequencing does not allow replication of individual positions – unless the genome is sequenced to multiple coverage. Therefore, errors caused by DNA damage cannot be detected by multiple amplification of the same position, which substantially elevates the error rate for the genomic sequences, especially with regard to C to T and G to A changes.[31,74] The second problem is that most ancient bone samples – apart from some permafrost remains[30] – contain only a small percentage of endogenous DNA.[31,57,73] Thus, substantially more effort is required for sequencing an entire ancient genome as compared to sequencing modern DNA. For example, with an extract containing only 5% endogenous DNA, 20 times as many sequencing runs are needed than when sequencing an extract of modern DNA, which typically contains 100% endogenous DNA, to obtain the same sequencing coverage.

Finally, we should say a few words about ancient human DNA analysis, which probably comprises the most controversial part of the field. Although the first ancient DNA amplified from bones came from humans,[75] ancient human DNA sequences have always been suspicious due to the problem of contamination. Human DNA is ubiquitous in the environment, and almost any fossil is contaminated with modern human DNA because of handling.[22,32,33] The amount of modern human DNA often exceeds the amount of endogenous ancient DNA by several orders of magnitude.[31] Because human sequences up to 50,000 years old are not expected to differ much from sequences in the current human gene pool, it is often almost impossible to distinguish contamination from endogenous ancient DNA. Even if all possible rules for working with ancient DNA are followed,[20] sample contamination still may cause false-positive results.[21] The genetic analysis of two 24,000-year-old human skeletons from Italy[76] illustrates this problem. Although the authors rigorously followed protocols for excluding contamination from their resulting sequences – and the sequences may well be correct – this publication sparked a protracted debate over whether their results are reliable[77,78] (see also references 8, 9, and 21 for reviews discussing this topic). Ultimately, it is impossible to rule out contamination as a possible source for human DNA sequences, which creates a severe bias – sequences that differ from modern human DNA are accepted as endogenous, whereas identical ones are rejected.[78] This bias may explain why researchers often prefer samples from geographical regions with sequences expected to be distant enough from modern European sequences as to be distinguishable, such as the Andaman islands[79,80] or the Americas.[81–84] Because such a view would clearly lead to biased sampling of sequences, researchers have been looking for ways to circumvent this problem. Probably the best solution is to work with complete skeletons and to sample from various parts of the skeleton in the hope that sequences consistently recovered from different subsamplings are endogenous. This approach was recently used to obtain mitochondrial sequences from Neolithic Europeans.[34] Although the authors took great care to obtain authentic results, they were also lucky because they recovered a certain haplogroup that is almost absent from modern populations. Again, this study identified sequences that somehow differ from modern human DNA. More generally, it is almost impossible to authenticate an individual's sequence from human remains, but population analyses – if conducted correctly – may provide insights into the fate of past populations.

CONCLUSIONS

Like many fields within molecular biology, ancient DNA research became possible only with the invention of PCR. If one considers how unlikely it seems that a fragile molecule like DNA can survive for tens and even hundreds of thousands of years, the recent developments in ancient DNA analysis using PCR amplification are a real success story. However, the incredible sensitivity of PCR has caused spectacular errors that have raised questions regarding the credibility of the entire field. Contamination will probably stay with ancient DNA research forever, and some parts of ancient DNA research, such as the analysis of ancient human sequences, face intrinsic obstacles. Overall, however, ancient DNA sequences have provided a whole range of new biological insights and are likely to continue to do so in the future.

ACKNOWLEDGMENTS

We thank Knut Finstermeier and Sabine Giesser for help with the figure design and Christine Green and Viola Mittag for comments on a previous version of this chapter. This work was funded by the Max Planck Society and the Deutsche Forschungsgemeinschaft.

REFERENCES

1. Higuchi R, Bowman B, Freiberger M, Ryder OA, Wilson AC (1984) DNA sequences from the quagga, an extinct member of the horse family. *Nature* **312**: 282–284.
2. Saiki RK, Scharf S, Faloona F, Mullis KB, Horn GT, Erlich HA, et al. (1985) Enzymatic amplification of beta-globin genomic sequences and restriction site analysis for diagnosis of sickle cell anemia. *Science* **230**: 1350–1354.
3. Pääbo S, Wilson AC (1988) Polymerase chain reaction reveals cloning artefacts. *Nature* **334**: 387–388.
4. Pääbo S, Gifford J, Wilson A (1988) Mitochondrial DNA sequences from a 7000-year-old brain. *Nucleic Acids Research* **16**: 9775–9787.
5. Rollo F, Amici A, Salvi R, Garbuglia A (1988) Short but faithful pieces of ancient DNA. *Nature* **335**: 774.
6. Hofreiter M, Jaenicke V, Serre D, Haeseler Av A, Pääbo S (2001) DNA sequences from multiple amplifications reveal artifacts induced by cytosine deamination in ancient DNA. *Nucleic Acids Research* **29**: 4793–4799.
7. Saiki RK, Gelfand DH, Stoffel S, Scharf SJ, Higuchi R, Horn GT, et al. (1988) Primer-directed enzymatic amplification of DNA with a thermostable DNA polymerase. *Science* **239**: 487–491.
8. Pääbo S, Poinar H, Serre D, Jaenicke-Despres V, Hebler J, Rohland N, et al. (2004) Genetic analyses from ancient DNA. *Annual Review of Genetics* **38**: 645–679.
9. Willerslev E, Cooper A (2005) Ancient DNA. *Proceedings. Biological Sciences* **272**: 3–16.
10. Pääbo S (1985) Molecular cloning of ancient Egyptian mummy DNA. *Nature* **314**: 644–645.
11. Del Pozzo G, Guardiola J (1989) Mummy DNA fragment identified. *Nature* **339**: 431–432.
12. Thomas RH, Schaffner W, Wilson AC, Pääbo S (1989) DNA phylogeny of the extinct marsupial wolf. *Nature* **340**: 465–467.
13. Golenberg EM, Giannasi DE, Clegg MT, Smiley CJ, Durbin M, Henderson D, et al. (1990) Chloroplast DNA sequence from a miocene Magnolia species. *Nature* **344**: 656–658.
14. Cano RJ, Poinar HN, Pieniazek NJ, Acra A, Poinar GO (1993) Amplification and sequencing of DNA from a 120–135-million-year-old weevil. *Nature* **363**: 536–538.
15. Woodward SR, Weyand NJ, Bunnell M (1994) DNA sequence from Cretaceous period bone fragments. *Science* **266**: 1229–1232.
16. Lindahl T (1993) Instability and decay of the primary structure of DNA. *Nature* **362**: 709–715.

17. Fish SA, Shepherd TJ, Mcgenity TJ, Grant WD (2002) Recovery of 16S ribosomal RNA gene fragments from ancient halite. *Nature* **417**: 432–436.
18. Hebsgaard MB, Phillips MJ, Willerslev E (2005) Geologically ancient DNA: fact or artefact? *Trends in Microbiology* **13**: 212–220.
19. Pääbo S, Wilson AC (1991) Miocene DNA sequences – a dream come true? *Current Biology* **1**: 45–46.
20. Cooper A, Poinar HN (2000) Ancient DNA: do it right or not at all. *Science* **289**: 1139.
21. Gilbert MTP, Bandelt HJ, Hofreiter M, Barnes I (2005) Assessing ancient DNA studies. *Trends in Ecology & Evolution* **20**: 541–544.
22. Hofreiter M, Serre D, Poinar HN, Kuch M, Pääbo S (2001) Ancient DNA. *Nature Reviews. Genetics* **2**: 353–359.
23. Smith CI, Chamberlain AT, Riley MS, Stringer C, Collins MJ (2003) The thermal history of human fossils and the likelihood of successful DNA amplification. *Journal of Human Evolution* **45**: 203–217.
24. Willerslev E, Cappellini E, Boomsma W, Nielsen R, Hebsgaard MB, Brand TB, et al. (2007) Ancient biomolecules from deep ice cores reveal a forested southern Greenland. *Science* **317**: 111–114.
25. Willerslev E, Hansen AJ, Binladen J, Brand TB, Gilbert MTP, Shapiro B, et al. (2003) Diverse plant and animal genetic records from Holocene and Pleistocene sediments. *Science* **300**: 791–795.
26. Römpler H, Dear PH, Krause J, Meyer M, Rohland N, Schöneberg T, et al. (2006) Multiplex amplification of ancient DNA. *Nature Protocols* **1**: 720–728.
27. Krause J, Dear PH, Pollack JL, Slatkin M, Spriggs H, Barnes I, et al. (2006) Multiplex amplification of the mammoth mitochondrial genome and the evolution of Elephantidae. *Nature* **439**: 724–727.
28. Römpler H, Rohland N, Lalueza-Fox C, Willerslev E, Kuznetsova T, Rabeder G, et al. (2006) Nuclear gene indicates coat-color polymorphism in mammoths. *Science* **313**: 62.
29. Margulies M, Egholm M, Altman WE, Attiya S, Bader JS, Bemben LA, et al. (2005) Genome sequencing in microfabricated high-density picolitre reactors. *Nature* **437**: 376–380.
30. Poinar HN, Schwarz C, Qi J, Shapiro B, Macphee RDE, Buigues B, et al. (2006) Metagenomics to paleogenomics: large-scale sequencing of mammoth DNA. *Science* **311**: 392–394.
31. Green RE, Krause J, Ptak SE, Briggs AW, Ronan MT, Simons JF, et al. (2006) Analysis of one million base pairs of Neanderthal DNA. *Nature* **444**: 330–336.
32. Serre D, Langaney A, Chech M, Teschler-Nicola M, Paunovic M, Mennecier P, et al. (2004) No evidence of Neandertal mtDNA contribution to early modern humans. *PLoS Biology* **2**: 313–317.
33. Wandeler P, Smith S, Morin PA, Pettifor RA, Funk SM (2003) Patterns of nuclear DNA degeneration over time – a case study in historic teeth samples. *Molecular Ecology* **12**: 1087–1093.
34. Haak W, Forster P, Bramanti B, Matsumura S, Brandt G, Tanzer M, et al. (2005) Ancient DNA from the first European farmers in 7500-year-old Neolithic sites. *Science* **310**: 1016–1018.
35. Leonard JA, Shanks O, Hofreiter M, Kreuz E, Hodges L, Ream W, et al. (2007) Animal DNA in PCR reagents plagues ancient DNA research. *Journal of Archaeological Science* **34**: 1361–1366.
36. Schmidt T, Hummel S, Herrmann B (1995) Evidence of contamination in PCR laboratory disposables. *Naturwissenschaften* **82**: 423–431.
37. Soltis PS, Soltis DE, Smiley CJ (1992) An Rbcl sequence from a Miocene Taxodium (Bald Cypress). *Proceedings of the National Academy of Sciences of the United States of America* **89**: 449–451.

38. Sidow A, Wilson AC, Pääbo S (1991) Bacterial DNA in Clarkia fossils. *Philosophical Transactions of the Royal Society of London. Series B, Biological Sciences* **333**: 429–432; discussion 432–433.
39. Gutierrez G, Marin A (1998) The most ancient DNA recovered from an amber-preserved specimen may not be as ancient as it seems. *Molecular Biology and Evolution* **15**: 926–929.
40. Zischler H, Höss M, Handt O, Von Haeseler A, Van Der Kuyl AC, Goudsmit J (1995) Detecting dinosaur DNA. *Science* **268**: 1192–1193; discussion 1194.
41. Vreeland RH, Rosenzweig WD, Powers DW (2000) Isolation of a 250 million-year-old halotolerant bacterium from a primary salt crystal. *Nature* **407**: 897–900.
42. Graur D, Pupko T (2001) The Permian bacterium that isn't. *Molecular Biology and Evolution* **18**: 1143–1146.
43. Nickle DC, Learn GH, Rain MW, Mullins JI, Mittler JE (2002) Curiously modern DNA for a "250 million-year-old" bacterium. *Journal of Molecular Evolution* **54**: 134–137.
44. Handt O, Höss M, Krings M, Pääbo S (1994) Ancient DNA: methodological challenges. *Experientia* **50**: 524–529.
45. Pääbo S (1989) Ancient DNA: extraction, characterization, molecular cloning, and enzymatic amplification. *Proceedings of the National Academy of Sciences of the United States of America* **86**: 1939–1943.
46. Burger J, Kirchner M, Bramanti B, Haak W, Thomas MG (2007) Absence of the lactase-persistence-associated allele in early Neolithic Europeans. *Proceedings of the National Academy of Sciences of the United States of America* **104**: 3736–3741.
47. Thomas WK, Pääbo S, Villablanca FX, Wilson AC (1990) Spatial and temporal continuity of kangaroo rat populations shown by sequencing mitochondrial DNA from museum specimens. *Journal of Molecular Evolution* **31**: 101–112.
48. Shapiro B, Drummond AJ, Rambaut A, Wilson MC, Matheus PE, Sher AV, et al. (2004) Rise and fall of the Beringian steppe bison. *Science* **306**: 1561–1565.
49. Cooper A, Mourer-Chauvire C, Chambers GK, Von Haeseler A, Wilson AC, Pääbo S (1992) Independent origins of New Zealand moas and kiwis. *Proceedings of the National Academy of Sciences of the United States of America* **89**: 8741–8744.
50. Höss M, Dilling A, Currant A, Pääbo S (1996) Molecular phylogeny of the extinct ground sloth Mylodon darwinii. *Proceedings of the National Academy of Sciences of the United States of America* **93**: 181–185.
51. Noro M, Masuda R, Dubrovo IA, Yoshida MC, Kato M (1998) Molecular phylogenetic inference of the woolly mammoth *Mammuthus primigenius*, based on complete sequences of mitochondrial cytochrome b and 12S ribosomal RNA genes. *Journal of Molecular Evolution* **46**: 314–326.
52. Ozawa T, Hayashi S, Mikhelson VM (1997) Phylogenetic position of mammoth and Steller's sea cow within Tethytheria demonstrated by mitochondrial DNA sequences. *Journal of Molecular Evolution* **44**: 406–413.
53. Cooper A, Lalueza-Fox C, Anderson S, Rambaut A, Austin J, Ward R (2001) Complete mitochondrial genome sequences of two extinct moas clarify ratite evolution. *Nature* **409**: 704–707.
54. Haddrath O, Baker AJ (2001) Complete mitochondrial DNA genome sequences of extinct birds: ratite phylogenetics and the vicariance biogeography hypothesis. *Proceedings of the Royal Society of London. Series B, Containing papers of a Biological character. Royal Society (Great Britain)* **268**: 939–945.
55. Holdaway RN, Jacomb C (2000) Rapid extinction of the moas (Aves: Dinornithiformes): model, test, and implications. *Science* **287**: 2250–2254.
56. Chamberlain JS, Gibbs RA, Ranier JE, Nguyen PN, Caskey CT (1988) Deletion screening of the Duchenne muscular dystrophy locus via multiplex DNA amplification. *Nucleic Acids Research* **16**: 11141–11156.
57. Noonan JP, Coop G, Kudaravalli S, Smith D, Krause J, Alessi J, et al. (2006) Sequencing and analysis of Neanderthal genomic DNA. *Science* **314**: 1113–1118.

58. Lambert DM, Ritchie PA, Millar CD, Holland B, Drummond AJ, Baroni C (2002) Rates of evolution in ancient DNA from Adelie penguins. *Science* **295**: 2270–2273.
59. Hagelberg E, Thomas MG, Cook CE Jr, Sher AV, Baryshnikov GF, Lister AM (1994) DNA from ancient mammoth bones. *Nature* **370**: 333–334.
60. Höss M, Pääbo S, Vereshchagin NK (1994) Mammoth DNA sequences. *Nature* **370**: 333.
61. Rogaev EI, Moliaka YK, Malyarchuk BA, Kondrashov FA, Derenko MV, Chumakov I, et al. (2006) Complete mitochondrial genome and phylogeny of Pleistocene mammoth *Mammuthus primigenius*. *PLoS Biology* **4**: e73.
62. Bunce M, Worthy TH, Ford T, Hoppitt W, Willerslev E, Drummond A, et al. (2003) Extreme reversed sexual size dimorphism in the extinct New Zealand moa Dinornis. *Nature* **425**: 172–175.
63. Huynen L, Millar CD, Scofield RP, Lambert DM (2003) Nuclear DNA sequences detect species limits in ancient moa. *Nature* **425**: 175–178.
64. Baker AJ, Huynen LJ, Haddrath O, Millar CD, Lambert DM (2005) Reconstructing the tempo and mode of evolution in an extinct clade of birds with ancient DNA: the giant moas of New Zealand. *Proceedings of the National Academy of Sciences of the United States of America* **102**: 8257–8262.
65. Hofreiter M, Rabeder G, Jaenicke-Despres V, Withalm G, Nagel D, Paunovic M, et al. (2004) Evidence for reproductive isolation between cave bear populations. *Current Biology* **14**: 40–43.
66. Rabeder G, Hofreiter M, Nagel D, Withalm G (2003) New taxa of Alpine cave bears (Ursidae, Carnivora). In: M Philippe, A, Argant, and J Argant (eds), 9th Cave Bear Symposium. Entremont-le-Vieux (Savoie, France), Museum Lyon.
67. Hofreiter M, Münzel S, Conard NJ, Pollack J, Slatkin M, Weiss G, et al. (2007) Sudden replacement of cave bear mitochondrial DNA in the late Pleistocene. *Current Biology* **17**: R122–R123.
68. Hofreiter M, Serre D, Rohland N, Rabeder G, Nagel D, Conard N, et al. (2004) Lack of phylogeography in European mammals before the last glaciation. *Proceedings of the National Academy of Sciences of the United States of America* **101**: 12963–12968.
69. Rohland N, Pollack JL, Nagel D, Beauval C, Airvaux J, Pääbo S, et al. (2005) The population history of extant and extinct hyenas. *Molecular Biology and Evolution* **22**: 2435–2443.
70. Poinar HN, Hofreiter M, Spaulding WG, Martin PS, Stankiewicz BA, Bland H, et al. (1998) Molecular coproscopy: dung and diet of the extinct ground sloth *Nothrotheriops shastensis*. *Science* **281**: 402–406.
71. Hofreiter M, Poinar HN, Spaulding WG, Bauer K, Martin PS, Possnert G, et al. (2000) A molecular analysis of ground sloth diet through the last glaciation. *Molecular Ecology* **9**: 1975–1984.
72. Hofreiter M, Mead JI, Martin P, Poinar HN (2003) Molecular caving. *Current Biology* **13**: R693–R695.
73. Noonan JP, Hofreiter M, Smith D, Priest JR, Rohland N, Rabeder G, et al. (2005) Genomic sequencing of Pleistocene cave bears. *Science* **309**: 597–599.
74. Stiller M, Green RE, Ronan M, Simons JF, Du L, He W, et al. (2006) Patterns of nucleotide misincorporations during enzymatic amplification and direct large-scale sequencing of ancient DNA. *Proceedings of the National Academy of Sciences of the United States of America* **103**: 13578–13584.
75. Hagelberg E, Sykes B, Hedges R (1989) Ancient bone DNA amplified. *Nature* **342**: 485.
76. Caramelli D, Lalueza-Fox C, Vernesi C, Lari M, Casoli A, Mallegni F, et al. (2003) Evidence for a genetic discontinuity between Neandertals and 24,000-year-old anatomically modern Europeans. *Proceedings of the National Academy of Sciences of the United States of America* **100**: 6593–6597.
77. Abbott A (2003) Anthropologists cast doubt on human DNA evidence. *Nature* **423**: 468.
78. Barbujani G, Bertorelle G (2003) Were Cro-Magnons too like us for DNA to tell? *Nature* **424**: 127.

79. Endicott P, Gilbert MTP, Stringer C, Lalueza-Fox C, Willerslev E, Hansen AJ, et al. (2003) The genetic origins of the Andaman Islanders. *American Journal of Human Genetics* **72**: 178–184.
80. Thangaraj K, Singh L, Reddy AG, Rao VR, Sehgal SC, Underhill PA, et al. (2003) Genetic affinities of the Andaman Islanders, a vanishing human population. *Current Biology* **13**: 86–93.
81. Kaestle FA, Smith DG (2001) Ancient mitochondrial DNA evidence for prehistoric population movement: the Numic expansion. *American Journal of Physical Anthropology* **115**: 1–12.
82. Lalueza-Fox C, Gilbert MTP, Martinez-Fuentes AJ, Calafell F, Bertranpetit J (2003) Mitochondrial DNA from pre-Columbian Ciboneys from Cuba and the prehistoric colonization of the Caribbean. *American Journal of Physical Anthropology* **121**: 97–108.
83. Poinar HN, Kuch M, Sobolik KD, Barnes I, Stankiewicz AB, Kuder T, et al. (2001) A molecular analysis of dietary diversity for three archaic Native Americans. *Proceedings of the National Academy of Sciences of the United States of America* **98**: 4317–4322.
84. Stone AC, Stoneking M (1998) mtDNA analysis of a prehistoric Oneota population: implications for the peopling of the new world. *American Journal of Human Genetics* **62**: 1153–1170.
85. Rohland N, Siedel H, Hofreiter M (2004) Nondestructive DNA extraction method for mitochondrial DNA analyses of museum specimens. *BioTechniques* **36**: 814–816, 818–821.
86. Boesch C (1994) Cooperative hunting in wild chimpanzees. *Animal Behaviour* **48**: 653–667.
87. Höss M, Pääbo S (1993) DNA extraction from Pleistocene bones by a silica-based purification method. *Nucleic Acids Research* **21**: 3913–3914.
88. Hänni C, Laudet V, Stehelin D, Taberlet P (1994) Tracking the origins of the cave bear (Ursus spelaeus) by mitochondrial DNA sequencing. *Proceedings of the National Academy of Sciences of the United States of America* **91**: 12336–12340.
89. Loreille O, Orlando L, Patou-Mathis M, Philippe M, Taberlet P, Hanni C (2001) Ancient DNA analysis reveals divergence of the cave bear, *Ursus spelaeus*, and brown bear, *Ursus arctos*, lineages. *Current Biology* **11**: 200–203.

Index

Printed in the United States
By Bookmasters